Design and Analysis of Centrifugal Compressors

WILEY-ASME PRESS SERIES LIST

Design and Analysis of Centrifugal Compressors

René Van den Braembussche
von Karman Institute
Belgium

This Work is a co-publication between ASME Press and John Wiley & Sons Ltd

Registered Offices
John Wiley & Sons, Inc., 111 River Street, Hoboken, NJ 07030, USA
John Wiley & Sons Ltd, The Atrium, Southern Gate, Chichester, West Sussex, PO19 8SQ, UK

Editorial Office
The Atrium, Southern Gate, Chichester, West Sussex, PO19 8SQ, UK

For details of our global editorial offices, customer services, and more information about Wiley products visit us at www.wiley.com.

Wiley also publishes its books in a variety of electronic formats and by print-on-demand. Some content that appears in standard print versions of this book may not be available in other formats.

Library of Congress Cataloging-in-Publication Data applied for

ISBN: 9781119424093

Cover Design: Wiley
Cover Images: Abstract Background © derrrek/Getty Images; Centrifugal Compressor: Courtesy of Koen Hillewaert & René Van den Braembussche

Set in 10/12pt WarnockPro by SPi Global, Chennai, India

Printed and bound by CPI Group (UK) Ltd, Croydon, CR0 4YY

10 9 8 7 6 5 4 3 2 1

to Leen

Contents

Preface

The growing awareness of the need for energy savings and the increase of efficiency of centrifugal compressors over the last decades has resulted in an increasing field of applications. The compactness, small weight and simplicity of the components allow an efficient replacement of multistage axial compressors by a single stage radial one. The absence of mechanical friction, lower life time cost and high reliability makes centrifugal compressors also superior to reciprocal ones. All this has lead to a revival of centrifugal compressor research.

Centrifugal compressors are very different from axial ones and require a specific approach. This book intends to respond to that. Extensive reference is made to the experimental results and analytical flow models that have been developed during the last (pre-computer) century and published in the open literature. This is complemented by the research conducted in the context of the PhD. thesis of Drs. Paul Frigne, George Verdonk, Marios Sideris, Antonios Fatsis, Erkan Ayder, Koen Hillewaert, Alain Demeulenaere, Olivier Léonard, Stephane Pierret, Tom Verstraete, Alberto Di Sante and the research projects of the many Master students that I had the pleasure to supervise.

The book does not provide the recipe to design "the optimal compressor" but rather insight into the flow structure. The purpose is to help remediating problems, finding a compromise between the different design targets and restrictions and help for a better reading and hence a more efficient use of Navier-Stokes results.

Numerical techniques are not described in detail but attention is given to their application, in particular to the correct operating conditions and restrictions of the different approaches and to their use in the modern computational design and optimization techniques developed during the last two decades.

The book is based on the "Advanced Course Centrifugal Compressors" that is taught by the author in the "Research Master Program" at the von Karman Institute. It intends to be a reference for engineers involved in the design and analysis of centrifugal compressors as well as teachers and students specializing in this field.

I am indebted to my former colleagues at the von Karman Institute, Profs. Frans Breugelmans, Claus Sieverding, Tony Arts and my successor Tom Verstraete. Working with them has been a very enriching and motivating experience. Thanks also to Dr. Z. Alsalihi and ir. J. Prinsier for the many years of fruitfull collaboration and CFD support including the preparation of figures, and to the VKI librarians Christelle De Beer and Evelyne Crochard for their logistic help in preparing this book.

This book would not have been realized without the understanding, encouragement and unlimited support of Leen, my wife and soul-mate for more than fifty years. Special thanks for that.

Alsemberg, February 25, 2018

René Van den Braembussche,
Hon. Professor von Karman Institute

Acknowledgements

The author wishes to thank the Forschungsvereinigung Verbrennungskraftmaschinen (FVV) for permission to include results in this book and in particular the members of the working group Vorleitschaufeln for the many fruitful discussions.

Permissions from the following to reproduce copyrighted material is gratefully recognized:

The American Society of Mechanical Engineers for: figs. 5 and 6 from Abdelhamid (1980) as fig. 8.33, fig. 3 from Abdelhamid (1982) as fig. 9.39, figs. 2 and 5 from Abdelhamid (1987) as fig. 9.40, fig. 9b from Abdelwahab (2010) as fig. 7.8, figs. 4 and 6 from Agostinelli et al. (1960) as figs. 7.25 and 7.26, figs. 8 and 9 from Amann et al. (1975) as fig. 9.27, figs. 15 and 16 from Ariga et al. (1987) as fig. 8.43, fig. 16 from Arnulfi et al. (1999b) as fig. 9.7, figs. 1 and 9 from Baljé (1970) as fig. 3.15, fig. 3 from Bhinder and Ingham (1974) as fig. 3.12, fig. 6b from Bonaiuti et al. (2002) as fig. 4.27, fig. 11b from Bowermann and Acosta (1957) as fig. 6.22, fig. 1 from Casey (1985) as fig. 1.20, figs. 2 and 13 from Chen and Lei (2013) as fig. 9.23, figs. 7 and 14 from Chen et al. (1993) as fig. 9.39, figs. 1 and 2 from Childs and Noronha (1999) as fig. 1.17, fig. 8d from Conrad et al. (1980) as fig. 4.36, fig. 6 from Davis and Dussourd (1970) as fig. 3.2, fig. 7 from Dean and Senoo (1960) as fig. 4.13, fig. 6 from Dean and Young (1977) as fig. 8.55, fig. 3 from Dickmann et al. (2005) as fig. 9.22, fig. 3 from Dussourd and Putman (1960) as fig. 9.20, figs. 5, 11, and 12 from Dussourd et al. (1977) as figs. 9.10 and 9.9, figs. 7 to 11 from Eckardt (1976) as figs. 3.34 and 8.60, fig. 9 from Elder and Gill (1984) as fig. 8.47, figs. 10, 11, 13, 15, and 17 from Ellis (1964) as figs. 3.16, 4.20, and 4.21, figs. 7 and 96 from Everitt and Spakovszky (2013) as figs. 7.11 and 8.45, fig. 8 from Fink et al. (1992) as fig. 8.64, figs. 4, 10, and 11 from Flathers et al. (1996) as figs. 6.10 and 6.14, figs. 16 and 17 from Flynn Weber (1979) as fig. 9.16, fig. 4 from Fowler (1966) as fig. 1.10, fig. 7b from Greitzer (1976) as fig. 8.67, fig. 16 from Gyarmathy (1996) as figs. 8.50 and 8.51, fig. 3 from Hayami et al. (1990) as fig. 9.42, fig. 11 from Hunziker and Gyarmathy (1994) as fig. 8.44, figs. 12 and 13 from Ishida et al. (2000) as fig. 9.38, figs. 8 and 14 from Jansen (1964b) as fig. 8.11, fig. 2 from Kammer and Rautenberg (1986) as fig. 8.61, figs. 4 and 6 from Kang Jeong-seek et al. (2000) as fig. 4.32, fig. 9 from Kinoshita and Senoo (1985) as fig. 8.14, figs. 5 and 10 from Koch et al. (1995) as figs. 6.12 and 6.13, fig. 1 from Krain (1981) as fig. 4.2, fig. 12 from Kramer et al. (1960) as fig. 3.11, figs. 4 and 6 from Lennemann and Howard (1970) as figs. 8.1 and 8.37, figs. 7.1 and 7.2 from Lüdtke (1983) as fig. 9.37, figs. 16 and 26 from Lüdtke (1985) as figs. 6.16, 6.8, and 6.9, figs. 15 and 16 from Mishina and Gyobu (1978) as figs. 6.20 and 6.21, figs. 6a and 9 from Mizuki et al. (1978) as figs. 8.34 and 8.38, fig. 5 from Morris et al. (1972) as fig. 2.3, figs. 1b, 2c, and 3c from Mukkavilli et al. (2002) as figs. 9.43, 9.44, and 9.45, fig. 3 from Nece and Daily (1960) as fig. 3.53, fig. 4 from Pampreen (1989) as fig. 4.25, figs. 4b and 4d from Reneau et al. (1967) as fig. 4.41, fig. 10 from Rodgers (1968) as fig. 9.33, figs. 2b and 5 from Rodgers (1977) as figs. 2.10 and 9.15, figs. 5 and 6 from Rodgers (1978) as fig. 8.40, fig. 3 from Rodgers (1962) as fig. 2.23, fig. 4 from Rodgers (1998) as fig. 2.31, fig. 7 from Rodgers (1991) as fig. 1.15, fig. 2 from Rodgers (1980), as fig. 1.14, figs. 4, 14, and 17 from Sapiro (1983) as figs.

9.46 and 9.47, fig. 1 from Strub et al. (1987) as fig. 1.16, fig. 1 from Rothe and Runstadler (1978) as fig. 8.63, figs. 5, 23, 25, 26, and 27 from Runstadler Dean (1969) as figs. 4.1 and 4.42, figs. 4 and 2 from Salvage (1998) as figs. 4.39 and 9.35, fig. 12 from Senoo Ishida (1975) as fig. 4.12, fig. 1 from Senoo et al. (1977) as fig. 4.16, figs. 7, 4a, 5a, and 6 from Senoo and Kinoshita (1977) as figs. 8.15 to 8.19, fig. 6 from Senoo and Kinoshita (1978) as fig. 8.18, fig. 3 from Senoo et al. (1983a) as fig. 4.26, figs. 7 and 13 from Simon et al. (1986) as figs. 2.6 and 9.34, fig. 1 from Simon et al. (1993) as fig. 9.5, fig. 15 from Sugimura et al. (2008) as fig. 5.30, fig. 19 from Tamaki et al. (2012) as fig. 9.25, figs. 6 and 7 from Toyama et al. (1977) as figs. 8.65 and 8.66, fig. 11 from Trébinjac et al. (2008) as fig. 4.31, fig. 8 and 5b from Tsujimoto et al. 1994 as figs. 8.35 and 8.9, fig. 126 from Wiesner (1967) as fig. 3.49, figs. 3 and 15 from Yoon et al. (2012) as fig. 9.8, fig. 4 from Yoshida et al. (1991) as fig. 9.29, figs. 7 and 8 from Yoshinaga et al. (1985) as fig. 9.45, figs. 5b and 8a and c from Tsujimoto et al. (1994) as figs. 8.9 and 8.35, fig. 7 from Kosuge et al. (1982) as fig. 8.41, fig. 14 from Marsan et al. (2012) as fig. 9.30.

ConceptsNREC for: fig. 1.7 from Japikse and Baines (1994) as fig. 1.13, fig. 6.9 from Japikse (1996) as fig. 8.42.

Dr. Heinrich for fig. 11 from Heinrich and Schwarze (2017) as fig. 6.59.

Gas Turbine Society of Japan for: figs. 2 and 4 from Krain at al. (2007) as fig. 4.29.

Institution of Mechanical Engineering for: figs. 14 and 15 from Japikse (1982) as fig. 6.62, fig. 1 from Sideris et al. (1986) as fig. 4.2.

International Association of Hydraulic Research for: figs. 2, 3, 4, 5, 6, and 7 from Matthias (1966) as fig. 6.4, fig. 1 from Rebernik (1972) as fig. 4.19.

Japan Society of Mechanical Engineering for: fig. 11 from Hasegawa et al. (1990) as fig. 7.23, fig. 13 from Aoki et al. (1984) as fig. 7.27, figs. 13, 6, and 7 from Nishida et al. (1991) as figs. 8.27 and 9.41, figs. 2 and 8 from Nishida et al. (1988) as figs. 8.21 and 8.23, figs. 2, 4, 5, 6, 7, 11, and 13 from Kobayashi et al. (1990) as figs. 8.22, 8.23, 8.24, 8.25, and 8.28.

J. Wiley and Sons for: figs. 6.66 and 5.133 from Baljé (1981) as figs. 9.35 and 2.24, figs. 2.12 and 2.15 from Neumann (1991) as figs. 6.11 and 6.15.

NATO Science and Technology Organization for: fig. 6a from Benvenuti et al. (1980) as fig. 2.27, fig. 7 from Walitt (1980) as fig. 3.1, fig. 3c2 from Poulain and Janssens (1980) as fig. 3.33b, figs. 14 and 15 from Vavra (1970) as figs. 3.37 and 3.54, figs. 3b, 10, 4, and 112 from Jansen et al. (1980) as figs. 9.21 and 9.26, fig. 1a from Japikse (1980) as fig. 8.48, fig. 13 from Vinau et al. (1987) as fig. 9.31, figs. 25 and 37 from Kenny (1970) as figs. 4.33 and 8.46.

Prof. M. Rautenberg for: figs. 8, 9, 10, and 11 from Rautenberg et al. (1983) as figs. 1.24a and 1.24b.

Sage Publications for: fig. 8 from Casey and Robinson (2011) as fig. 1.18, fig. 22 from Peck (1951) as fig. 6.57.

Solar Turbines Incorporated for: figs. 7, 15, and 16 from White and Kurz (2006) as figs. 9.13 and 9.14.

Springer Verlag for fig. 6.16.6 from Traupel (1966) as fig. 3.50, fig. 381 from Eckert and Schnell (1961) as fig. 3.51, figs. 12, 15, and 17 from Pfau (1967) as fig. 7.2.

The Academic Computer center in Gdansk TASC for: fig. 9 from Dick et al. (2001) as fig. 7.21.

Toyota Central Laboratory for: fig. 4 from Uchida et al. (1987) as fig. 7.3.

Tsukasa Yoshinaka for: figs. 3, 4, 5, and 9 from Yoshinaka (1977) as figs. 8.68, 8.69, 8.70, and 8.71.

von Karman Institute for: figs. 3, 15, and 17 from Benvenuti (1977) as figs. 1.3 and 1.4, fig. 4 from Breugelmans (1972) as fig. 2.4, figs. 2, 3b, 4, 7, and 25 from Dean (1972) as figs. 1.1, 2.1, 3.41, 3.42, and 4.43, fig. 11 from Senoo (1984) as fig. 4.17, figs. 20 and 22 from Stiefel (1972) as fig. 6.19, fig. 26 from Schmallfuss (1972) as fig. 9.17.

List of Symbols

A	cross section area
$A(U(\vec{X}), \vec{X})$	performance constraint function
AIRS	abrupt impeller rotating stall
AR	area ratio
AS	aspect ratio (b/O_{th})
a	speed of sound
$\vec{a}$	acceleration
A_σ	real part of growth rate S
b	impeller outlet or diffuser width
B^2	Greitzer B^2 factor (Equation 8.43)
B_f	distortion factor
bl	relative blockage
c	chord length
C	impeller outlet jet flow area at zero wake velocity
C_d	dissipation coefficient
C_f	Darcy friction coefficient
CDF	cumulative density function
CFD	computational fluid dynamics
CFL	Courant-Friedrichs-Lewy
C_m	momentum or torque coefficient
C_M	jet-wake friction coefficient
CP	static pressure rise coefficient
C_p	specific heat coefficient
D	diameter
DH	hydraulic diameter
DOE	design of experiment
DR	diffusion ratio (W_1/W_{SEP})
dS	control surface
EL	equivalent channel length
ESD	emergency shut down
EM	emergency shut-off valve
f	frequency of unsteadiness
F	force
FEA	finite element stress analysis
g	gravity acceleration
G	controller gain
$G_k(\vec{X})$	geometric constraint function

GPM	gallons per minute
h	static enthalpy
h_b	blade to blade distance
H	total enthalpy
	manometric height
i	incidence
J	moment of inertia
k_s	equivalent sand grain size of roughness
K	radial force coefficient (eqn. 7.14)
k_b	blade blockage
L	length of channel
LH	hydraulic length
LSD	low solidity diffuser
LWR	length over width ratio
m	meridional distance
$\dot{m}$	mass flow
M	Mach number
Mo	momentum or torque
M_R	radial momentum
M_u	tangential momentum
M_x	axial momentum
NACA	National Advisory Committee for Aerodynamics
NS	specific speed
$NUEL$	number of circumferential positions
n	distance perpendicular to axisymmetric streamsurface
n_D	number of design parameters
N	number of rotations (RPM)
	number of individuals in a population
NPSHR	net positive suction head required
O	opening or throat width
$OF(U(\vec{X}), \vec{X})$	objective function
P	pressure
	amplitude of power spectrum
	penalty
PDF	probability density function
PIRS	progressive impeller rotating stall
Pw	power (W)
$\dot{Q}$	volumetric flow
q	heat flux per unit mass
Q	heat flux (W)
	dynamic pressure ($P^o - P$)
R	radius measured from impeller axis
$R(U(\vec{X}), \vec{X})$	performance evaluator
r	radius measured from the volute cross section center
	degree of reaction
$\mathfrak{R}$	curvature radius
	diffuser inlet round-off radius
Re	Reynolds number
R_f	relaxation factor

R_G	gas constant
RHS	rigth hand side
Ro	rothalpy
RPM	rotations per minute
RV	hub/shroud radius ratio (R_{1H}/R_{1S})
s	distance along streamline
S	surface
	exponential growth rate of perturbation
	entropy
S_r	acoustic Strouhal number
S_x	axial gap between impeller backplate and casing
t	time
	pitch
T	temperature
u	non-dimensionalized meridional length
U	peripheral velocity
$U(\vec{X})$	output of performance evaluator
v	absolute velocity in the boundary layer
V	free stream absolute velocity
VDRS	vaneless diffuser rotating stall
$\mathcal{V}_C$	compressor volume
$\mathcal{V}_P$	plenum volume
w	relative velocity in the boundary layer
W	free stream relative velocity
x	axial or longitudinal distance
$\vec{X}$	geometry
y	distance in pitchwise direction
	direction perpendicular to x and z
Z	number of blades or vanes
	controller transfer function
Z_p, Z_u	parameters defining diffuser inlet conditions (Equations 8.9 and 8.10)
z	direction perpendicular to x and y
α	absolute flow angle measured from meridional plane
β	relative flow angle measured from meridional plane
β_σ	phase shift of controller
γ	angle between meridional streamsurface and axial direction
δ	boundary layer thickness
	ratio of inlet pressures (Equation 1.106)
δ_{cl}	impeller - shroud clearance gap
δ_{bl}	blade thickness perpendicular to camber
ε	skewness angle between wall streamline and main flow direction
ϵ	relative wake width
ϵ_{kb}	relative blade blockage
η	isentropic efficiency
η_W	wheel diffusion efficiency (Equation 3.40)
θ	angular coordinate (measured from the tongue)
	half diffuser opening angle
	ratio of inlet total temperatures (Equation 1.103)
κ	isentropic exponent

λ	number of stall cells or rotating waves
	ratio of wake mass flow/total mass flow
μ	work reduction factor
	dynamic viscosity
ν	wake/jet velocity ratio
	kinematic viscosity μ/ρ
ω	total pressure loss coefficient
Ω	impeller rotational speed (rad/sec)
Ω_R	reduced frequency (Equation 7.1)
ω_m	m^{th} modal frequency of the impeller
ω_s	streamwise vorticity
ω_σ	rotational speed of stall cell
	imaginary part of S
π	pressure ratio
ϕ	flow coefficient (V_m/U)
ψ	non dimensional pressure rise coefficient
Ψ	streamfunction
ρ	density
σ	slip factor
	solidity (chord/pitch)
	stress (MPa)
τ	time for one impeller rotation
	period of perturbation
	shear stress
∇	$\vec{i}_x \frac{\partial}{\partial x} + \vec{i}_y \frac{\partial}{\partial x} + \vec{i}_z \frac{\partial}{\partial x}$
$\vee$	vector product
∇^2	Laplace operator

Subscripts

0	upstream of IGV or inlet volute
01	downstream IGV
1	impeller inlet
2	impeller outlet
3	vaned diffuser leading edge
4	diffuser outlet
5	volute exit
6	compressor outlet - return channel exit
11	at the inner radius of the impeller backplate
a	absolute frame of reference
ad	adiabatic
b	in blade to blade direction
bl	of the blade
$b2$	based on the impeller outlet width
C	of the compressor
c	critical value
	at center of volute cross section
ce	due to centrifugal forces
ch	at choking

cl	due to clearance
Cor	due to Coriolis forces
curv	due to curvature
D	of the diffuser
	deterministic solution
d	downstream
des	design value
dia	diabatic
EC	of the exit cone
F	of the force
fl	of the flow
fr	due to friction
H	at the hub
i, j, k	indices in meridional, tangential and normal direction
IGV	inlet guide vane setting angle
inc	incompressible
	due to incidence
inl	at the inlet
iw	at the inner wall
j	in the jet
	index of circumferential position
kb	due to blade blockage
LE	leading edge value
m	meridional component
max	maximum value
mech	mechanical
min	minimum value
MC	corresponding to remaining swirl
MVDL	due to meridional velocity dump losses
n	normal component
N	nominal value
o	at the outlet
opt	optimum value
ow	at the outer wall
p	of the pipe
	polytropic
P	due to pressure
	of the plenum
PS	on the pressure side
r	of the rotor (relative frame)
R	radial component
	at resonance
	robust solution
ref	reference value, reference gas
ret	at return flow
Ro	corresponding to rothalpy / corrected for rotation
s	streamwise component
S	at the shroud
	swirl component

SS	on the suction side
$S - S$	static to static
SEP	at separation point
T	trough flow or tangential component of the throttle device
TE	trailing edge value
th	at the throat section
$T - S$	total to static
$T - T$	total to total
$TVDL$	due to tangential velocity dump losses
u	peripheral component
	upstream
un	uncontrolled
V	based on absolute velocity
w	in the wake
	on the wall
W	based on relative velocity
x	axial component
∞	free stream value
	at high Reynolds number

Superscripts

i	isentropic
k	number of the time step
nr	non rotating
o	stagnation conditions
t	at next time step or generation
$\wedge$	perturbation component
$\sim$	average
$\rightarrow$	vector
∞	assuming an infinite number of blades
$\star$	target value
$'$	$\partial../\partial\phi$

1

Introduction

A radial compressor can be divided into different parts, as shown in Figure 1.1. The flow is aspirated from the **inlet plenum** and after being deflected by the **inlet guide vanes** (IGV), it enters the **inducer**. From there on the flow is decelerated and turned into the axial and radial directions before leaving the impeller in the **exducer**. The presence of a radial velocity component is responsible for Coriolis forces, which, together with the blade curvature effect, tends to stabilize the boundary layer at the shroud and suction side of the inducer (Johnston 1974; Koyama et al. 1978). The boundary layer becomes less turbulent and will more easily separate under the influence of an adverse pressure gradient.

Two different flow zones can be observed inside the impeller resulting from flow separation and secondary flows (Carrad 1923; Dean 1972):

- A highly energetic zone with a high relative Mach number, commonly called the **jet**. The flow in this zone is considered quasi isentropic.
- A lower energetic zone with a low relative Mach number where the flow is highly influenced by losses. This zone, commonly called the **wake**, is fed by the boundary layers and influenced by secondary flows.

After leaving the impeller, rapid mixing takes place between the two zones due to the difference in angular momentum (**mixing region**). This intensive energy exchange results in a fast uniformization of the flow.

The flow is further decelerated by an area increase corresponding to the radius increase of the **vaneless diffuser** and influenced by friction on the lateral walls.

In case of a **vaned diffuser**, the flow, after a short vaneless space, enters the **semi-vaneless space**, i.e. the diffuser entry region between the leading edge and the throat section where a rapid adjustment rearranges the isobar pattern from nearly circumferential to perpendicular to the main flow direction. If the Mach number is higher than one, a shock system may decelerate the flow such that the **throat section** becomes subsonic.

A further decrease in the velocity in the divergent **diffuser channel** downstream of the throat realizes an additional increase in the static pressure. Depending on the throat flow conditions, the boundary layers in this channel will thicken or even separate, which limits the static pressure rise.

The flow may exit the compressor by a **volute** or plenum, or can be guided into the next stage by a **return channel**.

The following chapters describe the flow in the different parts (IGV, impeller, diffuser, etc.) together with the equations governing the flow in these components. A first objective is to provide insight into the flow structure to allow a better understanding of numerical and experimental results. A second objective is the characterization of the compressor components based on a limited number of geometrical parameters, experimental correlations, and flow

Design and Analysis of Centrifugal Compressors, First Edition. René Van den Braembussche.
© 2019, The American Society of Mechanical Engineers (ASME), 2 Park Avenue, New York, NY, 10016, USA (www.asme.org).
Published 2019 by John Wiley & Sons Ltd.

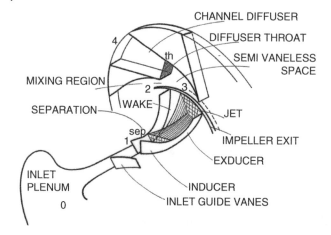

Figure 1.1 Schematic view of the radial compressor components and flow (from Dean 1972).

parameters such as the diffusion ratio (DR), the jet wake mass flow ratio (λ) for the impeller flow, the pressure recovery (CP) for the diffuser, etc.

The ultimate purpose is to provide input for the design of compressors that better satisfy the design requirements in terms of pressure ratio, efficiency, mass flow, and stable operating range.

1.1 Application of Centrifugal Compressors

Experience has shown that the specific speed NS is a valuable parameter in the selection of the type of compressor (axial, centrifugal or volumetric) that is best suited for a given application.

The specific speed is defined by

$$NS = \frac{RPM\sqrt{\dot{Q}}}{\Delta H^{3/4}} \tag{1.1}$$

This is a non-dimensional parameter only if coherent units are used (m^3/s for the volume flow $\dot{Q}$, m^2/s^2 for the enthalpy rise ΔH). However, a commonly used definition of specific speed for compressors

$$NS_C = \frac{RPM\sqrt{ft^3/s}}{ft^{3/4}} \tag{1.2}$$

does not use SI units and is not non-dimensional.

A common definition for pumps is

$$NS_P = \frac{RPM\sqrt{GPM}}{ft^{3/4}} \tag{1.3}$$

where GPM = US gallon/min and the manometric head is in ft.

The following definitions in SI units are non-dimensional:

$$NS_1 = \frac{\Omega\sqrt{m^3/s}}{\Delta H^{3/4}} \qquad or \qquad NS_2 = \frac{RPS\sqrt{m^3/s}}{\Delta H^{3/4}} \tag{1.4}$$

Previous definitions are linked by:

$$NS_C = 129.01NS_1 \qquad NS_C = 2\pi 129.01NS_2 \qquad NS_P = 21.22NS_C \qquad (1.5)$$

Radial compressors can achieve high pressure ratios and the inlet volume flow can be very different from the one at the outlet. We should therefore verify which one of the two has been used in the definition of NS. Rodgers (1980) proposes using an average value of the inlet and outlet volumetric flow:

$$\tilde{Q} = \frac{\dot{Q}_1 + \dot{Q}_6}{2}$$

The variation of efficiency as a function of specific speed for axial, centrifugal, and volumetric compressors is shown in Figure 1.2. Test results for numerous compressors lie within the shaded areas and the full lines envelop the data corresponding to the different types. The meridional cross section of the corresponding type of compressor geometry is shown on top. The limiting curves on the figure intend only to show the trend in compressor efficiency as a function of specific speed. They should not be used for prediction purposes because the information dates from a period when the flow in radial impellers was not yet fully understood (Baljé 1961). Great improvements have been made since then, thanks to the information obtained by CFD and optical measurement techniques. More recent results are shown in Figure 1.14.

Centrifugal compressors can also be designed for specific speed values away from the optimum indicated on Figure 1.2 but this does not facilitate the job. Positive displacement (volumetric) compressors are often replaced by less efficient very low specific speed centrifugal compressors for operational and maintenance reasons.

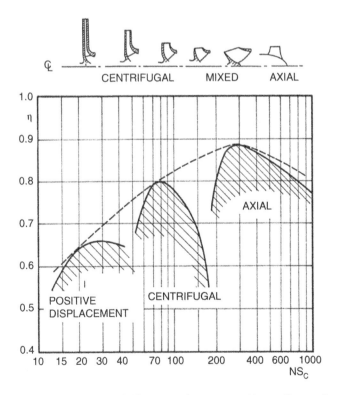

Figure 1.2 Variation of efficiency and geometry with specific speed.

Figure 1.3 Industrial centrifugal impellers (from Benvenuti 1977).

Centrifugal compressors are used at lower NS than axial compressors. The low NS may result from:

- operation at low RPM: this is often the case with industrial compressors (Figure 1.3) for reasons of maximizing lifetime
- small volume flow as occurring in last stages (Figure 1.4a) of multicorps industrial compressors (Figure 1.4b)
- a high pressure ratio per stage in combination with a small volume flow (Figure 1.5) or even large volume flow in combination with very large pressure ratios (Figure 1.6) as occurs in turbochargers
- a high pressure ratio and small volume flow as in small gas turbines for automotive applications (Figure 1.7), in the last compressor stages of small gas turbines, turboprop or jet engines (Figure 1.8), and in micro gas turbines (Figure 1.9).

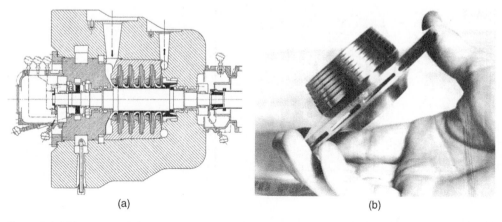

(a)   (b)

Figure 1.4 (a) Last corps of a high pressure industrial centrifugal compressor with (b) very low specific speed impeller (from Benvenuti 1977).

Figure 1.5 Cross section of a turbocharger (courtesy of MHI).

Figure 1.6 Large turbocharger for ship diesel engine (courtesy of ABB Turbo Systems Ltd).

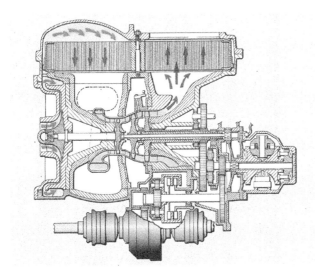

Figure 1.7 Layout of a gas turbine for automotive applications (courtesy of Volvo Group Trucks Technology).

1.2 Achievable Efficiency

Figure 1.2 shows a much lower maximum efficiency for radial compressors than for axial ones. As already mentioned, this figure dates from the time that the flow in radial compressors was not yet well understood and they were designed by simple rules and intuition, complemented by analytical considerations and empiricism. The relative flow in radial compressors being

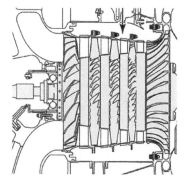

Figure 1.8 Compressor of a turboprop engine with radial endstage (courtesy of Pratt & Whitney Canada Corp.).

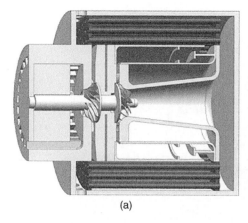

(a)   (b)

Figure 1.9 (a) Cross section of a micro gas turbine and (b) view of generator and impellers (diameter of 20 mm).

rotational, it is not possible to study the flow experimentally in a stationary (non-rotating) facility, as was done by NACA for axial compressors (Herrig et al. 1957). The heroic experimental campaign of Fowler (1966, 1968) was the start of a better understanding of the real three-dimensional (3D) flow in radial impellers (Figure 1.10). It has been complemented by advanced optical measurements (Eckardt 1976).

Before starting to discuss the maximum value of achievable efficiencies, one should first clarify the different definitions of efficiency (Figure 1.11). The temperature and enthalpy are related by the specific heat coefficient C_p. The following theoretical considerations assume constant C_p. Hence the T, S diagram is interchangeable with the H, S diagram.

The flow entering the compressor has a static temperature T_1 and a total temperature T_1^o. The difference is the kinetic energy at the impeller inlet $V_1^2/2C_p$. The static pressure at the impeller exit P_2 is achieved with a static temperature rise to T_2 and a total temperature T_2^o. An isentropic compression to the same static pressure would have resulted in an outlet static temperature T_2^i and total temperature $T_2^{o,i}$.

Considering only the static pressure at the *impeller* exit (2), the ratio of the minimum required energy over the real added one is called total to static efficiency, and is defined by:

$$\eta_{2,T-S} = \frac{T_2^i - T_1^o}{T_2^o - T_1^o} = \frac{(P_2/P_1^o)^{\frac{\kappa-1}{\kappa}} - 1}{(T_2^o - T_1^o)/T_1^o} \tag{1.6}$$

In most cases this will be a low value because in this definition the kinetic energy at the impeller exit is considered lost or useless.

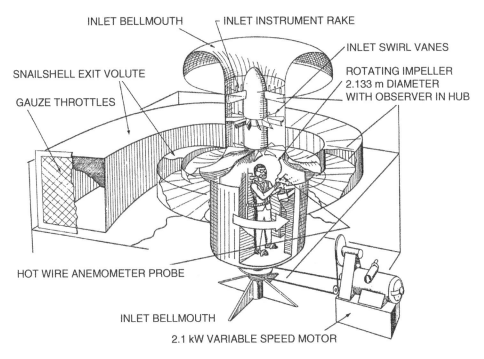

INLET BELLMOUTH

INLET INSTRUMENT RAKE

INLET SWIRL VANES

SNAILSHELL EXIT VOLUTE

ROTATING IMPELLER
2.133 m DIAMETER
WITH OBSERVER IN HUB

GAUZE THROTTLES

HOT WIRE ANEMOMETER PROBE

INLET BELLMOUTH

2.1 kW VARIABLE SPEED MOTOR

Figure 1.10 Measurements of the relative flow in a rotating impeller (from Fowler 1966).

Figure 1.11 Definition of efficiencies.

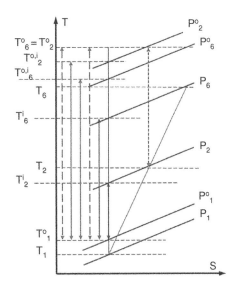

The total to total efficiency, defined by:

$$\eta_{2,T-T} = \frac{T_2^{o,i} - T_1^o}{T_2^o - T_1^o} = \frac{(P_2^o/P_1^o)^{\frac{\kappa-1}{\kappa}} - 1}{(T_2^o - T_1^o)/T_1^o} \tag{1.7}$$

is much higher (Figure 1.11) because it considers that the kinetic energy at the impeller exit ($V_2^2 = 2C_p(T_2^o - T_2)$) is also useful.

Considering the static and total flow conditions at the *compressor* exit (6), the previous definitions become:

$$\eta_{6,T-S} = \frac{T_6^i - T_1^o}{T_6^o - T_1^o} = \frac{(P_6/P_1^o)^{\frac{\kappa-1}{\kappa}} - 1}{(T_6^o - T_1^o)/T_1^o} \tag{1.8}$$

$$\eta_{6,T-T} = \frac{T_6^{o,i} - T_1^o}{T_6^o - T_1^o} = \frac{(P_6^o/P_1^o)^{\frac{\kappa-1}{\kappa}} - 1}{(T_6^o - T_1^o)/T_1^o} \tag{1.9}$$

The total to static efficiency is now much larger than at the impeller exit because part of the kinetic energy available at the impeller exit has been transformed into pressure by the stator/diffuser.

However, the total to total efficiency at the compressor exit is lower than at the impeller exit because the $T_6^{o,i}$ is smaller than $T_2^{o,i}$, due to the stator/diffuser losses. The total temperature rise $T_6^o - T_1^o = T_2^o - T_1^o$ because no energy is added in an adiabatic non-rotating diffuser. When comparing the efficiency of different compressors one should therefore verify if the same definition of the efficiency has been used.

The polytropic efficiency is commonly used for multistage and high pressure ratio compressors to correct for the divergence of the iso-pressure lines. Polytropic efficiency compares the real enthalpy rise with the hypothetical one of an infinite number of compressor stages each with an infinitesimal small pressure rise producing the same overall pressure and temperature rise of the complete compressor (Figure 1.12). Hence:

$$\eta_p = \frac{\sum \Delta T^i}{\sum \Delta T} = \frac{\sum \Delta T^i}{T_n - T_1} > \frac{T_n^i - T_1}{T_n - T_1} = \eta \tag{1.10}$$

This is therefore called small stage efficiency and can also be written as:

$$\eta_p = \frac{\kappa - 1}{\kappa} \frac{\ln(\frac{P_n}{P_1})}{\ln(\frac{T_n}{T_1})} \qquad \text{or} \qquad \frac{T_2}{T_1} = \frac{P_2}{P_1}^{\frac{\kappa-1}{\eta_p \kappa}} \tag{1.11}$$

Figure 1.12 illustrates how this definition results in an efficiency than is higher than the isentropic one because, due to the divergence of the iso-pressure lines, $\sum \Delta T^i > T_n^i - T_1$. The use of polytropic efficiency is recommended for multistage compressors because it results in a value of the overall efficiency that is closer to the efficiency of the individual stages.

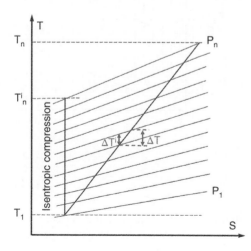

Figure 1.12 Definition of polytropic efficiency.

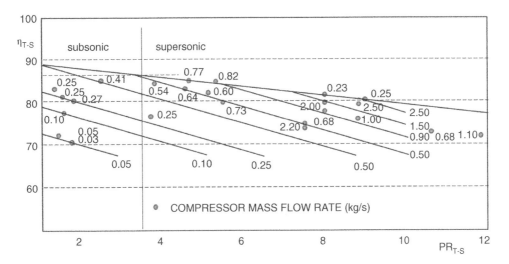

Figure 1.13 Variation of efficiency with pressure ratio and mass flow (from Japikse and Baines 1994).

Figure 1.13 shows an estimation of the achievable total to static efficiency of radial compressors in function of the mass flow and pressure ratio. Black dots indicate experimental data. The number next to them specifies the mass flow in kg/s at which that efficiency has been obtained. The lines on the figure define trends based on average values and may help in estimating the achievable performance of new designs. Very high values of maximum total to static efficiencies (up to 89%) are predicted at low pressure ratio. They are the result of an extrapolation of the high pressure ratio values. They are also not confirmed by experimental data and seem too optimistic.

The trends on Figure 1.13 can be explained by means of a model based on correlations available in the literature. It starts from the maximum impeller efficiency curve corresponding to large compressors operating at high Reynolds number and optimal specific speed (Figure 1.14) (Rodgers 1980). We observe maximum efficiencies that are much higher than the ones on Figure 1.2. This is a consequence of the improved understanding of the flow in radial compressors, obtained from more detailed experimental results (Fowler 1966; Eckardt 1976) and

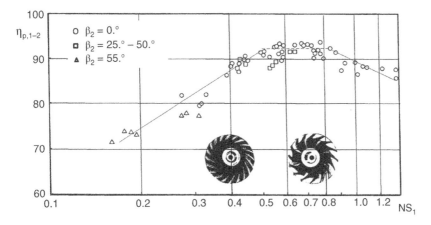

Figure 1.14 Variation of impeller polytropic efficiency (T–T) with non-dimensional specific speed (from Rodgers 1980).

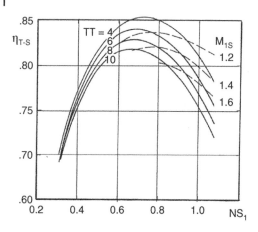

Figure 1.15 Variation of compressor *T–S* efficiency with non-dimensional specific speed and inlet Mach number (from Rodgers 1991).

full 3D Navier–Stokes analyses that have become possible on modern computers. The top line on this figure fixes the maximum polytropic impeller efficiency at 92%. This is in agreement with a maximum *T–S* stage efficiency of 86.5% for pressure ratios below 3 on Figure 1.13.

Two corrections to this reference value have to be implemented. The first one accounts for Mach number effects. It results in a decrease in the efficiency for pressure ratios larger than 3, when the impeller inlet flow becomes transonic. This is illustrated in Figure 1.15 and the impact on efficiency is estimated at

$$\Delta\eta_C = \lambda_M(M_{1,S} - 1.0) \tag{1.12}$$

where $\lambda_M = -0.1$ (Rodgers 1991). This explains the interest in designing impellers for minimum relative inlet Mach number at the shroud in order to postpone transonic flows to higher pressure ratios.

The second correction makes use of the Reynolds number to account for the change in efficiency with compressor size and operating conditions:

$$\frac{1-\eta}{1-\eta_{ref}} = a + (1-a)\left(\frac{Re_{b2,ref}}{Re_{b2}}\right)^n \tag{1.13}$$

where η_{ref} corresponds to the efficiency at a known reference Reynolds number $Re_{b2,ref}$, Re_{b2} is the Reynolds number at the operating point, a expresses the fraction of the compressor losses that do not scale with viscosity, such as clearance and leakage losses and therefore independent of Reynolds number, and n is an empirical factor, typically between 0.16 and 0.50, that depends on Reynolds number, roughness, and geometry.

The Reynolds number used in this correlation

$$Re_{b2} = \frac{U_2\, b_2}{\mu}\, \rho_2 \tag{1.14}$$

is based on the impeller outlet width b_2 because $2b_2 \approx DH$, the hydraulic diameter of an impeller flow passage near the exit, where the friction is dominant.

Previous correction accounts only implicitly for the impact of roughness on compressor losses by a change in the exponent n. A more explicit estimation has been proposed by Simon and Bulskamper (1984), Casey (1985), and Strub et al. (1987). They scale the losses by the friction coefficient instead of Reynolds number. This allows more explicit accounting for changes in both viscosity and roughness:

$$\frac{1-\eta}{1-\eta_{ref}} = \frac{a + (1-a)C_f/C_{f,\infty}}{a + (1-a)C_{f,ref}/C_{f,\infty}} \tag{1.15}$$

C_f is the Darcy friction coefficient, a function of the Reynolds number, wall roughness specified by the equivalent sand grain size k_s, and the hydraulic diameter DH. It is defined by the implicit formula of Colebrook (1939):

$$\frac{1}{\sqrt{C_f}} = -2\log_{10}\left(\frac{k_s}{3.7DH} + \frac{2.51}{Re\sqrt{C_f}}\right) \tag{1.16}$$

In explicit form it reads:

$$C_f = \frac{0.0625}{\left\{\log\left[\frac{k_s}{3.7DH} - \frac{5}{Re}\log\left(\frac{k_s}{3.7DH} - \frac{5}{Re}\log\left(\frac{k_s}{3.7DH}\right)\right)\right]\right\}^2} \tag{1.17}$$

$C_{f,\infty}$ is the friction coefficient on hydraulically smooth walls at high Reynolds number. $C_{f,ref}$ is the friction coefficient at the flow conditions at which η_{ref} has been defined.

This relation is shown in Figure 1.16. We observe that an increase in Reynolds number will not result in a decrease in the friction coefficient unless the surface is sufficiently smooth. This figure also shows that smoothing of the surface is useful only if the Reynolds number is larger than a critical value function of the relative roughness. Hence Equation (1.15) allows the impact of a change in the roughness of a given geometry at constant Reynolds number to be evaluated. It turns out that smoothing of the surfaces is useful only if the Reynolds number based on the sand grain size k_s:

$$Re_{k_s} = \frac{U\,k_s}{\mu}\rho < 100 \tag{1.18}$$

Childs and Noronha (1999) pointed out that the effect of roughness depends on the shape of the roughness, which in term depends on the manufacturing technique. Casting results in an unstructured sand grain type roughness (Figure 1.17a) whereas machining gives rise to a structured pattern composed of cusp heights and cutter path roughness in between (Figure 1.17b). In the first case the effect of roughness on friction is independent of the flow direction. This is not the case on machined surfaces where an alignment of the machine's cutter path to the flow

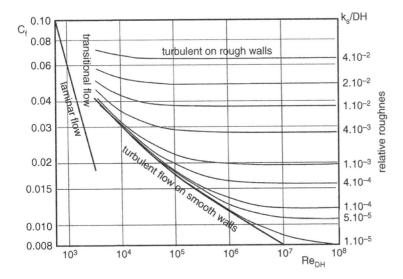

Figure 1.16 Variation of friction coefficient with Reynolds number and roughness (from Strub et al. 1987).

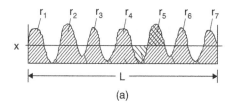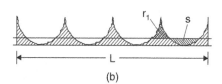

Figure 1.17 Centerline average roughness definition for (a) casted and (b) machined surfaces (from Childs and Noronha 1999).

direction may reduce the apparent roughness to the one inside each cutting path and may even have a favorable effect on the performance by a kind of alignment of the boundary layer to the main flow.

The width of the cusps and cutter path roughness depend on the size of the cutting tool and cutter speed, which in turn have an important impact on manufacturing cost. As stated by Childs and Noronha (1999), cutter marks may also affect the fatigue life of the blades, tend to retain deposits, and accelerate stress corrosion on the substrate metal.

Depending on the geometry, the fraction of viscous losses a in (1.13) and (1.15) at peak efficiency can vary between 0.0 and 0.57 (Wiesner 1979). Casey and Robinson (2011) tried to eliminate this dependence by calculating the change in efficiency directly:

$$\Delta\eta = -B_{ref}\frac{\Delta C_f}{C_{f\infty}} \tag{1.19}$$

This expression is nothing other than Equation (1.15), but written in a different way:

$$\Delta\eta = -\frac{a+(1-a)}{a+(1-a)C_{f,ref}/C_{f,\infty}}(1-\eta_{ref})\frac{\Delta C_f}{C_{f,\infty}} \tag{1.20}$$

Hence also B_{ref} depends on a and the authors provide a very useful correlation defining B_{ref} as a function of specific speed NS (Figure 1.18):

$$B_{ref} = 0.05 + \frac{0.025}{(NS_1 + 0.2)^3} \tag{1.21}$$

where NS_1 is the specific speed based on the flow at reference conditions.

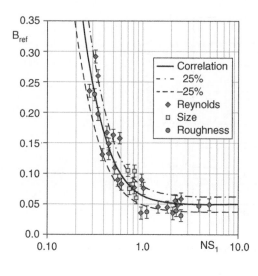

Figure 1.18 Variation of parameter B_{ref} as a function of non-dimensional specific speed (from Casey and Robinson 2011).

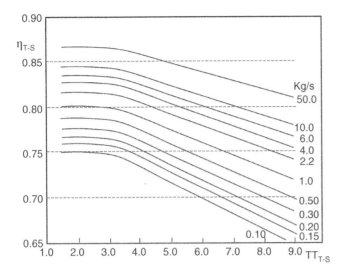

Figure 1.19 Achievable compressor efficiency as a function of pressure ratio and mass flow.

The Mach number effect (Equation (1.12)) defines the change in maximum efficiency as a function of pressure ratio. Combining the correction for Mach number and Reynolds number results in a variation of the maximum achievable efficiency, as shown in Figure 1.19. This figure is based on Equation (1.13) with $a = 0.5$ and $n = 0.9879/Re^{0.24335}$, with atmospheric inlet flow conditions and, strictly speaking, valid only for a change in Reynolds number at unchanged relative roughness $k_s/DH = C^{te}$. The following less relevant definition of the Reynolds number is used because the impeller outlet width may not be known when estimating the maximum efficiency at the early stage of a design,

$$Re = \frac{U_2 R_2}{\mu} \rho_2 \tag{1.22}$$

Re_{ref} is the Reynolds number corresponding to an impeller with 50 kg/s mass flow at atmospheric inlet conditions and sufficiently smooth surfaces to eliminate roughness effects. The upper curve on Figure 1.19 corresponds to such impellers. The surface roughness is assumed to be sufficiently small as to have no influence.

The decrease in efficiency for pressure ratios above three is due to increasing transonic flow losses in the inducer. The decrease in efficiency at a fixed pressure ratio (abscissa) is the consequence of a decreasing Reynolds number with decreasing volume flow or size. The final efficiency may be lower because not all compressors are designed at optimum specific speed and sufficiently small relative roughness and not necessarily for maximum efficiency.

A typical variation of the compressor efficiency curves with Reynolds number is shown in Figure 1.20. The corresponding change in the pressure rise and volume flow curve is similar to the one resulting from a small change in RPM. At unchanged throttle setting the change in volume flow $\dot{Q}$ is defined by:

$$\frac{\dot{Q}}{\dot{Q}_{ref}} = \sqrt{\frac{\Delta H^i}{\Delta H^i_{ref}}} \tag{1.23}$$

ΔH^i is the enthalpy rise corresponding to the pressure rise ΔP. According to Strub et al. (1987) only half of the decrease/increase in efficiency appears as an decrease/increase in isentropic head (ΔH^i) because the change in flow also results in an opposite change (increase/decrease)

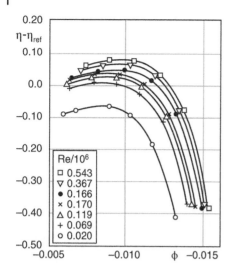

Figure 1.20 Variation of efficiency and volume flow with Reynolds number (from Casey 1985).

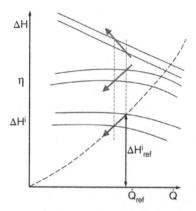

Figure 1.21 Variation of work input and isentropic head with efficiency.

in the work input (ΔH) as defined by the following relations, illustrated in Figure (1.21):

$$\frac{\Delta H^i}{\Delta H^i_{ref}} = 0.5 + 0.5\frac{\eta}{\eta_{ref}} \tag{1.24}$$

$$\frac{\Delta H}{\Delta H_{ref}} = 0.5 + 0.5\frac{\eta_{ref}}{\eta} \tag{1.25}$$

1.3 Diabatic Flows

Most compression and expansion processes are treated as adiabatic, neglecting the heat exchange with the external world. However, large amounts of heat transfer may take place in turbochargers and small gas turbines between the hot turbine and colder compressor, and between the compressor impeller and an external shroud. The amount of heat exchange depends on the temperature of the heat source and the geometry. It is largest in an overhang layout where the compressor is next to the turbine (Figure 1.9b). In a more traditional layout with a central bearing (Figure 1.9a) the heat transfer may be smaller and influenced by the oil temperature.

This heat loss in the turbine decreases the amount of available energy in the gas but increases the polytropic efficiency because of a reduction in the reheat effect. The heat addition in the compressor increases the shaft power needed to compress the gas because the compression takes place at a higher temperature. The main consequence of the internal heat transfer is a modification of the operating point of the turbocharger or gas turbine (Van den Braembussche 2005).

A further consequence of an increase in the compressor fluid temperature is a lower gas density at the impeller outlet and hence a reduced diffusion, resulting in a lower pressure rise and a change in the velocity triangles at the diffuser inlet. At unchanged shaft power, the compressor pressure ratio will be lower (Gong et al. 2004). Experimental results by Rautenberg et al. (1983) and Sirakov and Casey (2011), however, indicate that at unchanged pressure ratio more power is needed to drive the compressor. In what follows one will neglect these changes and assume that the velocity is unchanged along a streamline and that the friction losses can be evaluated from the polytropic efficiency of an adiabatic compression.

The measured exit temperatures are no longer representative for the mechanical power consumption of the compressor. They will lead to erroneous values of the efficiency if the non-adiabatic effects are not taken into consideration. Adiabatic efficiencies have to be used when calculating the turbocharger efficiency:

$$\eta_{TC} = \eta_{C,ad}\eta_{T,ad}\eta_{mech} \tag{1.26}$$

The effect of a non-adiabatic compression or expansion is illustrated in Figure 1.22.

Heating the flow during compression has a negative effect on the efficiency because the enthalpy dH needed for an elementary isentropic compression dP increases with temperature:

$$dH = \frac{dP}{\rho} = \frac{dP}{P}R_G T \tag{1.27}$$

Cooling the flow during the expansion in a turbine has also a negative effect on the power output because the energy dH obtained from an isentropic pressure drop dP decreases with decreasing temperature. Neglecting the heat loss may result in an apparent efficiency in excess of 100%.

The second law of thermodynamics provides the relation for non-isentropic diabatic compression (Equation (1.57)):

$$dH = \frac{dP}{\rho} + T\,dS = \frac{dP}{\rho} + dH_{fr} + dq \tag{1.28}$$

where dH_{fr} is the heat produced by the internal friction losses and dq is the amount of heat per unit mass transmitted through the walls. Distributing the losses and heat addition uniformly

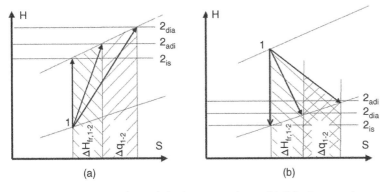

Figure 1.22 *H, S* diagram for (a) diabatic compression or (b) diabatic expansion.

over the enthalpy rise

$$dH_{fr} = \frac{\Delta H_{fr,1-2}}{H_2 - H_1} dH \qquad\qquad dq = \frac{\Delta q_{12}}{H_2 - H_1} dH \tag{1.29}$$

and substituting it into Equation (1.28) provides the following relation:

$$\left(1 - \frac{\Delta H_{fr,1-2}}{H_2 - H_1} - \frac{\Delta q_{1-2}}{H_2 - H_1}\right) dH = \frac{dP}{\rho} \tag{1.30}$$

Using the perfect gas relation to calculate the density and expressing the enthalpy as a function of the temperature change and constant specific heat coefficient C_p one obtains

$$\left(1 - \frac{\Delta H_{fr,1-2}}{C_p(T_2 - T_1)} - \frac{\Delta q_{1-2}}{C_p(T_2 - T_1)}\right) dT = \frac{\kappa - 1}{\kappa} T \frac{dP}{P} \tag{1.31}$$

Integrating this relation from inlet to outlet results in

$$\ln \frac{T_2}{T_1} \left(1 - \frac{\Delta H_{fr,12} + \Delta q_{1-2}}{C_p(T_2 - T_1)}\right) = \frac{\kappa - 1}{\kappa} \ln \frac{P_2}{P_1} \tag{1.32}$$

or

$$\frac{T_2}{T_1} = \frac{P_2}{P_1}^\mu \qquad \mu = \frac{\dfrac{\kappa - 1}{\kappa}}{1 - \dfrac{\Delta H_{fr,1-2} + \Delta q_{1-2}}{C_p(T_2 - T_1)}} \tag{1.33}$$

This equation is similar to the definition of the polytropic efficiency of an adiabatic compression (Equation (1.11)). Hence

$$\eta_{p,dia} = 1 - \frac{\Delta H_{fr,1-2} + \Delta q_{1-2}}{C_p(T_2 - T_1)} \tag{1.34}$$

which means that $\eta_{p,dia}$ decreases with positive values of Δq_{1-2} because any heat addition increases both the nominator and denominator of the RHS by the same amount. Similar equations can be derived for turbines.

The variation in compressor efficiency as a function of the adiabatic efficiency and heat flux is shown in Figure 1.23. Δq_{1-2} is non-dimensionalized by the adiabatic compressor energy input at $\eta_p = 0.7$.

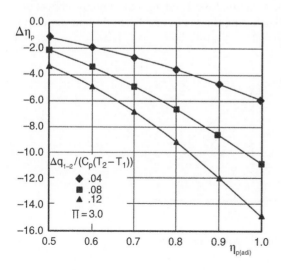

Figure 1.23 Decrease in the compressor polytropic efficiency as a function of adiabatic efficiency for different values of the non-dimensionalized heat flux.

The measured diabatic outlet temperature cannot be directly used to calculate the power absorbed by the compressor or delivered by the turbine. The heat transfer from the turbine to the compressor ($\Delta Q_{1-2} \neq 0$) is required to correctly assess the compressor aerodynamic performance map. The shaft power needed to drive the compressor is obtained by subtracting the heat flux from the total energy transfer, defined from the measured inlet and outlet temperature.

$$Pw_{dia,C} = \dot{m}C_p(T_{2,dia} - T_1) - \Delta Q_{1-2} \tag{1.35}$$

The change in polytropic efficiency depends on the amount of heat that is added (Equation (1.34)). The value of $\Delta H_{fr,1-2}$ can be directly derived from the efficiency of an adiabatic compression ($\Delta Q_{1-2} = 0$):

$$\Delta H_{fr,1-2} = (1 - \eta_{p,ad})C_p(T_{2,ad} - T_1) \tag{1.36}$$

Comparing the adiabatic and diabatic efficiency allows estimation of the amount of heat transfer. Cold tests are not representative for the performance under real operating conditions but comparing the results with a hot test allows the amount of heat flux from the hot parts to the compressor to be assessed.

Measurements by Rautenberg et al. (1983) show the impact of the heat transfer on efficiency in a classical turbocharger geometry with a central bearing (Figure 1.24a) and in an overhang geometry with the compressor close to the hot turbine (Figure 1.24b). Solid lines are for a given amount of heat flux defined by

$$\epsilon_{dia} = \frac{\eta_{ad} - \eta_{dia}}{\eta_{ad}} = \frac{\Delta Q_{1-2}}{\dot{m}(H_2 - H_1)_{dia}} \tag{1.37}$$

As the heat transfer ΔQ_{1-2} depends mainly on the wall temperatures and less on the compressor operating conditions, it is obvious that the impact on efficiency will be largest at low mass flows and low pressure ratios because ($\dot{m}(H_2 - H_1)$) is smaller.

The heat transfer inside a micro gas turbine with central bearing has been numerically evaluated by Verstraete et al. (2007) for a compressor with seven full and seven splitter blades operating at atmospheric inlet conditions and a turbine inlet temperature of 1200 K. The results

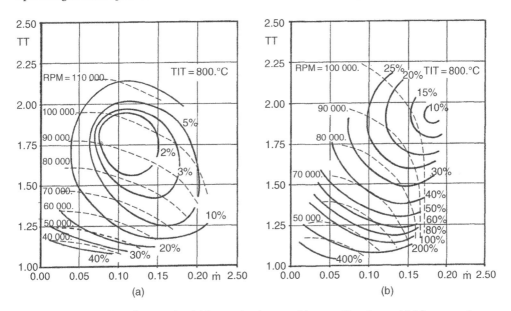

Figure 1.24 Typical heat flux number (a) for a turbocharger with central bearing and (b) for an overhang geometry with the compressor close to the turbine (from Rautenberg et al. 1983).

Table 1.1 Variation of the heat transfer with compressor size, conductivity, and geometry.

	$\lambda = 28$ W/(m.K)			Gap in house	$\lambda = 50$ W/(m.K)	$Q_{shroud} = 0$
	Geo. 1	Geo. 2	Geo. 3	Geo. 2M	Geo. 2	Geo. 2
2R (mm)	10	20	40	20	20	20
Pw (W)	900	3 570	14 250	3 570	3 570	3 570
$\dot{m}$ (g)	5.37	21.50	86.00	21.50	21.50	21.50
ΔQ_{1-2} (W)	11.6	34.4	124.0	26.0	37.0	76.0
ΔQ_{2-4} (W)	50	140	382	56	160	134
$\Delta\eta_{T-T}$ (%)	1.7	0.7	0.7			3.2

for impeller diameters between 10 and 40 mm with different shapes of bearing house and conductivity of the material are summarized in Table 1.1. All dimensions scale with the impeller diameter. The compressor and turbine shroud wall temperatures are fixed at 300 and 1000 K except for the data in the last column obtained with an adiabatic compressor shroud. The geo. 2M differs from the other by a 2-mm wide cavity in the bearing house to reduced the heat transfer from the hot turbine side to the cooler compressor side (Figure 1.25).

The total amount of heat flux ΔQ_{1-2} transmitted to the compressor fluid by the impeller hub and blades, minus the heat loss through the shroud, varies between 11.60 W for geo. 1 and 124 W for geo. 3. This is the value to be used in Equation (1.34). Only 25 W passes through the shaft of geo. 2 when it is made of the lower conductivity material. This value increases with the cross section (R^2) and decreases with the length L, hence scales with R^1. The rest of the heat enters through the impeller hub cavity by heat transfer from the bearing wall and by disk friction. The first one is proportional to the surface, hence scales with R^2. The disk friction losses are proportional to $U_2^3 R_2^2$. Identical pressure ratio in the three geometries requires the same value for U_2, so that this contribution to the heat transfer scales with R^2. The predicted values are in line with the experimental ones in Figure 1.24.

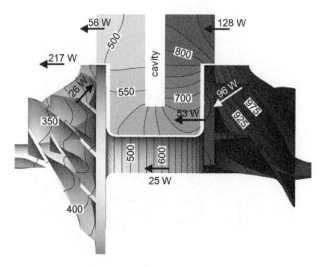

Figure 1.25 Temperature and heat transfer inside a compressor–hot turbine combination with the compressor shroud wall at 300 K (from Verstraete et al. 2007).

Only a small drop in efficiency $\Delta\eta_{T-T}$ is observed when imposing a 300 K wall temperature on the shroud as a consequence of the large heat loss through that wall. Assuming an adiabatic shroud wall for geo. 2 (the last column in Table 1.1), the total amount of heat transmitted to the impeller is transmitted to the fluid, resulting in a much larger efficiency drop (3.2%). The thermal insulation of the compressor shroud wall enhances the impact of the internal heat transfer on efficiency.

A large amount of heat enters the fluid at the vaneless diffuser hub (140 W in geo. 2 and 56 W in geo. 2M). This increases the compressor outlet temperature but has no direct impact on the real compressor efficiency.

A more detailed split of the heat transfer in geo. 2M is shown in Figure 1.25. The imbalance between the heating and cooling of the bearing house shown on this figure is due to the leakage flow transporting heat from the compressor hub cavity to the turbine inlet along the air bearing. The 217 W heat flux on the impeller and diffuser shroud is also partially due to the temperature increase of the fluid during compression.

The impact of the wall temperature on the compressor efficiency has also been studied by Isomura et al. (2001) and Sirakov et al. (2004).

1.4 Transformation of Energy in Radial Compressors

The main dimensions of the impeller and diffuser are defined in Figure 1.26.

The flow entering the impeller has a meridional velocity component V_{m1} and in case of preswirl also a tangential component V_{u1} (Figure 1.27). These velocity components together

Figure 1.26 Definition of geometrical parameters.

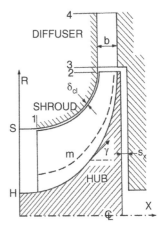

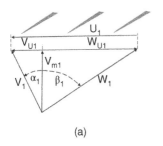

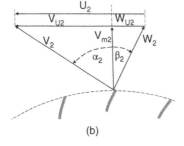

(a)   (b)

Figure 1.27 Velocity triangle at impeller (a) inlet and (b) outlet.

with the peripheral speed $\vec{U}_1$ define the velocity triangle at the rotor inlet. The angle convention for the relative flow is:

- $\beta_1 > 0$ is as shown in Figure 1.27a
- $\beta_2 > 0$ for backward leaning blades is as shown in Figure 1.27b. This is a logical choice because it does not require a change of sign for the relative flow angle between inlet and outlet of backward leaning blades.

For forward leaning blades ($\beta_2 < 0$) an obvious position for the change of sign is where the blade becomes radial:

$$\vec{V}_1 = \vec{V}_{u1} + \vec{V}_{m1} \tag{1.38}$$

The spanwise distribution of the inlet axial velocity depends on the shape of the inlet channel. In the case of an axial inlet a uniform axial velocity can be assumed. In the case of a radial to axial inlet section the spanwise variation will depend on the meridional curvature.

The relative velocity and flow angle vary over the blade height at the impeller inlet because of the change in peripheral and eventually axial velocity with radius:

$$\vec{V}_1 = \vec{W}_1 + \vec{U}_1 \tag{1.39}$$

Optimum impeller performance is expected when the relative inlet velocity is nearly parallel to the blade at the leading edge. Mass flow variation around this point is limited by the increasing diffusion or choking losses at positive or negative incidence, respectively.

The absolute velocity at the impeller outlet is defined from the relative outlet velocity $\vec{W}_2$ and the local circumferential velocity $\vec{U}_2$ (Figure 1.27b):

$$\vec{V}_2 = \vec{U}_2 + \vec{W}_2 \tag{1.40}$$

In a first approach one assumes that the outlet relative velocity (magnitude and direction) is uniform in the spanwise direction. The large peripheral velocity results in a large absolute velocity and hence large kinetic energy at the diffuser inlet. This energy is then transformed into potential energy (pressure) by decelerating the flow from $\vec{V}_2$ to the diffuser exit velocity $\vec{V}_3$.

The fluid enters the impeller with an angular momentum

$$\int_{R_H}^{R_S} R_1 V_{u1} d\dot{m} \tag{1.41}$$

and leaves the impeller with an angular momentum

$$\int_0^b R_2 V_{u2} d\dot{m} \tag{1.42}$$

The difference results from the forces exerted by the impeller on the fluid. Assuming a spanwise uniform flow at the exit, the equation of angular momentum results in

$$Mo = R_2 V_{u2}\, \dot{m} - \int_{R_H}^{R_S} R_1 V_{u1} d\dot{m} = \dot{m}(R_2 V_{u2} - \tilde{R}_1 \tilde{V}_{u1}) = \iint R(F_{u,PS} - F_{u,SS}) dS \tag{1.43}$$

where $\tilde{R}_1$ and $\tilde{V}_{u1}$ are mass averaged values at the rotor inlet. $F_{u,PS}$ and $F_{u,SS}$ are the tangential force components on the suction and pressure side of the blades, respectively.

The energy per unit mass flow transmitted by the impeller *along a streamline* does not require an averaging at the rotor inlet:

$$\frac{\Omega F_u \tilde{R}}{\dot{m}} = \frac{P_W}{\dot{m}} = U_2 V_{u2} - U_1 V_{u1} = \Delta H \tag{1.44}$$

This equation is valid along any streamline for isentropic and non-isentropic flows as well, as long as the flow is adiabatic. Adiabatic must be interpreted here in a strict way, i.e. not only no heat but also no work addition/subtraction from outside the control volume by friction on walls that are moving relative to the rotor (i.e. the fixed shroud in open impellers) (Lyman 1993). A model to estimate this energy dissipation is described by Sikarov et al. (2004). The torque on the shaft contributes to the enthalpy rise of the impeller fluid but part of it is needed to overcome the drag generated by the non-rotating shroud:

$$Pw_{shaft} - \Omega Mo_{fr} = \dot{m}(U_2 V_{u2} - \tilde{U}_1 \tilde{V}_{u1}) \tag{1.45}$$

Gong et al. (2004) consider this removal of energy from the rotor fluid as casing drag loss. They have estimated it in the extreme case of a 4-mm diameter micro gas turbine impeller with central bearing at 13% of the total power transmitted by the shaft. Except for the disk friction losses on the impeller backplate this non-adiabatic energy exchange is normally neglected in one-dimensional (1D) prediction methods. In shrouded impellers the shroud drag losses are replaced by disk friction losses on the outer wall of the shroud.

The following relations can be derived from the Figure 1.27:

$$W_1^2 = V_1^2 + U_1^2 - 2U_1 V_{u1} \tag{1.46}$$

$$W_2^2 = V_2^2 + U_2^2 - 2U_2 V_{u2} \tag{1.47}$$

Substituting them in Equation (1.44) one obtains

$$\Delta H = \frac{1}{2}(V_2^2 - V_1^2 + W_1^2 - W_2^2 + U_2^2 - U_1^2) \tag{1.48}$$

This equation is different from the one derived for axial compressors by the term $\frac{1}{2}(U_2^2 - U_1^2)$ and explains why:

- a much larger enthalpy rise can be achieved in radial compressors ($U_2 > U_1$) than in axial ones
- the diffusion from $W_1 \rightarrow W_2$ must be much larger at the shroud than at the hub to obtain the same work input on both sides and as a consequence flow separation is more common at the shroud than at the hub
- radial compressors can be very efficient in spite of an eventual flow separation in the impeller. The diffusion from W_1 to W_2 is only part of the energy transfer in the impeller and the impact of an inefficient deceleration of the relative flow can be minimized by the large extra work input $U_2^2 - U_1^2$.

The same equation can also be derived from the equations of motion and energy. Although its derivation is more complicated, it is also worth looking at because it provides a better physical understanding of the mechanisms of energy transfer and allows a further clarification of the conditions of validity.

The **equation of motion in non-rotating systems** (absolute motion) expresses the equilibrium between the acceleration $\vec{a}$ and:

- the local pressure gradients $\frac{\nabla P}{\rho}$
- the gravity forces/unit mass $\nabla(gz)$
- the friction forces/unit mass $\vec{f}_{fr}$

$$\vec{a} = -\frac{\nabla P}{\rho} - \nabla(gz) + \vec{f}_{fr} \tag{1.49}$$

whereby the acceleration is the change in velocity along a streamline, defined by:

$$\vec{a} = \frac{DV}{dt} = \frac{\partial V}{\partial t} + \vec{V}.\nabla \vec{V} \tag{1.50}$$

Combining Equations (1.49) and (1.50) for steady flow ($\frac{\partial V}{\partial t} = 0$) provides the equation of motion

$$\vec{V}.\nabla \vec{V} = -\frac{\nabla P}{\rho} - \nabla(gz) + \vec{f}_{fr} \tag{1.51}$$

Making use of the following vector identity (Vavra 1974)

$$\vec{V}.\nabla \vec{V} = \nabla\left(\frac{V^2}{2}\right) - \vec{V} \vee (\nabla \vee \vec{V}) \tag{1.52}$$

it can also be written as:

$$\vec{V} \vee (\nabla \vee \vec{V}) = \frac{\nabla P}{\rho} + \frac{\nabla V^2}{2} + \nabla(gz) - \vec{f}_{fr} \tag{1.53}$$

The **energy equation in the absolute frame of reference** is obtained by integrating the scalar product of the equation of motion (1.51) with a displacement $d\vec{s} = \vec{V}dt$ along a streamline (energy input is force times displacement):

$$\vec{V}.\nabla \vec{V}d\vec{s} = -\frac{\nabla P}{\rho}.d\vec{s} + \vec{f}_{fr}.d\vec{s} - \nabla gz.d\vec{s} \tag{1.54}$$

Taking into account that in the absence of heat transfer the entropy increase along a streamline is due only to friction,

$$T\nabla S\vec{V}dt = -\vec{f}_{fr}.\vec{V}dt \tag{1.55}$$

Equation (1.54) can be written as:

$$\left(\nabla\frac{V^2}{2} + \frac{\nabla P}{\rho} + T\nabla S + \nabla(gz)\right)d\vec{s} = 0 \tag{1.56}$$

Considering the second law of thermodynamics, illustrated in Figure 1.28,

$$\frac{\nabla P}{\rho} + T\nabla S = \nabla h \tag{1.57}$$

Equation (1.56) reduces to

$$\left(\nabla(h + \frac{V^2}{2} + gz)\right)d\vec{s} = 0 \tag{1.58}$$

This equation states that the total enthalpy

$$H = h + \frac{V^2}{2} + gz \tag{1.59}$$

Figure 1.28 Variation of pressure, enthalpy, and entropy.

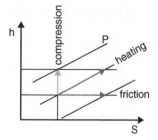

is constant along a streamline of steady adiabatic flows in the absence of body forces, with or without losses. However, any energy addition by friction on a moving wall violates the adiabatic condition.

Substituting Equation (1.57) into (1.53) results in

$$\vec{V} \vee (\nabla \vee \vec{V}) = \nabla \left(h + \frac{V^2}{2} + gz \right) - T\nabla S - \vec{f}_{fr} \tag{1.60}$$

Although in adiabatic flows the local entropy increase along a streamline is due to friction (Equation (1.55)), this does not mean that the local entropy gradient $T\nabla S$ is everywhere equal to the local friction forces. The entropy can be different from streamline to streamline depending on the friction forces on the upstream flow path. Hence the last two terms of Equation (1.60) do not cancel.

Isentropic ($\nabla S = 0$), hence frictionless ($\vec{f}_{fr} = 0$), and adiabatic flows with uniform H over the inlet section have constant H over the whole flow field according to Equation (1.59) and are called Beltrami flows. For absolute flows Equation (1.53) then reduces to

$$\vec{V} \vee (\nabla \vee \vec{V}) = 0 \tag{1.61}$$

This equation is satisfied if $\vec{V} = 0$, or $\vec{V}$ is parallel to $(\nabla \vee \vec{V})$ or when the curl

$$\nabla \vee \vec{V} = 0 \tag{1.62}$$

The first condition corresponds to a trivial solution and the second one requests that the velocity is parallel to its curl. Hence only Equation (1.62) is a relevant condition.

Two additional accelerations must be added to Equation (1.50) to obtain the **equation of motion for relative flows**:

- one corresponding to the centrifugal force $\vec{a}_{ce} = \vec{\Omega} \vee (\vec{\Omega} \vee \vec{R}) = \Omega^2 \vec{R}$
- one corresponding to the Coriolis force $a_{Co} = 2\,(\vec{\Omega} \vee \vec{W})$.

Hence the acceleration is

$$\vec{a} = \frac{D\vec{V}}{Dt} = \frac{\partial \vec{W}}{\partial t} + \vec{W}\nabla\vec{W} + 2(\vec{\Omega} \vee \vec{W}) + \Omega^2 \vec{R} \tag{1.63}$$

$\frac{\partial \vec{W}}{\partial t} = 0$ in steady relative flows (i.e. flows in a rotor at constant RPM with no inlet or outlet circumferential distortions other than the ones rotating with the rotor).

Combining Equations (1.49) and (1.63) results in the following equation of motion for the steady relative flow:

$$\vec{W}.\nabla\vec{W} + 2(\vec{\Omega} \vee \vec{W}) + \Omega^2 \vec{R} = -\frac{\nabla P}{\rho} + \vec{f}_{fr} - \nabla gz \tag{1.64}$$

Substituting again the vector identity Equation (1.52) but for the relative velocity W one obtains

$$\vec{W} \vee (\nabla \vee \vec{W}) + 2\vec{\Omega} \vee \vec{W} = \frac{\nabla P}{\rho} - \Omega^2 \vec{R} + \frac{\nabla W^2}{2} - \vec{f}_{fr} + \nabla gz \tag{1.65}$$

The corresponding **energy equation for relative flows** is obtained by integrating Equation (1.64) along a relative streamline starting at the impeller inlet. Considering that $d\vec{s} = \vec{W}.dt$ and everywhere perpendicular to the Coriolis force $\vec{\Omega} \vee \vec{W}$ (Figure 1.29) the integral of the second term is identical to zero:

$$\int_1 2(\vec{\Omega} \vee \vec{W})\vec{W}dt \equiv 0 \tag{1.66}$$

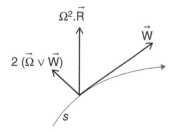

Figure 1.29 Forces in the relative plane.

The integrated value of the centrifugal term is:

$$-\int_1 \Omega^2 \vec{R} dR = -\frac{1}{2}(U^2 - U_1^2) \tag{1.67}$$

The other terms are unchanged so that with Equations (1.55) and (1.56), and after replacing V by W in Equation (1.58) one obtains

$$h + \frac{W^2}{2} - \frac{U^2}{2} + gz = h_1 + \frac{W_1^2}{2} - \frac{U_1^2}{2} + gz_1 = C^{te} = Ro \tag{1.68}$$

In analogy with the total enthalpy, one can commonly define rothalpy Ro for rotating systems which, according to Equation (1.68), is constant along a streamline in an adiabatic rotating system. One should keep in mind that the rothalpy is different from the relative total enthalpy defined by $H_r = h + \frac{W^2}{2}$.

As explained in more detail in Lyman (1993), $Ro = C^{te}$ is valid for steady adiabatic flow (with or without friction) along a streamline in a rotor at constant RPM, in absence of body forces. Adiabatic must again be considered in the most strict way: "without energy transfer by friction forces on the non-rotating walls". Equation (1.68) is very useful to estimate local flow conditions inside an impeller.

Considering the second law of thermodynamics Equation (1.57) and substituting constant rothalpy into Equation (1.65) one obtains the following equation of motion, again for isentropic and frictionless flows with constant rothalpy at the inlet (Beltrami flows):

$$\vec{W} \vee (\nabla \vee \vec{W} + 2\vec{\Omega}) = 0 \tag{1.69}$$

In the case of prerotation, rothalpy relates to total enthalpy at the inlet by

$$Ro_1 = H_1 + \Omega R V_{u1} \tag{1.70}$$

Hence spanwise constant total enthalpy at the inlet corresponds to constant rothalpy only in the case of a free vortex swirl distribution $(RV_{u1} = C^{te})$.

Equation (1.69) is satisfied if $\vec{W} = 0$, $\vec{W}$ parallel to $(\nabla \vee \vec{W} + 2\vec{\Omega})$ or when the curl

$$\nabla \vee \vec{W} = -2\vec{\Omega} \tag{1.71}$$

The first condition is trivial and the second one requests that the velocity is parallel to its curl. Hence only Equation (1.71) is relevant.

In what follows the gravity terms are neglected because they are irrelevant for gas flows. The total enthalpy by definition being $H = h + V^2$, constant Ro along a streamline (Equation (1.68)) results in the following expression for the enthalpy change between the inlet and the outlet of an impeller:

$$\Delta H = H_2 - H_1 = \frac{1}{2}(V_2^2 - V_1^2 + W_1^2 - W_2^2 + U_2^2 - U_1^2) \tag{1.72}$$

which is identical to Equation (1.48). For non-rotating systems $(U = 0$ and $W = V)$ Equation (1.72) reduces to $H_2 = H_1$ as specified in Equation (1.59).

1.5 Performance Map

A performance map describes the variation of the pressure ratio, mass flow, and efficiency at different operating conditions. The curves can be obtained by changing either:

- the speed of rotation at constant outlet pressure
- the outlet pressure at constant speed of rotation.

The latter curves are most commonly used. They are obtained by changing the resistance (pressure drop) of the throttling mechanism (valve or turbine) at constant RPM. A typical performance map is shown in Figure 1.30.

1.5.1 Theoretical Performance Curve

A simple 1D prediction method is presented in this section. We will assume here, for simplicity, that the compressor operates without preswirl ($V_{u1} = 0$).

The flow in an impeller with an infinite number of infinitely thin blades has an exit flow direction tangent to the blade ($\beta_2^\infty = \beta_{2,bl}$). Hence the tangential velocity components at the impeller exit are readily defined by (Figure 1.27):

$$W_{u2}^\infty = V_{m2} \tan \beta_{2,bl} \qquad\qquad V_{u2}^\infty = U_2 - V_{m2} \tan \beta_{2,bl} \qquad (1.73)$$

The total enthalpy rise for adiabatic flow without preswirl ($V_{u1} = 0$) in such an impeller is

$$\Delta H^\infty = U_2 V_{u2}^\infty = U_2^2 - U_2 V_{m2} \tan \beta_{2,bl} \qquad (1.74)$$

The meridional component of the outlet velocity (V_{m2}) is a function of mass flow and outlet geometry. Starting with an estimation of ρ_2 the outlet meridional velocity can be defined from:

$$\dot{m} = \rho_2 V_{m2} 2\pi R_2 b_2 \qquad (1.75)$$

Eliminating the dependence of ΔH and $\dot{m}$ on the peripheral velocity U_2 results in the following non-dimensional work and flow coefficients:

$$\psi^\infty = 2\frac{\Delta H^\infty}{U_2^2} = 2\frac{V_{u2}^\infty}{U_2} \qquad (1.76)$$

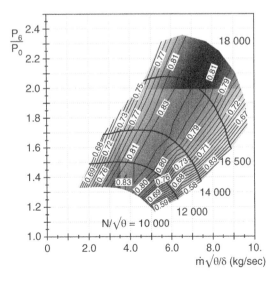

Figure 1.30 Typical centrifugal compressor performance map (from Steglich et al. 2008).

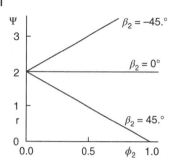

Figure 1.31 Theoretical impeller performance variation for an infinite number of blades.

$$\phi_2 = \frac{\dot{m}}{2\pi R_2 b_2 \rho_2 U_2} = \frac{V_{m2}}{U_2} \tag{1.77}$$

The inlet meridional velocity is linked to the outlet one V_{m2} by continuity. In the absence of compressibility effects (low Mach number flow) it scales with the outlet meridional velocity. The inlet peripheral velocity scales with the outlet peripheral velocity. When operating at different RPM but at the same ϕ and ψ, the inlet and outlet flow angles are conserved and hence also the impeller incidence and diffuser inlet absolute flow angle. Neglecting further the effect of a change in Reynolds number one can conclude that the efficiency will also be unchanged.

The non-dimensionalized parameters ϕ and ψ are useful to define the performance in similar operating points when scaling up or down the geometry or changing rotational speed. The velocity components at the impeller inlet and outlet, corresponding to a given RPM or scale factor, are readily obtained by multiplying ψ and ϕ, respectively, by the corresponding value of U_2^2 and U_2.

Dividing all terms in Equation (1.74) by $U_2^2/2$, one obtains

$$\psi^\infty = 2(1 - \phi_2 \tan \beta_{2,bl}) \tag{1.78}$$

This relation is plotted in Figure 1.31 for different values of $\beta_{2,bl}$.

Forward leaning blades ($\beta_{2,bl} < 0$) are normally used only for ventilators in air-conditioning systems (squirrel-cage fans). The increasing work input with increasing mass flow compensates for the increase in friction losses in the ducts of the air conditioning system ($\approx \phi^2$). This type of impeller is not used for high pressure ratio compressors because of stability problems.

Radial blades have lower bending stresses and allow higher RPM. They are used for very high pressure ratio compressors and the work input is independent of volume flow.

Backward leaning blades ($\beta_{2,bl} > 0$) show a more stable operation over a large mass flow range at nearly constant power input. They were traditionally used for industrial compressors running at moderate RPM. However, the increased reliability of the stress predictions allows the stress levels to be controlled and the geometry can be adapted accordingly. Backward leaning blades are now commonly used, including in impellers running at large peripheral speeds.

1.5.2 Finite Number of Blades

Previous performance curves are for flows that are parallel to the blades, which is hypothetical as this would require an infinite number of blades. We must correct the curves on Figure 1.31 for real flow effects to find the real enthalpy rise and what part of it is transformed into pressure.

Similar to what is called the deviation in axial compressors, the flow direction at the trailing edge is influenced by the blade shape and viscous effects, including an eventual flow separation in the impeller. This deviation is larger in radial impellers than in axial ones because it is enhanced by the rotationality of the flow resulting from the Coriolis forces.

The impact of this difference in flow direction on the work input is quantified by the work reduction factor, commonly called the slip factor:

$$\mu = \frac{\Delta H}{\Delta H^\infty} \tag{1.79}$$

Its value is normally obtained from empirical correlations. The following theoretical evaluation of μ is due to Stodola (1924) and is theoretically valid only for Beltrami flows.

As will be demonstrated in Section 3.1.2, Equation (1.71) defines the suction to pressure side velocity difference as a function of the blade curvature and impeller rotation. Assuming a straight two-dimensional (2D) rotating channel with incompressible flow and zero blade thickness the suction to pressure side velocity difference (Equation (3.18)) reduces to

$$\frac{W_{SS} - W_{PS}}{h_b} = 2\Omega \tag{1.80}$$

where h_b is the distance from suction to pressure side normal to the streamlines (Figure 1.32). Hence the velocity on the suction side W_{SS} and pressure side W_{PS} can be calculated as a function of a mean velocity $\widetilde{W}$ and the passage vortex -2Ω:

$$W_{SS} = \widetilde{W} + 2\Omega\frac{h_b}{2} \qquad W_{PS} = \widetilde{W} - 2\Omega\frac{h_b}{2} \tag{1.81}$$

This passage vortex is at the origin of the slip factor. Stodola (1924) assumed that ΔW_{u2} is equal to the mean value of the tangential velocity component created by the passage vortex in the trailing edge plane (Figure 1.32). Hence

$$\Delta \widetilde{W}_{u2} = 2\Omega\frac{h_b}{2} \times 0.5 \tag{1.82}$$

The impact of this change in tangential velocity (Figure 1.33) on the work input or enthalpy rise is calculated from the Euler momentum equation:

$$\mu = 1 - \frac{\Delta V_{u2}}{V_{u2}^\infty} = 1 - \frac{\Delta W_{u2}}{V_{u2}^\infty} \tag{1.83}$$

where $\Delta V_{u2} = V_{u2}^\infty - V_{u2}$ and $\Delta W_{u2} = W_{u2} - W_{u2}^\infty$.

This definition of the slip factor is common in German literature (Minderleistungsfaktor) and in what follows it is called the work reduction factor.

Another parameter that characterizes the impact of the vorticity on the outlet flow relates the change in tangential velocity to the peripheral velocity:

$$\sigma = 1 - \frac{\Delta V_{u2}}{U_2} = 1 - \frac{\Delta W_{u2}}{U_2} \tag{1.84}$$

Figure 1.32 Variation of the velocity from suction to pressure side in a straight rotating channel.

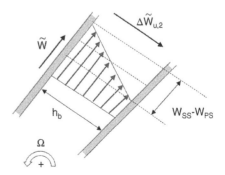

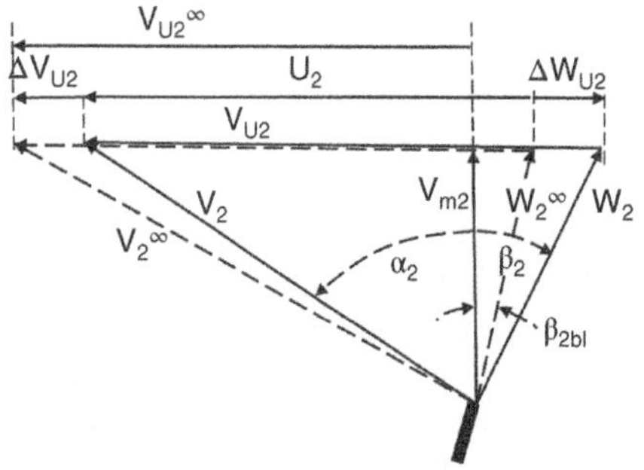

Figure 1.33 Impeller exit velocity triangles with and without slip.

This definition is common in American literature (Wiesner 1967) and in what follows is called the slip factor. Both definitions are identical for compressors with radial ending blades when $V_{u2}^\infty = U_2$.

The channel height at the impeller exit h_b can be approximated by

$$h_b = \frac{2\pi R_2 \cos \beta_{2,bl}}{Z_r} \tag{1.85}$$

where Z_r is the number of rotor blades, so that (Figure 1.33)

$$\Delta W_{u2} = \frac{U_2 \pi \cos \beta_{2,bl}}{Z_r} \tag{1.86}$$

From the definition of ψ^∞ (Equation (1.76)) one has

$$V_{u2}^\infty = \frac{\psi^\infty U_2}{2} \tag{1.87}$$

After substituting Equations (1.86) and (1.87) into Equation (1.83) one obtains the following expression for the work reduction factor:

$$\mu = 1 - \frac{2\pi \cos \beta_2}{Z_r \psi^\infty} \tag{1.88}$$

which relates the hypothetical work coefficient, corresponding to an infinite number of blades ψ^∞, to ψ, the real work input taking into account the limited number of blades.

μ is independent of ϕ and the work reduction

$$\Delta \psi_\mu = \psi^\infty(\mu - 1) = \frac{-2\pi \cos \beta_2}{Z_r} \tag{1.89}$$

is independent of the flow coefficients. The work reduction $\Delta \psi_\mu$ is not a loss but quantifies an amount of energy that is not added to the fluid. The corresponding shift of the performance curve is shown in Figure 1.34.

The power to drive the compressor is given by

$$P_w = \Delta H \dot{m} = \psi \phi_2 \, \pi R_2 b_2 \rho_2 U_2^3 \tag{1.90}$$

1.5.3 Real Performance Curve

The real performance curve for a finite number of blades differs from the one described previously because of a decrease in pressure rise due to losses, as there are friction losses, incidence losses, etc.

Friction losses are proportional to W^2, or $\Delta \psi_{fr} \approx C_{fr} \phi^2$ where C_{fr} is a friction loss coefficient depending on the geometry, Reynolds number, etc.

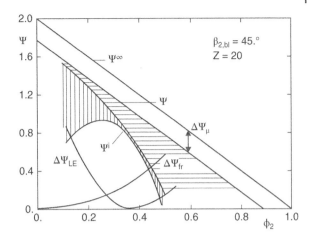

Figure 1.34 Influence of limited blade number and losses on performance curve.

Although there are no shocks in low speed compressors, the name shock losses is often used to account for the incidence losses at off-design mass flow. In analogy to axial compressors, these losses $\Delta\psi_{LE}$ increase on both sides of the ϕ value corresponding to optimum incidence.

An approximation of both losses is shown in Figure 1.34. After subtracting them from the theoretical performance curve we obtain the real performance curve (ψ^i, ϕ_2). It is clear that the shape of this curve is strongly dependent on the blade trailing edge angle β_2.

The non-dimensional performance curve allows the performance map to be calculated at different operating conditions as a function of the parameters ψ^i, ϕ:

$$\frac{P_6}{P_0} = 1 + \frac{\Delta P}{P_0} \approx 1 + \frac{\rho\Delta H^i}{P_0} = 1 + \frac{\rho\psi^i \frac{U_2^2}{2}}{P_0} \tag{1.91}$$

Substituting $\rho = \frac{P_0}{R_G T_0}$ the pressure ratio can be defined by

$$\frac{P_6}{P_0} = 1 + \frac{\psi^i \Omega^2 R_2^2}{2R_G T_0} \tag{1.92}$$

The mass flow is given by

$$\dot{m} = \rho\phi U_2 (2\pi R_2 b_2) \tag{1.93}$$

This allows the pressure ratio and mass flow to be calculated as a function of the operational conditions (pressure, temperature, rotational speed) and geometrical scale factor.

1.6 Degree of Reaction

The degree of reaction is the ratio of the static enthalpy rise in the impeller over the stage total enthalpy rise (Figure 1.35):

$$r = \frac{\Delta h_r}{\Delta H} = \frac{T_2 - T_1}{T_6^o - T_1^o} \tag{1.94}$$

An alternative definition is the impeller static pressure rise over the stage pressure rise. Expressing the enthalpy change as a function of the velocities

$$\Delta h_r = \frac{1}{2}(W_1^2 - W_2^2 + U_2^2 - U_1^2) \tag{1.95}$$

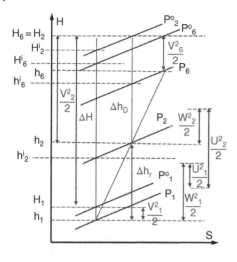

Figure 1.35 Impeller pressure rise and degree of reaction.

in combination with the Euler momentum equation

$$\Delta H = U_2 V_{u2} - U_1 V_{u1} \tag{1.96}$$

and the geometrical relations

$$V^2 = V_m^2 + V_u^2$$

$$W^2 - U^2 = V^2 - 2UV_u \tag{1.97}$$

with $V_{u1} = 0$ and assuming $V_1 = V_{m2}$, Equation (1.94) can be written as

$$r \approx 1 - \frac{V_{u2}}{2U_2} = 1 - \frac{\psi}{4} \tag{1.98}$$

This relation is illustrated in Figure 1.36.

The rotor static head coefficient is

$$\psi_r = \frac{\Delta h_r}{U_2^2/2} = r\psi = 2r(1 - \phi_2 \tan \beta_{2,bl}) \tag{1.99}$$

This relation between the flow coefficient, blade exit angle β_2, degree of reaction, and ψ_r is graphically represented in Figure 1.37. It illustrates that a compressor with 70° backward lean has a degree of reaction of 0.87 and the rotor has a pressure rise coefficient of 0.44 compared

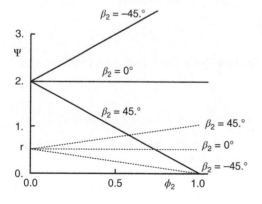

Figure 1.36 Variation of the degree of reaction as a function of the flow coefficient and impeller exit angle.

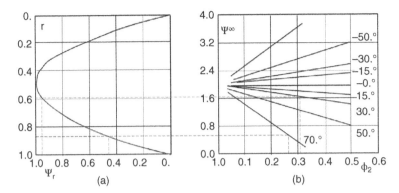

Figure 1.37 Variation of (a) the rotor static head coefficient and (b) degree of reaction as a function of the flow coefficient and impeller exit angle.

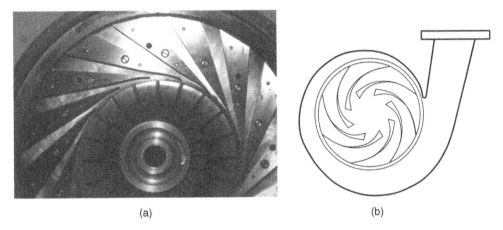

Figure 1.38 (a) A compressor with a degree of reaction of 0.5 and (b) a pump with a high degree of reaction.

to a total pressure rise coefficient of 0.51. This means that only a small fraction of the static pressure rise can be realized in the diffuser. This type of compressor or pump often operates with a very small or no diffuser (Figure 1.38). Centrifugal compressors with radial ending blades have a degree of reaction of ≈ 0.5. Hence the pressure rise is equally split between the impeller and the diffuser.

Each component (impeller and diffuser) impacts on the stage efficiency:

$$\eta_{stage} = \frac{H_6^i - H_0}{H_6 - H_0} = \frac{H_6^i - H_0}{\Delta H} \tag{1.100}$$

H_1 being equal to H_0, the numerator can be written as

$$H_6^i - H_1 = \left(h_2^i - h_1 - \frac{V_1^2}{2} \right) + \left(h_6^i - h_2 + \frac{V_6^2}{2} \right)$$

Assuming that the compressor inlet velocity V_1 equals the stage outlet velocity V_6 this reduces to

$$H_6^i - H_1 \approx \Delta h_r^i + \Delta h_D^i$$

where

$$\Delta h_r^i = \eta_r \Delta h_r = \eta_r r \Delta H$$

$$\Delta h_D^i = \eta_D \Delta h_D = \eta_D \left(\frac{V_2^2}{2} - \frac{V_6^2}{2} \right) = \eta_D \left(\Delta H - \Delta h_r \right) + \left(\frac{V_1^2}{2} - \frac{V_6^2}{2} \right)$$

Assuming again that the compressor inlet velocity V_1 equals the stage outlet velocity V_6 the last term becomes

$$\Delta h_D^i \approx \eta_D (\Delta H - \Delta h_r) = \eta_D (1 - r) \Delta H$$

and after substitution in Equation (1.100) we obtain:

$$\frac{H_6^i - H_1}{\Delta H} = \eta_{stage} \approx r\eta_r + (1 - r)\eta_D \tag{1.101}$$

This explains why impellers with radial blades ($r \approx 0.5$) profit from a good diffuser whereas impellers with very backward leaning blades ($r \gg 0.5$), such as pump impellers, often operate without a diffuser (Figure 1.38).

1.7 Operating Conditions

Inlet pressure and temperature can change with atmospheric conditions during the test period. In order to obtain a compressor map with coherent data or to compare the performance of compressors tested at different inlet conditions it is common to scale all results to reference conditions (typically 293.3 K and 101 kPa). The following describes such a transformation that can also be used to compare experimental data obtained with different gases.

A variation of the inlet temperature has a direct impact on the speed of sound and hence on the Mach number. Adjusting the operating conditions (RPM and mass flow) such that in two tests both the inlet relative Mach number and the flow angles are unchanged, eliminates the impact of compressibility and incidence change on performance. For a given geometry, the conservation of Mach number relates the RPM at test conditions to the one at reference conditions,

$$RPM = RPM_{ref} \frac{\sqrt{(\kappa R_G T_1)}}{\sqrt{(\kappa R_G T_1)_{ref}}} \tag{1.102}$$

$T_1/T_{1,ref} = T_1^o/T_{1,ref}^o$ when testing at unchanged inlet Mach numbers with the same gas, and Equation (1.102) reduces to

$$RPM = RPM_{ref} \sqrt{\theta} \tag{1.103}$$

where $\theta = T_1^o/T_{1,ref}^o$.

A change in the inlet pressure impacts on the fluid density and hence on the mass flow. The latter is defined by

$$\dot{m} = A_1 \rho_1 V_{m,1} = A_1 \frac{P_1}{R_G T_1} M_{V_{m,1}} \sqrt{\kappa R_G T_1} = A_1 P_1 M_{V_{m,1}} \sqrt{\frac{\kappa}{R_G T_1}} \tag{1.104}$$

Imposing the same similarity conditions as for the RPM (i.e. the same value of $M_{V_{m,1}}$) defines the mass flow at reference conditions as a function of the measured one:

$$\dot{m}_{ref} = \dot{m}\frac{P_{ref}\sqrt{\left(\frac{\kappa}{R_G T_1}\right)_{ref}}}{P_1\sqrt{\left(\frac{\kappa}{R_G T_1}\right)}} \tag{1.105}$$

When testing with the same gas, this reduces to

$$\dot{m}_{ref} = \dot{m}\frac{P_{ref}}{P_1}\sqrt{\frac{T_1^o}{T_{ref}^o}} = \dot{m}\frac{\sqrt{\theta}}{\delta} \tag{1.106}$$

where $\delta = P_1/P_{ref}$. Under these conditions the pressure ratio is the same at both operating conditions.

Under some circumstances it may be easier or more economical to test a compressor in air instead of an expensive or toxic gas. Even when scaling the RPM according to previous rules, the performance will be different because the difference in isentropic exponent κ modifies the ratio between the pressure and density:

$$P/\rho^\kappa = C^{te} \tag{1.107}$$

At a given pressure ratio, the impeller inlet to outlet density ratio and hence the corresponding meridional velocity ratio will vary with κ. The corresponding change in diffusion ratio will not only have an impact on the impeller efficiency but also on the work input coefficient by changing the outlet velocity triangles. The original outlet over inlet velocity ratio can be reestablished by trimming the impeller outlet and diffuser inlet width to compensate for the difference in outlet over inlet density ratio.

Assume that the outlet width of the compressor, tested in air, has been adjusted to reestablish the initial velocity ratio such that the outlet velocity and hence also the velocity triangles are the same as in the compressor operating in a gas. At unchanged Mach number, the total temperature rise at both test conditions is then related by

$$\frac{(C_p\Delta T^o)_{air}}{(C_p\Delta T^o)_{gas}} = \frac{(U_2 V_{u2} - U_1 V_{u1})_{air}}{(U_2 V_{u2} - U_1 V_{u1})_{gas}} = \frac{(\kappa R_G T_1)_{air}}{(\kappa R_G T_1)_{gas}} = \frac{RPM_{air}^2}{RPM_{gas}^2} \tag{1.108}$$

After rearranging the terms and assuming the same T_1^o we obtain

$$\frac{RPM_{air}^2}{RPM_{gas}^2} = \frac{(\kappa R_G)_{air}}{(\kappa R_G)_{gas}}\frac{F(\kappa, M_{V_{m,1}})_{gas}}{F(\kappa, M_{V_{m,1}})_{air}} \tag{1.109}$$

where $M_{V_{m,1}}$ is the Mach number corresponding to the meridional component of the inlet velocity

$$F(\kappa, M_{V1}) = \frac{T_1^o}{T_1} = 1 + \frac{\kappa - 1}{2}M_{V1}^2$$

Equation (1.108) can now be written as

$$\frac{\Delta T_{air}^o}{\Delta T_{gas}^o} = \frac{C_{p,gas}}{C_{p,air}}\frac{RPM_{air}^2}{RPM_{gas}^2} = \frac{(\kappa - 1)_{air}}{(\kappa - 1)_{gas}}\frac{F(\kappa, M_{Vm1})_{gas}}{F(\kappa, M_{Vm1})_{air}} \tag{1.110}$$

Except for the eventual influence of a difference in Reynolds number, the same diffusion, Mach number, and flow angles allow us to conclude that the isentropic efficiency (Equation (1.7)) will be the same. Together with Equation (1.110) this provides the following relation between the stage pressure ratio of the compressors when the modified compressor is tested with a gas of different isentropic exponent:

$$\frac{\left(\pi^{\frac{\kappa-1}{\kappa}} - 1\right)_{air}}{\left(\pi^{\frac{\kappa-1}{\kappa}} - 1\right)_{gas}} = \frac{(\kappa-1)_{air}}{(\kappa-1)_{gas}} \frac{F(M_{Vm1})_{gas}}{F(M_{Vm1})_{air}} \tag{1.111}$$

For an unchanged inlet geometry, the inlet to outlet velocity ratio and hence also the degree of reaction and outlet flow angles will be the same if the outlet width is trimmed by the inverse of the respective density ratios. Relating the latter to the pressure ratio by (Equation (1.11)) and considering that the impeller pressure ratio is related to the stage pressure ratio by the degree of reaction r:

$$\frac{b_{2air}}{b_{2gas}} = \frac{\rho_{2gas}}{\rho_{2air}} = \frac{(1 + r(\pi - 1))_{air}^{\frac{\kappa(1-\eta_p)-1}{\kappa\eta_p}}}{(1 + r(\pi - 1))_{gas}^{\frac{\kappa(1-\eta_p)-1}{\kappa\eta_p}}} \tag{1.112}$$

The procedure is as follows: starting from an estimation of the isentropic efficiency η_{T-T}, total pressure ratio π_{T-T}, and given inlet temperature T_1^o we can estimate the total temperature rise ΔT_{air}^o of the trimmed impeller from the definition of efficiency (Equation (1.7)). Imposing for the testing in gas the same Mach number as in air, the corresponding ΔT_{gas}^o is easily obtained from Equation (1.110). Assuming further the same polytropic efficiency and degree of reaction for both operating conditions the required change of impeller outlet/diffuser inlet width when testing in a different gas is obtained from Equation (1.112).

Listed in Table 1.2 are the required changes of b_2 when testing a compressor designed for air ($\kappa = 1.4$ and $C_p = 1004.8$) in a gas ($\kappa = 1.14$ and $C_p = 612$) for different values of the total pressure ratio when Mach number and flow angle similarity are imposed. It is assumed that the impeller polytropic efficiency $\eta_p = 1.0$, $\beta_1 = 60°$ and that the inlet Mach number changes with pressure ratio, as defined in Figure 2.1. For moderate pressure ratios (up to $\pi = 1.5$) the required change of impeller outlet width is less than 1%. This means that the assumption of identical flow angles and unchanged efficiency may still be acceptable and no change of geometry is required. The required change, however, increases with increasing pressure ratio to reach 10.5% at a pressure ratio of 4. The ratio of the rotational speed only slightly changes between 0.462 and 0.469 for increasing inlet Mach number.

An alternative way to reestablish the original diffusion without modification of the geometry is to adjust the pressure ratio to conserve the original outlet over inlet density ratio. The corresponding pressure ratio is obtained by imposing $(b_2)_{air}/(b_2)_{gas} = 1$ in Equation (1.112).

Table 1.2 Variation of pressure ratio, RPM and outlet width when testing compressors in different gases ($\kappa_{gas} = 1.14$ and $\kappa_{air} = 1.4$, $T_1^o = 293.3$ K, $\eta_{T-T} = 0.80$, $r = 0.6$, $\eta_{p,r} = 1.0$).

π_{air}	$M_{W,1}$	RPM (gas/air)	b_2 (gas/air)	π_{gas}
1.2	0.347	0.462	0.997	1.16
1.5	0.526	0.463	0.990	1.41
2.0	0.702	0.465	0.974	1.84
3.0	0.911	0.467	0.935	2.75
4.0	1.046	0.469	0.894	3.74

Table 1.3 Variation of the performance and flow characteristics for an unchanged density ratio in an impeller with unchanged geometry ($\kappa_{gas} = 1.14$ and $\kappa_{air} = 1.4$, $T_1^o = 293.3$ K, $\eta_{T-T} = 0.80, r = 0.6, \eta_{p,r} = 1.0$).

π_{air}	$M_{W1}(air)$	RPM (gas/air)	π_{gas}	$M_{W1}(gas)$	RPM(gas/gas)
1.2	0.347	0.459	1.16	0.308	0.979
1.5	0.526	0.456	1.38	0.466	0.970
2.0	7.020	0.451	1.75	0.623	0.956
3.0	0.911	0.444	2.44	0.809	0.935
4.0	1.046	0.438	3.09	0.928	0.919

Neglecting an eventual change of isentropic stage efficiency, the total temperature rise ΔT^o can be derived from its definition (Equation (1.7)) and the ratio between the respective RPM results from Equation (1.108).

We should be aware, however, that when adjusting the RPM to obtain the same density ratio, the Mach number and Reynolds number will change, which will have an impact on efficiency and incidence range and hence modify the surge and choking limits. The Reynolds number effect can be accounted for as explained in Section 1.2. Results are shown in Table 1.3. The relative inlet Mach number in air, corresponding with the pressure ratio in the first column, is listed in the second column. The third column lists the RPM ratio required to obtain the same impeller inlet to outlet density ratio. The corresponding pressure ratio is listed in the fourth column. The main drawback of this approach is that the Mach numbers will no longer be conserved and the efficiency might be different, in particular when approaching transonic inlet flow. The last column shows the ratio between the RPM for the unchanged density ratio and impeller outlet width over the RPM for the unchanged Mach number with adjusted impeller outlet width, both operating in the gas.

Gases with a low speed of sound have been used to test transonic and supersonic impellers at lower RPM than in air. The main disadvantage is the required adjustment of the impeller outlet width. Research has been carried out to find gas mixtures that have a lower speed of sound but the same isentropic exponent κ so that no modification of the geometry would be required (Chapman 1954; Block et al. 1972). Problems encountered are the uncertainty about the exact composition of the test gas and an eventual contamination by air in the test facility, leading to reduced accuracy.

More information and examples about testing in different gases have been reported by Roberts and Sjolander (2005), von Backström (2008), and Hartmann and Wilcox (1957). The influence of gas characteristics and how to handle real gas effects are discussed by Lüdtke (2004).

2

Compressor Inlets

2.1 Inlet Guide Vanes

When imposing constant specific speed, an increase in the pressure ratio will require a proportional increase in ΔH and hence in RPM or size. The corresponding increase in the inducer shroud relative Mach number M_{W1} has been estimated by Dean (1972) for different values of NS_C and pressure ratios (Figure 2.1). It can be seen that for optimal values of NS_C (103 for dimensional NS_C or 0.8 for the non-dimensional NS_1) the critical value $M_{W1} = 1$ will already be reached at $\pi = 4$ and that for a compressor with $\pi = 9$ the shroud inlet relative Mach number will be as high as 1.4.

The supersonic inlet Mach number will not only give rise to additional shock losses, but the shock boundary layer interaction will increase the throat blockage and may induce early flow separation, resulting in even higher losses. As illustrated in Figure 2.23, the operating range between surge and choking will also be very small. A possible solution for this problem is to reduce the specific speed by lowering the RPM (Figure 2.1).

A decrease in the RPM must be compensated for by an increase in the impeller exit radius to maintain the same value of U_2. The specific speed is no longer the optimal one. The flow channel will be longer and narrower, involving additional shroud leakage, friction, and secondary flow losses.

Another possibility to lower the leading edge Mach number is to introduce preswirl vanes, as illustrated by the velocity triangles in Figure 2.2. A turning of the flow in the direction of

Figure 2.1 Influence of specific speed and pressure ratio on $M_{W,1S}$ (from Dean (1972) ($NS_C = 129.01NS_1$).

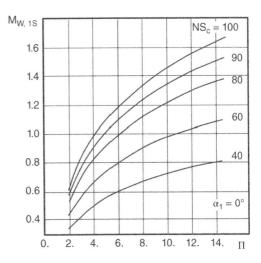

Design and Analysis of Centrifugal Compressors, First Edition. René Van den Braembussche.
© 2019, The American Society of Mechanical Engineers (ASME), 2 Park Avenue, New York, NY, 10016, USA (www.asme.org).
Published 2019 by John Wiley & Sons Ltd.

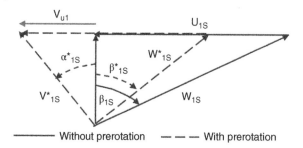

Figure 2.2 Influence of prerotation on W_{1S}.

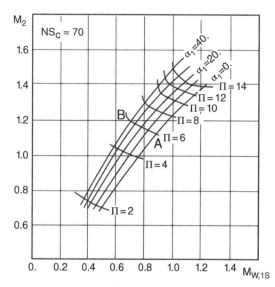

Figure 2.3 Influence of prerotation and pressure ratio on $M_{W,1S}$ and M_2 (from Morris and Kenny 1972).

rotation (positive preswirl) results in a noticeable decrease in the relative velocity component and relative flow angle. The Euler momentum equation (Equation (1.44)) $\Delta H = U_2 V_{u2} - \tilde{U}_1 \tilde{V}_{u1}$ indicates a decrease in the impeller energy input for positive values of the inlet tangential velocity $\tilde{V}_{u1}$. Consequently, an increase in the rotor exit swirl (V_{u2}) or peripheral velocity (U_2) will be necessary to maintain the same pressure ratio. Hence, the impeller outlet Mach absolute number will increase. The consequence of this approach is illustrated in Figure 2.3. It can be seen that for $\pi = 6$ and an increase in the inlet flow prerotation angle α_1 from 0° to 40°, the inducer tip Mach number decreases from 0.9 (point A) to 0.7 (point B), while the impeller outlet absolute Mach number increases from 1.12 to 1.2.

It has been observed by Kenny (1972) that a supersonic flow at the vaned diffuser inlet has no detrimental effect on performance. The supersonic flow at the impeller outlet can eventually be decelerated in a vaneless diffuser to subsonic at the vaned diffuser leading edge if the outlet-to-inlet radius ratio of the vaneless space is sufficiently large. This, however, will result in increased friction losses and, as explained in Section 8.2, can limit the stable operating range.

It is known from supersonic compressor research (Breugelmans 1972) that a normal shock at moderate Mach number is not an inefficient diffusion system. This is illustrated in Figure 2.4, where the downstream over upstream total pressure ratio due to a normal shock at different upstream Mach numbers is compared with the total pressure ratio corresponding to different total pressure loss coefficients ω:

$$\omega = \frac{P_u^o - P_d^o}{P_u^o - P_u}$$

Figure 2.4 Comparison between shock losses and diffusion losses (from Breugelmans 1972).

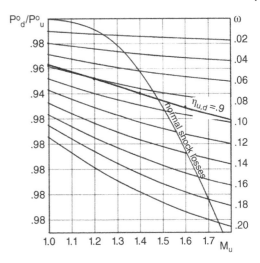

Up to Mach 1.4 the losses are less than 6% of the upstream dynamic pressure and hence are comparable to those of a well-designed stator. In conclusion, the amount of preswirl created by the inlet guide vane (IGV) should be defined by a compromise between the Mach-number dependent losses in the impeller and the diffuser, and on operating range considerations.

In addition to a change in the impeller inlet Mach number, variable IGVs can also be used to adjust the mass flow in order to increase the operating range. Optimum impeller performance is expected when the relative inlet velocity is nearly parallel to the blades at the leading edge. Mass flow variation around this point is limited by the increasing diffusion and separation losses or choking at positive or negative incidence, respectively.

Figure 2.5 shows how zero incidence can be obtained for different values of the mass flow by changing the rotational speed (a) or by means of prerotation (b, c) at constant RPM.

Changing the mass flow by adjusting the RPM (Figure 2.5a) requires a variable speed motor and has a direct influence on both U_2 and V_{u2} and therefore also a large impact on the energy input and pressure rise (Equation (1.44)). U_1 and U_2 remain unchanged in the case of prerotation and, as shown in next paragraph, it is possible to design impellers in such a way that the energy input and hence also the pressure ratio is nearly independent of the prerotation.

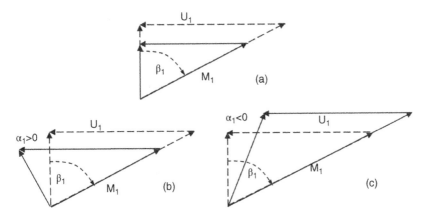

Figure 2.5 Variation of mass flow at zero incidence with (a) changing RPM, (b) positive prerotation, and (c) negative prerotation.

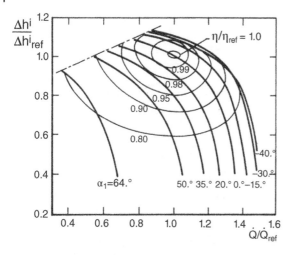

Figure 2.6 Influence of prerotation on the performance of a compressor with vaneless diffuser (from Simon et al. 1986).

The typical impact of prerotation on the operating curve of a centrifugal compressor is illustrated in Figure 2.6. Both mass flow and pressure ratio decrease with positive prerotation (Figure 2.5b). In addition to the decrease in energy input, specified by the Euler momentum equation, the pressure ratio may be limited by the diffuser instability resulting from a more tangential inflow velocity. As a consequence the point of maximum pressure rise may no longer be reachable. At negative prerotation, the increase in mass flow and pressure ratio is limited by choking due to the increase in the inlet relative Mach number (Figure 2.5c).

2.1.1 Influence of Prerotation on Pressure Ratio

It is possible to obtain a nearly constant pressure ratio for different values of the prerotation if an increase in diffuser losses at decreasing mass flow can be avoided and by defining the impeller in such a way that the energy input per unit mass flow increases with decreasing mass flow. In impellers with backward leaning blades ($\beta_2 > 0$) and prerotation vanes, a change in mass flow not only changes V_{u1} but also V_{u2} (Figure 2.7). We can design the impeller in such a way that the latter compensates the effect of prerotation on work input.

Assume that at all values of prerotation, we always operate at the best efficiency point, i.e. the relative inlet velocity is parallel to the leading edge (Figure 1.27). Starting from the following relation between prerotation and mass flow

$$V_{u1} = U_1 - V_{m1} \tan \beta_1 \tag{2.1}$$

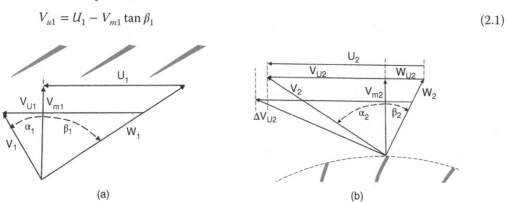

Figure 2.7 Variation of (a) inlet and (b) outlet velocity triangles with prerotation and mass flow.

and assuming further that the impeller relative outlet flow angle β_2 is not influenced by prerotation, provides the outlet tangential velocity

$$V_{u2} = U_2 - V_{m2} \tan \beta_{2,fl} \qquad (2.2)$$

Substituting both expressions into the Euler momentum equation (Equation (1.44)) and correlating the inlet and outlet meridional velocity by the continuity relation

$$\rho_1 V_{m1} A_1 = \rho_2 V_{m2} A_2 \qquad (2.3)$$

results in

$$\Delta H = U_2 \left(U_2 - \frac{\rho_1 A_1}{\rho_2 A_2} V_{m1} \tan \beta_{2,fl} \right) - U_1 \left(U_1 - V_{m1} \tan \beta_1 \right) \qquad (2.4)$$

The specific energy input (and at unchanged efficiency also the pressure rise) will be independent of the mass flow if:

$$\frac{d(\Delta H)}{d V_{m1}} = 0 \qquad (2.5)$$

hence when

$$U_2 \frac{\rho_1 A_1}{\rho_2 A_2} \tan \beta_{2,fl} - U_1 \tan \beta_1 = 0 \qquad (2.6)$$

Using Equation (2.3) we obtain

$$U_2 V_{m2} \tan \beta_{2,fl} = U_1 V_{m1} \tan \beta_1 \qquad (2.7)$$

as a condition for the impeller flow to produce a constant specific work input at different mass flows. The results are shown in Table 2.1.

Constant energy input will result in a constant pressure ratio if the modified flow conditions in the vaneless or vaned diffuser (Figure 2.7) do not create extra losses or destabilize the flow. In some cases this may require a variable vaned diffuser to adapt the geometry to the modified inlet flow (Simon et al. 1986) and results in operating curves like those shown in Figure 9.34.

2.1.2 Design of IGVs

A first type of IGV is similar to the ones used in axial compressors and turbines. Blades in the axial inlet duct deflect the flow in the tangential direction. These IGVs can be fixed or rotatable around radial axes (variable IGV), as shown in Figure 2.8. The shroud contour where the vanes are located should be shaped as part of a sphere to allow vane rotation with minimum tip clearance.

Table 2.1 Outlet flow angles for constant energy input at typical values of U_1/U_2 and V_{m1}/V_{m2}.

| $U_1/U_2 = 0.5$ | | $U_1/U_2 = 0.5$ | | $U_1/U_2 = 0.3$ | |
| $V_{m1}/V_{m2} = 1.0$ | | $V_{m1}/V_{m2} = 1.5$ | | $V_{m1}/V_{m2} = 1.0$ | |
β_1	$\beta_{2,fl}$	β_1	$\beta_{2,fl}$	β_1	$\beta_{2,fl}$
40	22.7	40	32.2	40	20.6
50	30.7	50	41.8	50	28.2
60	40.9	60	52.4	60	37.9

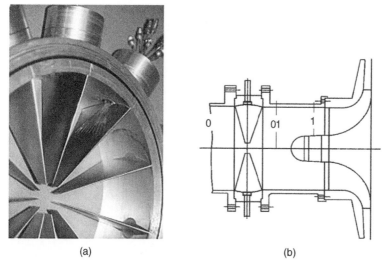

(a) (b)

Figure 2.8 (a) Geometry of variable IGV and (b) inlet cross section.

Different spanwise swirl distributions are possible (constant prerotation angle, forced vortex, free vortex, etc.). Typical variable IGV geometries have straight symmetric blades with a decreasing chord towards the hub such that the inlet channel can be closed by setting the vanes at a 90° stagger. Blocking the inlet channel during startup creates an underpressure and facilitates a quick acceleration of the impeller by limiting the power required. The vanes are then progressively opened once the impeller approaches nominal rotational speed.

The swirl velocity at the impeller inlet depends on the blade shape as well as on the shape of the inlet channel (Figures 2.8 and 2.9). The impeller inlet meridional velocity is defined by continuity:

$$\int_{R_{H,1}}^{R_{S,1}} \rho_1 V_{m,1} dR = \int_{R_{H,01}}^{R_{S,01}} \rho_{01} V_{m,01} dR \tag{2.8}$$

Any contraction of the inlet duct gives rise to an increased meridional velocity. The swirl velocity is defined by the conservation of angular momentum along the streamlines:

$$R_1 V_{u,1} = R_{01} V_{u,01} \tag{2.9}$$

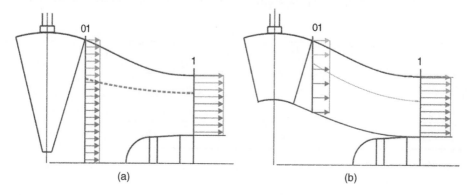

(a) (b)

Figure 2.9 Impact of (a) the meridional cross section and (b) the central body on the impeller inlet flow.

The increasing axial velocity towards the impeller inlet in the geometry without central core (Figure 2.8) in combination with a constant swirl velocity near the shroud results in a decrease in the prerotation angle between positions 01 and 1. An even larger reduction in the swirl velocity occurs near the hub where the small prerotation in section 01, because of poor guidance of the flow near the central hole, is further reduced by the increase in radius towards the impeller inlet according to the conservation of angular momentum RV_u = constant.

These phenomena are partially compensated for in geometries with a decreasing shroud radius (Figure 2.9a). The reduction in the swirl angle by the large increase in the meridional velocity is compensated for near the shroud by the increase in swirl velocity resulting from the reduction of the radius. The swirl at the hub is again very small and the outcome is an increased spanwise variation of the swirl at the impeller inlet.

The use of IGVs with a central core (Figure 2.9b) allows the conservation of or even an increase in prerotation angle between the IGV outlet and the impeller inlet. The change in meridional velocity depends on the change in flow area. The swirl velocity increases by reducing the radial position of the flow. The increase in the swirl angle is largest near the hub where the relative change of radius is the largest and the meridional velocity may be lowest because of the concave curvature of the meridional contour. A central core provides a lot of freedom in the definition of the spanwise swirl distribution at the impeller inlet. However, it requires some upstream struts to keep it in place, which results in some extra wake losses.

Centrifugal compressors with a radial inlet may have the IGV located in the radial part (Figure 2.10). The advantage is that only small deflections by the IGVs are required to achieve large tangential velocities at the impeller inlet. $V_{u,1}$ increases towards the impeller inlet because of the decreasing radius. The resulting increase in the swirl velocity is particularly large at the hub (because $R_{H,1} \ll R_{IGV}$) where the meridional velocity can also be very small because of the meridional curvature. One should therefore be careful in defining this type of IGV. The impeller hub prerotation angle can be reduced by shortening the chord of the vanes at the hub side of the inlet section. The larger pitch over chord radius results in a smaller deflection of the flow.

Losses in the IGV are very small compared to losses in other parts of the compressor. They are almost negligible except when separation occurs at high deflections. Most IGVs have a rather small thickness to limit the losses at the design point (zero swirl). The consequence is a rapidly increasing velocity peak at the leading edge suction side and a massive flow separation is likely to occur once the stagger angle exceeds 30°.

The design target is to postpone the suction side separation to larger vane setting angles. One way to achieve this is by means of variable geometry vanes (flexible vanes or tandem vanes with

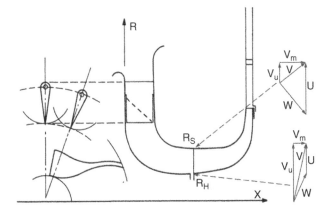

Figure 2.10 IGV in the radial part of the compressor inlet (from Rodgers 1977).

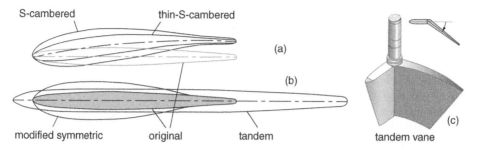

Figure 2.11 Typical IGV blade geometries: (a) and (b) fixed geometry vanes and (c) tandem vane (from Mohseni et al. 2012).

a rotating second part) (Figure 2.11c). This allows zero incidence to be conserved when turning the flow.

Negative preswirl (against the impeller rotational direction) results in an increase in the relative inlet Mach number (Figure 2.5c) and the expected increase in mass flow is often limited by choking problems. Hence variable IGVs are mostly used to create positive preswirl. In this case there is a possibility of postponing flow separation by using asymmetric blades (Figure 2.11a). These blades are cambered to make them better suited to create positive preswirl. The unfavorable effect this will have at negative prerotation is of less importance because IGVs are rarely used to increase the mass flow. However, one should take care that the asymmetry of the blades does not increase the losses at zero preswirl. S-shaped camberlines have been proposed for this purpose (Sanz et al. 1985; Doulgeris et al. 2003).

The iso-Mach surfaces at 10%, 50%, and 90% span of three different inlet guide vanes are shown in Figure 2.12 at selected stagger angles between −20° and +60°. The iso-Mach surfaces of the "original symmetric vane" show a massive flow separation at the hub for vane setting angles larger than 30°. The flow remains attached to the thicker S-cambered vane at all positive stagger angles. At −20° stagger, the flow shows a large separated flow zone at the hub and a small one at the tip.

The iso-Mach surfaces of the tandem blades show attached flow at all setting angles. The flow is identical at positive and negative setting angles. One reason for the absence of flow separation even at very high turning angles is the zero incidence at all operating points. The turning is entirely achieved inside the channel and in combination with the high solidity, due to the long blade chord, a good guidance is provided and the flow remains attached. The flow acceleration on the suction side is no longer followed by a deceleration because the downstream flow is at the same high velocity.

The pressure loss of the different blades, non-dimensionalized by the inlet kinetic energy, are compared in Figure 2.13. A magnification of the data near the zero setting angle illustrates the small difference in losses at this setting angle. The best performance is obtained with the thick S-cambered vane with optimized velocity distribution. These blades show only a small increase in losses at zero setting angle and the smallest increase in losses when increasing the stagger angle up to 60°.

The second best performance is obtained with the tandem vanes. However, one should take into account that this variable geometry blade has a much longer chord length and is more complex (expensive) than other ones. Furthermore, a rather large torque is needed to put the vanes at non-zero stagger because the pivot point is at the leading edge of the rotatable part.

The modified symmetric profile, an uncambered blade with the same (larger) thickness distribution as the optimized S-cambered blade, shows losses that are comparable to those of the

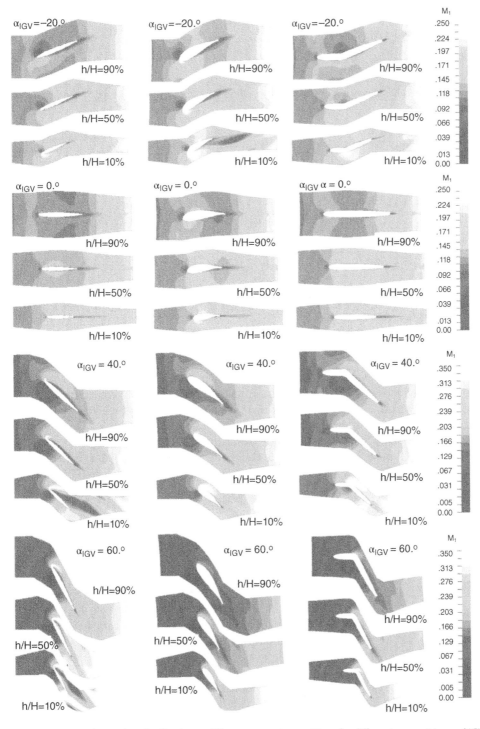

Figure 2.12 Mach number distribution at different spanwise positions for different geometries and IGV setting angles (from Mohseni et al. 2012).

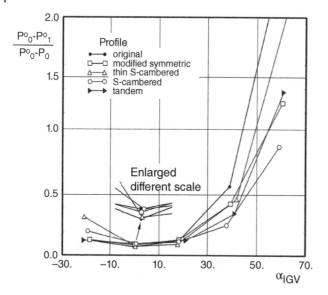

Figure 2.13 Variation in IGV losses as a function of turning angle and vane shape for the meridional contour of Figure 2.9b (from Mohseni et al. 2012).

tandem vanes. A decrease in the trailing edge thickness does not show a measurable effect on the losses.

A thin S-cambered vane, with the same camber line as the optimized one but with a smaller thickness, shows an increase in the losses to the same level as the thicker symmetric vane up to a 40° turning. Higher turning provokes a rapid increase in the losses because of flow separation. The zoom of the loss distribution around the zero stagger angle indicates a small decrease in the losses at zero stagger when using the thinner vanes.

A second aspect of flow separation is its impact on the spanwise flow distortion at the impeller inlet. The centrifugal forces, resulting from the swirling flow downstream of the IGV, generate a pressure rise towards the outer wall:

$$\frac{dP}{dR} = \rho \frac{V_u^2}{R} \tag{2.10}$$

As a consequence the low energy fluid in the separated flow zone, with a small V_u, is pushed towards the hub side of the inlet channel. The result is a transformation of the pitchwise wake distribution into a spanwise distortion of the impeller inlet flow with low energy fluid (low P^o) at the hub (Figure 2.14). This redistribution of the low energy fluid depends on the distance between the IGV and the rotor inlet, and the amount of flow separated from the IGV. The change in flow pattern between the IGV exit (section 01) and the impeller inlet (section 1) is illustrated in Figures 2.15 to 2.18. The low total pressure fluid of the wakes in section 01 has accumulated near the hub in section 1. The almost uniform static pressure in section 01 shows a radial pressure gradient in section 1, satisfying Equation (2.10). Fluid with a large tangential velocity has accumulated near the shroud in section 1. The high static and total pressure in combination with the large swirl velocity near the shroud results in a low through flow velocity defined by:

$$\frac{\rho V_m^2}{2} = P^o - P - \frac{\rho V_T^2}{2} \tag{2.11}$$

The axial velocity near the hub is also larger due to the acceleration around the impeller nose.

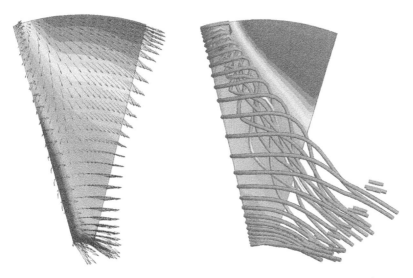

Figure 2.14 Velocity vectors and streamlines near the suction side at $\alpha_{IGV} = 75°$.

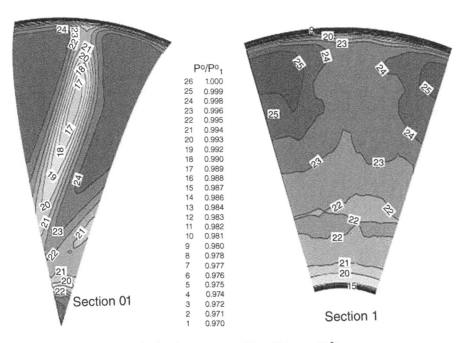

	Po/Po$_1$
26	1.000
25	0.999
24	0.998
23	0.996
22	0.995
21	0.994
20	0.993
19	0.992
18	0.990
17	0.989
16	0.988
15	0.987
14	0.986
13	0.984
12	0.983
11	0.982
10	0.981
9	0.980
8	0.978
7	0.977
6	0.976
5	0.975
4	0.974
3	0.972
2	0.971
1	0.970

Figure 2.15 Total pressure distribution at sections 01 and 1 ($\alpha_{IGV} = 20°$).

An interesting observation is that the Navier–Stokes calculations of the flow in the IGV and inlet duct are a good approximation of the full stage calculations if the radial equilibrium is used to define the spanwise static pressure variation at the exit boundary of the numerical domain. Imposing a uniform static pressure would result in an erroneous static pressure rise near the hub wall, resulting in very high total pressure losses because of return flow near the hub. In the absence of flow separation or return flow the upstream influence of the impeller is negligible, and a correct analysis of the IGV and the inlet duct is possible without considering also the

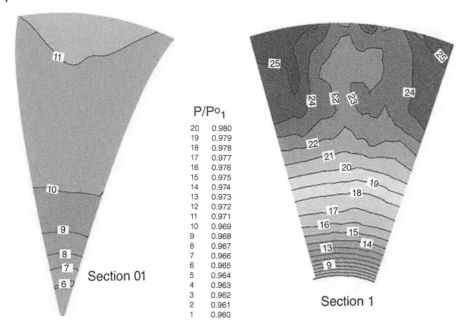

P/Po₁ legend:

20	0.980
19	0.979
18	0.978
17	0.977
16	0.976
15	0.975
14	0.974
13	0.973
12	0.972
11	0.971
10	0.969
9	0.968
8	0.967
7	0.966
6	0.965
5	0.964
4	0.963
3	0.962
2	0.961
1	0.960

Figure 2.16 Static pressure distribution at sections 01 and 1 ($\alpha_{IGV} = 20°$).

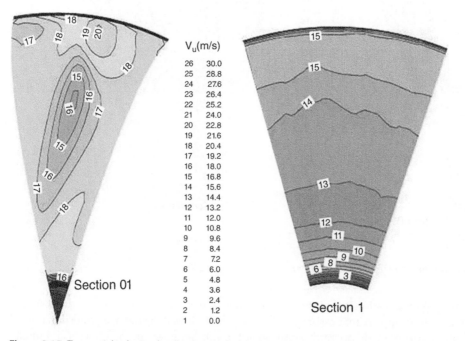

V_u(m/s) legend:

26	30.0
25	28.8
24	27.6
23	26.4
22	25.2
21	24.0
20	22.8
19	21.6
18	20.4
17	19.2
16	18.0
15	16.8
14	15.6
13	14.4
12	13.2
11	12.0
10	10.8
9	9.6
8	8.4
7	7.2
6	6.0
5	4.8
4	3.6
3	2.4
2	1.2
1	0.0

Figure 2.17 Tangential velocity distribution at sections 01 and 1 ($\alpha_{IGV} = 20°$).

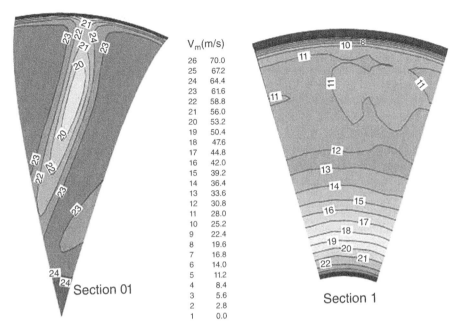

V_m (m/s)	
26	70.0
25	67.2
24	64.4
23	61.6
22	58.8
21	56.0
20	53.2
19	50.4
18	47.6
17	44.8
16	42.0
15	39.2
14	36.4
13	33.6
12	30.8
11	28.0
10	25.2
9	22.4
8	19.6
7	16.8
6	14.0
5	11.2
4	8.4
3	5.6
2	2.8
1	0.0

Figure 2.18 Axial velocity distribution at sections 01 and 1 ($\alpha_{IGV} = 20°$).

impeller geometry, at the condition that the exit boundary conditions are defined by the radial equilibrium between the hub and the shroud (Equation (2.10)).

2.2 The Inducer

The inducer has a large impact on the impeller performance. A non-optimal inducer will accelerate the fluid more than required and the larger subsequent deceleration will create extra losses and may even provoke an early flow separation. This gives rise to increased flow distortion and mixing losses downstream of the impeller.

Efficiency, however, is not the only criterion in the design of a compressor. Most applications also require a sufficiently large operating range between the surge and choking limits. Although the inducer inlet geometry is not the only parameter defining range, it certainly has an important influence on it. The incidence range decreases with Mach number and one of the design criteria for inducers is therefore to minimize the inlet relative Mach number and if possible to keep it subsonic.

The centrifugal impeller inducer differs from an axial compressor in several ways:

- a 60° flow turning between the inlet and axial parts is not exceptional and much more than what is achieved in axial compressors
- the blades do not terminate in the axial direction but continue in the radial direction so that the blade loading does not have to be zero at the end of the axial part (as required in axial blade rows by the Kutta conditions); loading can continue up to the blade trailing edge
- the blade solidity (chord length over pitch) is higher at the inlet and the blade thickness creates a higher blockage.

Typical radial compressor geometries (Figure 1.26) are within following limits:

- R_{1S}/R_2 is generally between 0.5 and 0.8.
  0.5 results in long channels with large friction and eventual tip leakage losses.
  0.8 results in an unsatisfactory meridional contour with an insufficient length to diffuse the flow along the shroud.
  The optimum is around 0.6 to 0.65.
- b_2/R_2 is ideally between 0.05 and 0.15 depending on the specific speed. It is a function of the mass flow and pressure ratio.
  0.05 is the lower limit to avoid too large friction and clearance losses in narrow channels. This value is typical for low *NS* impellers.
  0.15 is a maximum value to limit the diffusion in the impeller and to avoid large flow separation.
- R_{1H}/R_{1S} depends on the volume flow
  A too small value results in too much blockage at the hub section by the blade thickness. The latter is defined by stress and manufacturing considerations, and increases with decreasing value of R_{1H}/R_{1S}.
  The minimum hub radius may further be limited in multistage compressors by the minimum shaft diameter required for the torque transmission or by rotor dynamic considerations.
  Too high values reduce the inlet section and result in higher inlet velocities.
  The optimal value is between 0.3 and 0.7.
- β_{1S} should be below 70° to avoid too much blockage by the blade thickness and too much turning and losses in the inducer. Optimal values (between 50° and 60°) are related to minimizing the shroud leading edge Mach number.
- α_2 should be between 65° and 80°. The highest value is acceptable for high pressure ratio compressors with vaned diffusers but may result in high friction losses and instabilities in vaneless diffusers.
- M_{W1} is preferably less than 1, otherwise a special transonic inducer will be required.
- M_{V2} influences the compressor operating range when using vaned diffusers. The operating range decreases with increasing M_{V2} but values as high as 1.2 to 1.4 are acceptable in high pressure ratio vaned diffusers.
- $\tilde{W}_2/\tilde{W}_1$ is the deceleration ratio of the impeller relative velocity. Too low values will result in impeller flow separation. Too high values result in a decrease in performance because of a lack of pressure rise in the impeller.

2.2.1 Calculation of the Inlet

The fundamental relations used in the following chapters are for ideal gases. For real gas calculations, one is referred to Lüdtke (2004).

The impeller inlet flow conditions T_1^o, P_1^o, α_1, the required mass flow $\dot{m}$, and the dimensions of the inlet section R_{1H} and R_{1S} are given. Guessing a first value for V_1 allows the calculation of the following static conditions for a uniform inlet flow:

$$T_1 = T_1^o - V_1^2/(2C_p) \tag{2.12}$$

P_1 can be related to P_1^o assuming an isentropic change of state in an accelerating inlet flow:

$$\frac{P_1}{P_1^o} = \frac{T_1}{T_1^o}^{\frac{\kappa}{\kappa-1}} \tag{2.13}$$

The density is then given by the equation of state, which provides the following expression for the mass flow:

$$\dot{m}_{calc} = \pi(R_{1,S}^2 - R_{1,H}^2)\rho_1 V_1 \cos\alpha_1 \tag{2.14}$$

The difference between this mass flow and the required one allows a more accurate estimation of the inlet velocity V_1 by an iterative procedure. This system will only converge if the absolute inlet Mach number $M_{V1} < 1$. A reduction of M_{V1} may be achieved by increasing the shroud radius R_{1S} or decreasing the hub radius R_{1H}. Previous procedure defines the static inlet flow conditions on the H–S diagram (Figure 1.35).

Very important for the impeller performance predictions are the relative flow conditions at the inducer tip. They make the main contribution to the incidence losses, transonic flow shock losses, and, as will be shown later, influence separation and diffusion losses.

$$\tan \beta_{1S} = \frac{\Omega R_{1S} - V_1 \sin \alpha_1}{V_1 \cos \alpha_1} \tag{2.15}$$

and

$$M_{W_{1S}} = \frac{V_1 \cos \alpha_1}{\cos \beta_{1S} \sqrt{\kappa R_G T_1}} \tag{2.16}$$

Previous calculations assume a uniform inlet velocity V_1 corresponding to a straight axial inlet channel. In the case of a non-axial inlet one should account for the spanwise velocity variation resulting from the meridional curvature in a way similar to what is explained for the impeller flow in Section 3.1. In that case P_1 and T_1 will vary from hub to shroud.

2.2.1.1 Determination of the Inducer Shroud Radius

The minimum hub radius $R_{1H_{min}}$ is fixed by mechanical considerations or aerodynamic blockage. The determination of R_{1S} is more complicated because one wants to minimize the relative inducer shroud Mach number M_{W1S}. In what follows the relation between these two parameters is illustrated with a centrifugal compressor operating in a low speed of sound gas with the following specifications: $\dot{m} = 2.5$ kg/s, $R_G = 75.3$ J/(kg K), $\kappa = 1.136$, $\rho_1 = 4$ kg/m³, $T_1^o = 283$ K.

The rotational speed RPM, the prerotation angle α_1 and the inducer hub radius R_{1H} are input parameters. The inducer shroud radius is defined by the parameter $RV = R_{1H}/R_{1S}$. The inducer shroud relative Mach number $M_{W_{1S}}$ can be expressed as follows:

$$M_{W_{1S}} = \frac{1}{\sqrt{\kappa R_G T_1}} \sqrt{\left(\frac{\dot{m}}{k_{b1} \pi \rho_1 R_{1H}^2 (\frac{1}{RV^2} - 1)}\right)^2 + \left(\frac{2\pi RPM}{60} \frac{R_{1H}}{RV} - \frac{\dot{m}}{k_{b1} \pi \rho_1 R_{1H}^2} \frac{\tan \alpha_1}{\frac{1}{RV^2} - 1}\right)^2} \tag{2.17}$$

where k_{b1} is a correction for the boundary layer blockage in the inlet section. The relation between the shroud relative Mach number and the radius ratio RV is plotted in Figure 2.19 for $RPM = 14\,000$ to $18\,000$, $R_{1H} = 0.04$ m, and $\alpha_1 = 0°$.

At 16 000 RPM the minimum value of $M_{W_{1S}} = 0.85$ is observed at an inducer hub/shroud radius ratio $RV = 0.59$. At smaller values of RV, the Mach number $M_{W_{1S}}$ increases due to the increase in the circumferential velocity U_{1S}. The Mach number $M_{W_{1S}}$ also increases at higher than optimal values of RV because the inlet flow section is reduced, with a subsequent increase in the axial velocity V_{1m}.

An important conclusion is that for a given value of RPM, α_1, and R_{1H}, it will not always be possible to find an RV value (say R_{1S} value) satisfying a preset limit on the relative tip Mach number. The only way to overcome this difficulty is to adjust one of the other parameters (RPM, α_1 or R_{1H}). Figures 2.19 and 2.20 show that a reduction in RPM or a positive preswirl results in a decrease in the inducer shroud relative Mach number.

Figure 2.21 shows only a slight decrease in the minimum relative Mach number when reducing the hub radius. The reason is that a change in this radius has only a very small impact on the inlet section and hence on V_{m1}, and $V_{u,1S}$ remains unchanged.

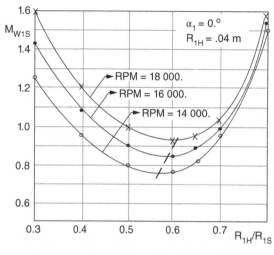

Figure 2.19 Influence of the RPM on the shroud relative inlet Mach number.

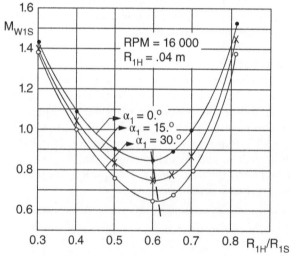

Figure 2.20 Variation of the shroud relative inlet Mach number with prerotation.

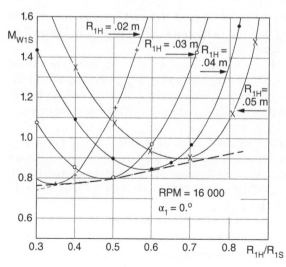

Figure 2.21 Influence of the hub radius on the shroud relative Mach number.

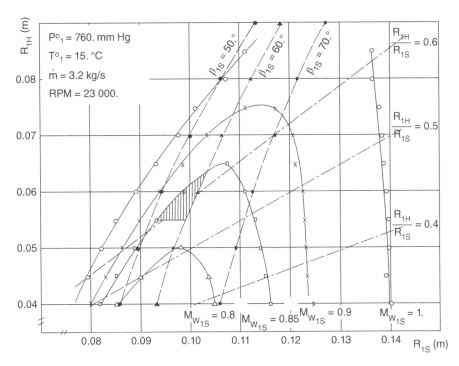

Figure 2.22 Influence of the hub and shroud radius and leading edge blade angle on the shroud relative Mach number.

The previous results are valid only for impellers with an axial inlet resulting in a uniform meridional velocity from hub to shroud. In multistage compressors, the flow enters radially and is turned to axial just upstream of the impeller. This results in a higher meridional velocity at the shroud and a lower one at the hub. This difference depends on the curvature of the inlet channel and the distance between shroud and hub. It can be calculated by the method explained in Chapter 3. Previous considerations about the optimum shroud radius should account for this hub to shroud variation.

Figure 2.22 summarizes the result of a systematic scanning of the design space for a compressor operating at the conditions shown on the graph. It presents a global view of the optimization of the impeller inlet for a minimum shroud relative Mach number. The hatched area defines the inlet geometry satisfying the following criteria: $R_H > 0.55$ m, $M_{1S} < 0.85$, and $\beta_{1,S} < 60°$.

2.2.2 Optimum Incidence Angle

Rodgers (1962) has evaluated the stall and choking incidence for three axial impellers with different leading edge blade angles (β_{1bl}). All three have an axial outlet ($\beta_2 = 0$) and hence different turning. The variation of the incidence limits with the relative inlet Mach number is plotted on Figure 2.23.

The operating range between choking and stall decreases significantly with increasing Mach number and depends on the leading edge blade angle. Smaller values of β_{1bl} show a larger range. However, one should keep in mind that the impellers used for Figure 2.23 turn the flow to the axial direction. Hence large β_1 means larger camber because of the axial outlet. The same figure also shows that the center of the operating range is between $0°$ and $15°$ positive incidence depending on the Mach number and blade angle. Turning the flow has a

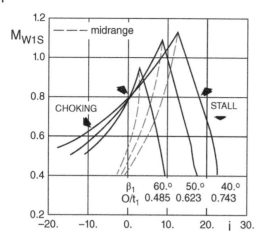

Figure 2.23 Inducer stall and choking incidence as a function of leading edge blade angle and inlet Mach number (from Rodgers 1962).

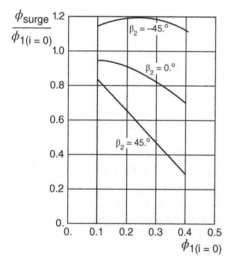

Figure 2.24 Ratio of surge flow over incidence-free inducer flow (from Baljé 1981).

potential effect upstream of the blades row. As shown by the NACA results and quantified in correlations (Lieblein 1960) a larger camber induces a more negative contribution to the optimum incidence. Blades with low inlet angles are less cambered, which explains the increase in the stall incidence for lower β_{1bl} values.

The trend of increasing operating range with smaller β_1 is confirmed by theoretical considerations by Baljé (1981) (Figure 2.24). The smallest values of $\phi_{surge}/\phi_{i=0}$ occur at the highest value of $\phi_{i=0}$, i.e. the mass flow range between the optimum incidence and surge is larger at low values of β_{1bl}. This trend is very pronounced for backward leaning blades ($\beta_{2bl} > 0$).

The positive mid-range incidence in Figure 2.23 results from blade blockage and can be very high at the inlet of centrifugal impellers. Blade blockage may amount to 10% at the shroud and 40% at the hub. Its influence on optimum incidence has been investigated by Stanitz (1953). The following calculations assume that the optimum incidence occurs at zero leading edge loading, i.e. with no change in absolute tangential velocity at the leading edge.

When the flow enters the impeller, the free frontal area is reduced by the presence of the blades with finite thickness. This contraction changes the velocity triangles (Figure 2.25) by increasing the meridional velocity component from V_{m1} to

$$V_{m1bl} = V_{m1}/(1 - \epsilon_{kb}/\cos\beta_{1bl})$$

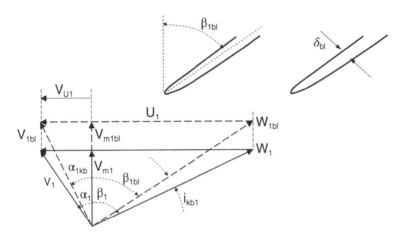

Figure 2.25 Influence of blade blockage on the inlet velocity triangle.

where

$$\epsilon_{kb} = \frac{\delta_{bl} Z_r}{2\pi R}$$

The relative flow will be turned over an angle i_{kb1} from β_1 to the value β_{1bl}. Imposing zero loading at the leading edge and neglecting the upstream influence of the blade camber, the zero-loading incidence i_{kb1} is defined by:

$$\tan i_{kb1} = \frac{\epsilon_{kb} \sin \beta_{1bl}}{1 - \epsilon_{kb} \cos \beta_{1bl}} \tag{2.18}$$

The variation of the zero-loading incidence as a function of β_{1bl} and blade blockage (ϵ_{kb}) is shown in Figure 2.26. The large values at the hub are due to the large blockage ϵ_{kb} resulting from the large blade thickness at the small radius. Lower blockage values (ϵ_{kb}) at the tip section result in large zero-loading incidence because of larger β_{1bl} values.

Figure 2.26 Zero leading edge loading incidence as a function of blade blockage.

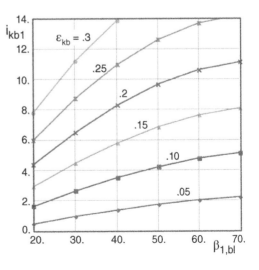

The combination of the potential upstream effect of the blade camber and the blade blockage defines the optimal incidence. The optimum blade leading edge angle is defined by

$$\beta_{1,bl} = \beta_{1,fl} - i_{kb1} \tag{2.19}$$

and the new relative velocity inside the impeller by

$$W_{1kb} = W_1 \sin \beta_1 / \sin \beta_{1bl} \tag{2.20}$$

The new axial velocity component is

$$V_{mkb} = V_m \tan \alpha_1 / \tan \alpha_{1bl} \tag{2.21}$$

The corresponding absolute flow angle α_{1kb} follows from:

$$\alpha_{1kb} = \arctan \left(\tan \alpha_1 \frac{\tan \beta_{1bl}}{\tan \beta_1} \right) = \arctan \left(\frac{U_1 + W_{1kb} \sin \beta_{1bl}}{W_{1kb} \cos \beta_{1bl}} \right) \tag{2.22}$$

The velocity V_{1kb} is :

$$V_{1kb} = \sqrt{W_{1kb}^2 + U_1^2 + 2W_{1kb}U_1 \sin \beta_{1bl}} \tag{2.23}$$

The new static temperature T_{1kb} can be derived from the energy equation, when accepting that no energy has been added to the fluid during the contraction:

$$T_{1kb} = T_1 + \frac{V_1^2 - V_{1kb}^2}{2C_p} \tag{2.24}$$

Supposing this is an isentropic process, the static pressure P_{1kb} becomes

$$P_{1kb} = P_1 \left(\frac{T_{1kb}}{T_1} \right)^{\frac{\kappa}{\kappa-1}} \tag{2.25}$$

The new density ρ_{1kb} follows from the ideal gas equation:

$$\rho_{1kb} = \frac{P_{1kb}}{R_G T_{1kb}} \tag{2.26}$$

It is important to notice in previous calculations that, since U_1 varies with the radius R, all quantities V_1, W_1, α_1, β_1, P_1, T_1, and ρ_1 are functions of R.

2.2.3 Inducer Choking Mass Flow

The maximum impeller inlet mass flow theoretically occurs when the flow is sonic in the inducer throat section. In reality, choking will already occur at lower mass flows because the incidence and blade curvature do not allow a uniform velocity in the throat section.

Boundary layer blockage on the hub and shroud contour and non-uniformity of the inlet flow (Figure 2.27), due to the wakes of the return vanes, further reduces the choking mass flow. The latter can be accounted for by a blockage coefficient.

The exact location of the throat section requires a detailed 3D knowledge of the inducer geometry and it is not feasible to do such a calculation in the context of a fast overall performance prediction. An approximate calculation of the throat section is possible by an estimation of the length O_{th} between the leading edge pressure side and the next blade suction side at different radii (Figure 2.28).

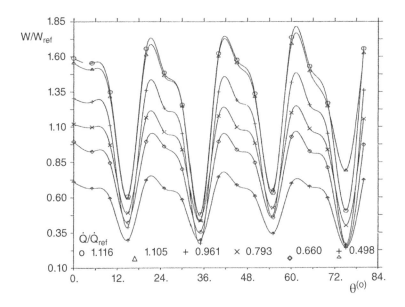

Figure 2.27 Circumferential variation of the velocity near the shroud wall downstream of return vanes for different mass flows (from Benvenuti et al. 1980).

Figure 2.28 Schematic view of the impeller throat section.

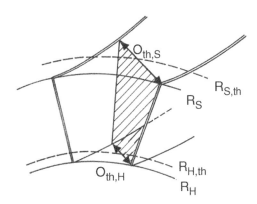

The mass flow in the throat is defined by following integral

$$\dot{m}_{th} = Z_r \int_{R_{H,th}}^{R_{S,th}} \rho_{th} W_{th} O_{th} dR_{th} \tag{2.27}$$

where both O_{th} and the flow conditions change with radius. This equation can also be written as:

$$\dot{m}_{th} = Z_r \int_{R_{H,th}}^{R_{S,th}} F(M_{th}) \frac{P_r^o(R_{th})}{\sqrt{T_r^o(R_{th})}} O(R_{th}) dR_{th} \tag{2.28}$$

with

$$F(M_{th}) = \sqrt{\frac{\kappa}{R_G}} M_{th} \frac{P(R_{th})}{P_r^o(R_{th})} \sqrt{\frac{T_r^o(R_{th})}{T(R_{th})}} = \sqrt{\frac{\kappa}{R_G}} \frac{M_{th}}{(1 + \frac{\kappa-1}{2} M_{th}^2)^{\frac{\kappa+1}{2(\kappa-1)}}} \tag{2.29}$$

Assuming that choking occurs at $M_{th} = 1$, Equation (2.29) reduces to:

$$F(M_{th} = 1) = \sqrt{\frac{\kappa}{R_G}} \left(\frac{2}{\kappa + 1} \right)^{\frac{\kappa+1}{2(\kappa-1)}} \tag{2.30}$$

Substituting $F(M_{th} = 1)$ in Equation (2.28) provides the theoretical choking mass flow whereby $T(R_{th})$ is calculated from constant rothalpy along the streamline between R_1 at the inlet and R_{th} at the throat section

$$T(R_{th}) = T(R_1) + \frac{W^2(R_1) - W^2(R_{th})}{2C_p} - \frac{\Omega^2}{2C_p}(R_1^2 - R_{th}^2) \tag{2.31}$$

The relative total temperature at the inlet and in the throat section are:

$$T_r^o(R_1) = T(R_1) + \frac{W^2(R_1)}{2C_p} \qquad\qquad T_r^o(R_{th}) = T(R_{th}) + \frac{W^2(R_{th})}{2C_p} \tag{2.32}$$

hence

$$T_r^o(R_{th}) = T_r^o(R_1) - \frac{\Omega^2}{2C_p}(R_1^2 - R_{th}^2) \tag{2.33}$$

$P_r^o(R_{th})$ is calculated assuming an isentropic change of state because the flow is accelerating between the inlet and throat section

$$P_r^o(R_{th}) = P_r^o(R_1)\left(\frac{T_r^o(R_{th})}{T_r^o(R_1)} \right)^{\frac{\kappa}{\kappa-1}} \tag{2.34}$$

The assumption of isentropic flow is acceptable if a boundary layer blockage coefficient accounts for the friction losses in the boundary layer and if no shocks occur upstream of the throat section.

$O(R_{th})$ depends on the blade number and local impeller geometry (blade curvature and thickness, and leading edge blade angle).

$$O(R_{th}) \approx \frac{2\pi R_{th}}{Z_r} \cos \beta_{1,bl} - \delta_{bl} \tag{2.35}$$

A decrease in the blade thickness δ_{bl} has a favorable effect on inducer choking but is limited by a minimum value required for mechanical integrity (stress and vibrations). A smaller value of $\beta_{1,bl}$ also results in larger values of O_{th} and can compensate for the unfavorable influence of blade thickness (Figure 2.29). However, this results in an increase in the incidence at all operating points and, if not compensated by an increase in the inlet meridional velocity, has an unfavorable impact on the surge limit.

Cambering the inducer suction side is a common way to increase the throat section without changing the incidence (Figure 2.29). However, it may not always be reflected in an increase in the choking mass flow because turning the flow from the inlet direction to the throat increases the velocity variation from suction to pressure side (Figure 2.30a). As a consequence most of the flow passes through the throat at a Mach number that is higher or lower than 1, as illustrated in Figure 2.30b. The local mass flux is everywhere lower than at $M = 1$ and the maximum mass flow is lower than the theoretical one. The non-uniformity increases with the suction side camber and a too large suction side curvature may annihilate the increase in the corresponding throat section.

At transonic inlet flow conditions the increase in the suction side Mach number will also lead to increased shock losses and boundary layer blockage. The latter has not only an unfavorable effect on the impeller efficiency but further limits the maximum mass flow. It leads to

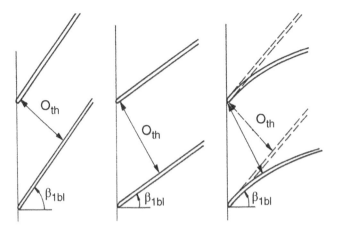

Figure 2.29 Influence of the blade angle and inlet curvature on the throat section.

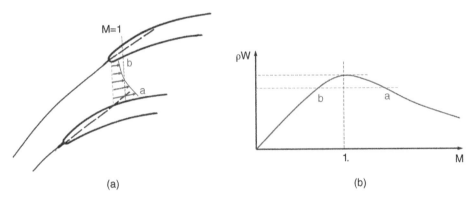

Figure 2.30 (a) Non-uniformity of the flow in the throat section and (b) the consequence on the mass flux.

the intriguing observation that an increase in the throat section by cambering the blade suction side may even result in a decrease in the choking mass flow because of increased boundary layer blockage and non-uniformity in the throat section.

The influence of blade suction side shape on choking mass flow has been studied by Rodgers (1961, 1998) on a radial impeller with parabolic inducer camber lines, staggered in a way to have constant incidence over the blade height. Figure 2.31 shows the difference between the experimental inlet Mach number, at choking, and the theoretical one corresponding to a uniform flow at $M_{th} = 1$ in the throat section. An increasing discrepancy is observed at inlet Mach numbers approaching one. It is obvious that this discrepancy is not universal but depends on the blade shape. Low suction side camber of the order of 1° to 2° between the leading edge and the throat seems to be optimal in transonic inducers. Negative camber may even lead to shock-free transonic inlet flow but makes it more difficult to maintain the minimum required blade thickness and to keep the choking mass flow constant (Figure 5.9) (Demeulenaere et al. 1998).

Rodgers (1961) further states that, for non-staggered inducers ($\beta_{1,bl} = C^{te}$), the discrepancy between the theoretical choking mass flow and the real one increases further because of the increasing incidence from shroud to hub, leading to flow separation at the hub.

The inducer choking mass flow will also increase with decreasing number of blades in the inducer. This may have an unfavorable effect on the performance because of the increased blade loading, in particular at the impeller outlet. The latter can be avoided by the use of splitter blades

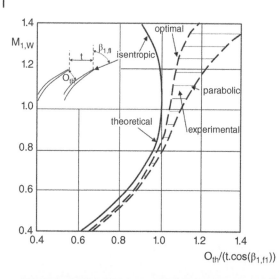

Figure 2.31 Theoretical versus measured choking Mach number for a radial impeller (from Rodgers 1998).

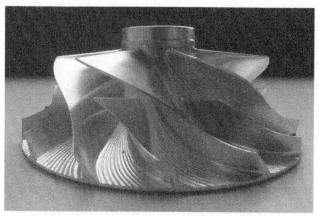

Figure 2.32 Impeller with splitter vanes.

(Figure 2.32). This allows a larger choking mass flow without increasing the blade loading in the exducer. However, this is often at the expense of slightly lower impeller efficiency (of the order of 1%) and more flow distortion at the impeller exit because of unequal flow on both sides of the splitter blade. Impeller optimizations, allowing a splitter vane geometry that is not a shorter version of the full blade, but has a leading edge shape that is locally adapted to the incoming flow, are discussed in Section 5.2.5. Furthermore, one should verify that choking does not occur in the second throat section at the splitter vane leading edge. Hence a detailed blade to blade calculation is required to define the influence of the splitter vanes on the velocity distribution and to optimize the splitter vane leading edge location.

3

Radial Impeller Flow Calculation

The flow in radial impellers is 3D, viscous, and unsteady. Unsteady 3D Navier–Stokes calculations require a very detailed geometry definition and considerable effort. The first Navier–Stokes steady flow calculations by Wallitt (1980) and Moore et al. (1983) nicely described the major flow phenomena such as secondary flows and suction side flow separation (Figure 3.1). However, excessive computational effort (6 weeks on a supercomputer for only 100 000 grid points in 1980) restricted their use to a final check of performance. The enormous increase in computational capacity and speed, and the improvement in the solvers, now allow a routine analysis of the 3D flow in impellers at design and off-design operation once the detailed geometry is defined.

The main advantage of 3D Navier–Stokes calculations is the availability of detailed information on the flow at a large number of points. This allows the designer to base his judgment and modifications of the geometry on quantities that are directly related to the flow. The interpretation of the Navier–Stokes results, however, requires a good understanding of the relations between geometry and flow in order to find out what geometry changes are required to achieve the target. This search may be supported by optimization techniques (Pierret 1999; Poloni 1999; Alsalihi et al. 2002; Verstraete et al. 2008; Thévenin and Janiga 2008; Giannakoglou et al. 2012; Gauger 2008; Saphar 2004). However, they require the analysis of a large number of geometries and are appropriate only in the final design phase after a first design by fast approximated techniques.

It is the purpose of the present chapter to provide insight into the flow structure based on theoretical considerations and on the results of numerical experiments. It describes and quantifies

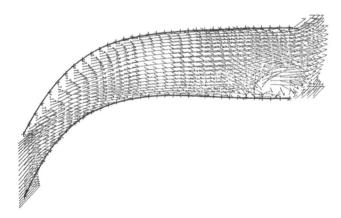

Figure 3.1 Blade to blade flow predicted by the Navier–Stokes solver (from Wallitt 1980).

Design and Analysis of Centrifugal Compressors, First Edition. René Van den Braembussche.
© 2019, The American Society of Mechanical Engineers (ASME), 2 Park Avenue, New York, NY, 10016, USA (www.asme.org).
Published 2019 by John Wiley & Sons Ltd.

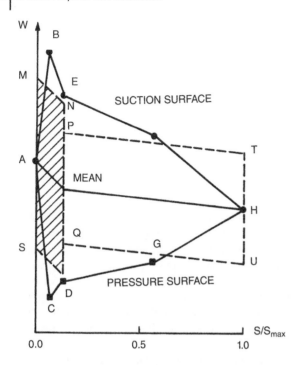

Figure 3.2 Approximate velocity distribution (from Davis and Dussourd 1970).

the various forces and losses that govern the 3D structure of the flow as a function of the impeller geometry and operating conditions.

The first type of methods that are suited for rapid performance predictions are the inviscid calculations. Quasi 3D and full 3D solutions of the Euler equations require much less computer effort than full 3D Navier–Stokes calculations. They make use of simplified equations applied to the real 3D geometry. Comparisons with experimental data have shown that inviscid flow calculations provide a useful approximation only if the influence of viscous effects does not result in flow separation. The distributed body force model proposed by Denton (1986) implemented into an Euler solver allows a rapid approximation of the viscous and secondary flow effects. The latter require imposing a boundary layer like total pressure distribution at the inlet section.

Other simplified models aim for a description of the real flow in simplified geometries. Typical methods are those by Herbert (1980), Davis and Dussourd (1970), Jansen (1970), and many others. The relative velocity distribution between inlet and outlet is approximated by a standardized variation. The difference between the suction and pressure side velocity, based on flow angle and radius variation along the blade, is superposed to the mean value (Figure 3.2). The blockage and losses are estimated from correlations or by calculating the boundary layer growth along the channel walls using a simple integral method. Those methods are very approximated but very useful in an early stage of the design, allow savings in computer effort, and provide insight into the way different flow mechanisms influence the performance.

In the first section of this chapter we will describe the interaction between the inviscid forces in the two surfaces used for the quasi 3D analysis of impellers, i.e. the blade to blade surface and the meridional plane. This will enlighten the various contributions to the blade loading and explain the difference between the flow at the hub and the shroud. The second section describes the full 3D flow and how 3D geometrical features may influence the flow and performance.

A method to predict the main characteristics of a centrifugal compressor and to investigate the influence of various design parameters on the performance is described in the third section. Such a tool has to be fast enough to allow an iterative use on a large number of geometries, but at the same time should be sufficiently elaborated to take into account the important flow phenomena, such as flow separation, compressibility, boundary layer blockage, and the different kind of losses. Design and off-design prediction methods have been described by Galvas (1973), Aungier (2000), Harley et al. (2012), and many others.

The model described here makes use of the actual knowledge of the real flow in centrifugal compressors as obtained from the numerous theoretical and experimental studies that are referenced in the text. This realistic flow model, known as the jet–wake model or the two-zone model, provides a fairly realistic picture of the flow, including flow separation, and is specific for performance predictions of centrifugal impellers.

3.1 Inviscid Impeller Flow Calculation

The quasi 3D model for unseparated flows, described in this section, allows a better insight into the way the velocity distribution is influenced by the impeller geometry. It is based on the simplified S1 S2 model of Wu (1952) splitting the 3D flow into two 2D flows, one in the meridional plane and one in the blade to blade plane. It is the basis for a discussion of the optimum velocity distribution and indicates what geometry modifications are needed to achieve it.

Figure 3.3 shows a quite different velocity distribution at the hub and shroud of a centrifugal impeller. The inlet Mach number is much larger at the shroud because of the larger peripheral velocity resulting from the larger radius. According to Equation (1.48) defining the energy input in the impeller, the larger diffusion at the shroud is needed to compensate for the smaller increase in the peripheral velocity at that location.

The hub and shroud velocity distribution can be split into an average velocity on which the suction to pressure side velocity difference is superposed. Although closely related to each other, the velocity variation in the meridional and in the blade to blade plane are discussed separately, considering the impact of the meridional shape on the average velocity variation and the influence of the blade to blade shape on the suction to pressure side velocity difference.

3.1.1 Meridional Velocity Calculation

The tangential $V_u = W_u - \Omega R$ and meridional component $\widetilde{V}_m = \widetilde{W}_m$ of the absolute velocity induce centrifugal forces in the meridional plane (Figure 3.4). The first one is due to the swirling motion at constant radius on the axisymmetric streamsurface, the second one results from turning the flow from axial to radial on a surface with curvature radius $\mathfrak{R}_n$. They are

Figure 3.3 Velocity distribution at hub (H) and shroud (S) suction (SS) and pressure side (PS).

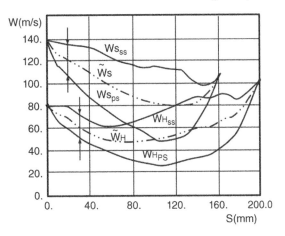

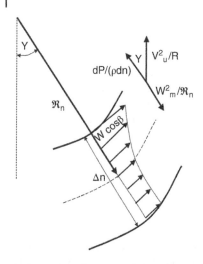

Figure 3.4 Meridional velocity distribution along quasi orthogonals.

balanced in the normal direction by the hub to shroud pressure gradient. Neglecting the forces due to the blade lean angle (assuming $\frac{\partial \theta_{blade}}{\partial n} = 0$), the equilibrium of forces is expressed by:

$$\frac{1}{\rho}\frac{\partial P}{\partial n} = \frac{(\Omega R - W_u)^2}{R}\cos\gamma - \frac{\widetilde{W}_m^2}{\mathfrak{R}_n} \tag{3.1}$$

Substituting $h = P/\rho$ along a streamline in the equation of constant rothalpy (Equation (1.68)) and assuming that the rothalpy is also constant across the streamlines (because it is uniform in the inlet section), the derivative in the n direction results in following expression for the pressure gradient:

$$\frac{1}{\rho}\frac{\partial P}{\partial n} = -\widetilde{W}\frac{\partial \widetilde{W}}{\partial n} + \Omega^2 R\frac{\partial R}{\partial n} \tag{3.2}$$

After elimination of $\frac{1}{\rho}\frac{\partial P}{\partial n}$ between Equations (3.1) and (3.2) we obtain

$$\frac{\partial \widetilde{W}}{\partial n} = \frac{\widetilde{W}\cos^2\beta}{\mathfrak{R}_n} + \cos\gamma\sin\beta\left(2\Omega - \frac{\widetilde{W}\sin\beta}{R}\right) \tag{3.3}$$

Katsanis (1964) used a finite difference technique to solve this equation for axisymmetric isentropic homoenthalpic flows. Solving it by a streamline curvature method may be less accurate but allows a better understanding of the link between the geometry and the flow. For this purpose a grid of intermediate streamsurfaces is defined by a linear interpolation along the quasi orthogonals connecting the hub and shroud contour (Figure 3.5).

Assuming pure axial-radial blades ($\beta = 0$) we obtain the following approximate expression for the pitchwise averaged meridional velocity variation along a quasi orthogonal:

$$\frac{\partial \widetilde{W}_m}{\partial_n} = \frac{\widetilde{W}_m}{\mathfrak{R}_n} \tag{3.4}$$

For small distances between the two grid lines (Figure 3.5) we can use a finite difference approximation to define the meridional velocity component $W_{m,j}$ in the successive locations between the hub and the shroud, as a function of the local curvature and the velocity in the previous location:

$$\widetilde{W}_{m,j} = \widetilde{W}_{m,j-1}(1 + \Delta n/\widetilde{\mathfrak{R}}_n) \tag{3.5}$$

where $\widetilde{\mathfrak{R}}_n$ is the average curvature over the distance Δn.

Figure 3.5 Quasi orthogonals and quasi streamsurfaces for the meridional flow calculation.

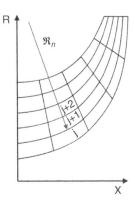

The starting value $W_{m,1}$ can be defined by applying continuity between the hub and the shroud. For compressible flow, the integration requires an iterative procedure because the density depends on the local velocity:

$$\sum_{j=1}^{j=j_{max}} \widetilde{W}_{m,j}\rho_{m,j}2\pi R\Delta n = \dot{m} \tag{3.6}$$

The conclusion is that the meridional velocities at the shroud and hub are related by the meridional curvature and the distance between the hub and the shroud. Increasing this distance decreases the average meridional velocity and, at unchanged curvature, increases the velocity difference between the hub and the shroud. This may result in a low or even negative meridional velocity at the hub. The latter is not a flow separation but a recirculation of inviscid nature. The difference between the shroud and hub meridional velocity can be reduced by reducing the curvature of the meridional contour or the distance between the hub and the shroud. The gradient (Equation (3.4)) also depends on W_m and hence changes with mass flow.

The mean relative velocity on the streamsurface between the suction and pressure side is defined by

$$\widetilde{W} = \frac{\widetilde{W_m}}{\cos(\beta_{fl})} \tag{3.7}$$

The flow angle (β_{fl}) is equal to the blade angle (β_{bl}) except near the leading and trailing edge, where it is influenced by incidence and slip. The radial position R^* where the flow starts to deviate from the blade because of slip has been estimated by Stanitz and Prian (1951):

$$\ln \frac{R^*}{R_2} = 0.71\frac{2\pi \cos \beta_{2bl}}{Z_r} \tag{3.8}$$

Downstream of this critical point, the flow angle can be approximated by a second degree polynomial:

$$\beta_{fl} = Am^2 + Bm + C \tag{3.9}$$

The three variables A, B, and C are defined by satisfying the following conditions:

$$\beta_{fl}(m^*) = \beta_{bl}(m^*) \tag{3.10}$$

$$\frac{d\beta_{fl}}{dm}\big|_{m^*} = \frac{d\beta_{bl}}{dm}\big|_{m^*} \tag{3.11}$$

to assure a first-order continuity between blade and flow angle at m^*, and

$$\beta_{2fl} = \beta_{2,slip} \tag{3.12}$$

The flow angle from the leading edge to the throat region can be approximated in the same way. The constants are defined by the inlet flow angle, the tangency to the blade, and the continuity of the streamline curvature at the throat section.

3.1.2 Blade to Blade Velocity Calculation

The difference between the blade suction and pressure side velocity on an axisymmetric stream-surface can be derived from the equation of motion (Equation (1.71)):

$$\nabla \vee \vec{W} = -2\vec{\Omega}$$

This equation must be satisfied in every point of the flow field. Hence the integral over any part of the flow surface must be zero. According to Stokes's theorem, the flux of the curl of the vector $\vec{W}$ through part of the blade to the blade streamsurface shown on Figure 3.6 can be replaced by the line integral of the vector along the contour of that surface:

$$\iint (\nabla \vee \vec{W} + 2\vec{\Omega}) d\vec{S} = \oint \vec{W}.d\vec{s} + \iint 2\vec{\Omega} d\vec{S} = 0.$$

which provides the following relation between the PS and SS velocity:

$$W_{SS} - W_{PS} = \left(\frac{2\pi}{Z_r} - \frac{\delta_{bl}}{R\cos\beta} \right) \frac{d}{ds} \left(\Omega R^2 - \widetilde{W}_m R \tan \beta_{fl} \right) \tag{3.13}$$

Superposing this velocity difference on the pitchwise averaged value $\widetilde{W}$ provides the SS and PS velocity distribution on the different axisymmetric streamsurfaces, as illustrated in Figure 3.3. Equation (3.13) shows how the suction to pressure side velocity difference depends on the change in radius in rotating systems:

$$\frac{d}{ds}(\Omega R^2) \tag{3.14}$$

and on the change in velocity and flow direction in the blade to blade plane:

$$\frac{d}{ds} \left(\widetilde{W}_m R \tan \beta_{fl} \right) = \widetilde{W}_m R \frac{d \tan \beta_{fl}}{ds} + \tan \beta_{fl} \frac{dR\widetilde{W}_m}{ds} \tag{3.15}$$

Considering that R and $\widetilde{W}_m$ are constant on a cylindrical streamsurface of an **axial compressor** and with $d\beta/ds = 1/\mathfrak{R}_b$, Equation (3.13) reduces to

$$W_{SS} - W_{PS} = \frac{\widetilde{W}}{\cos\beta_{fl}} \left(\frac{2\pi R}{Z_r} - \frac{\delta_{bl}}{\cos\beta_{bl}} \right) \frac{1}{\mathfrak{R}_b} \tag{3.16}$$

Figure 3.6 Zero absolute velocity circulation on a closed contour.

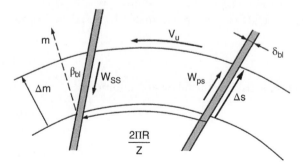

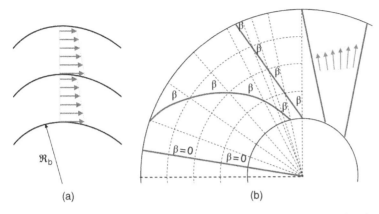

Figure 3.7 Relation between flow angle and curvature on (a) a cylindrical surface and (b) a 2D non-rotating radial plane.

In an axial turbomachine the suction to pressure side velocity difference increases with the average relative velocity, the distance between two successive blades $(2\pi R/Z_r - \frac{\delta_{bl}}{\cos \beta_{bl}})$, and is inversely proportional to the blade curvature radius $\Re_b$ (Figure 3.7a). As the flow is independent of the rotational speed, the blade to blade velocity distribution in axial compressors can be measured in non-rotating blade rows if the relevant inlet conditions are created (Herrig et al. 1957).

A similar result is obtained for a **2D non-rotating channel** in the radial plane (Figure 3.7b) where $R\widetilde{W}_m = C^{te}$ by continuity. Equation (3.13) then reduces to

$$W_{SS} - W_{PS} = \frac{\widetilde{W}}{\cos \beta_{fl}} \left(\frac{2\pi R}{Z_r} - \frac{\delta_{bl}}{\cos \beta} \right) \frac{d\beta}{ds} \tag{3.17}$$

The suction to pressure side velocity depends again on the change in β. However, the relation between a change in β and curvature is different from the one for axial turbomachines. As illustrated in Figure 3.7b any straight ($\Re_b = \infty$) but non-radial streamline has a varying β because the reference (radial) direction changes orientation along the streamline. Blade curvature is no longer the measure of blade loading.

In a **2D radial rotating channel** with incompressible flow ($\frac{d\widetilde{W}_m R}{dR} = 0$) and constant β (radial walls on Figure 3.8a), the blade loading depends on the blade pitch, rotational speed, and change

Figure 3.8 Blade to blade velocity variation for (a) straight rotating channels and (b) backward curved rotating channels.

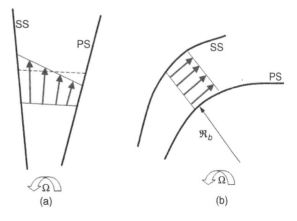

in radius:

$$W_{SS} - W_{PS} = \left(\frac{2\pi R}{Z_r} - \frac{\delta_{bl}}{\cos \beta} \right) 2\Omega \frac{dR}{ds} \tag{3.18}$$

Backward curvature has the opposite effect to rotation and can be used to decrease the blade loading (Figure 3.8b). The latter will be zero if the backward curvature is such that

$$\frac{d}{ds} \left(\Omega R^2 - \widetilde{W}_m R \tan \beta_{fl} \right) = 0 \tag{3.19}$$

which is equivalent to

$$\frac{d}{ds}(RV_u) = 0 \tag{3.20}$$

expressing the conservation of momentum in the absence of any blade force. Hence no energy will be transmitted by such blades to the fluid at that rotational speed and mass flow.

The hub to shroud (Equation (3.3)) and blade to blade velocity variation (Equation (3.13)) are strongly interconnected. The average velocities at the shroud and hub are related by the meridional curvature, the flow angle β, the distance between the hub and the shroud, and the rotational speed (Equation (3.3)). The pressure and suction side velocity distributions depend on the local average velocity, the blade number, the rotational speed, the blade curvature, and the change in radius of the meridional surface (Equation (3.13)). Any modification of the meridional contour has an impact on the blade loading. Any change in the blade to blade flow changes the balance of the forces in the meridional plane. The β distribution and meridional curvature have a distinct impact on the velocity distribution, but with a strong mutual interaction.

Once the velocities on the pressure and suction side have been defined by superimposing $W_{SS} - W_{PS}$ on $\widetilde{W}$, they can be corrected for viscous effects by introducing the boundary layer blockage along the hub, shroud, and blade surfaces (Figure 3.9). The three-dimensionality of the boundary layers, due to the secondary flows generated in the long curved channels between inlet and outlet, is neglected.

3.1.3 Optimal Velocity Distribution

The impact of the rotational speed on the meridional and blade to blade velocity distribution means that experiments can only be performed in rotating systems. Defining an optimal geometry for centrifugal impellers is therefore more complex than for axial ones.

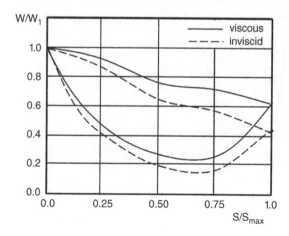

Figure 3.9 Calculated velocity distribution on a mean surface between the hub and the shroud: correction for boundary layer blockage.

Before discussing the optimal velocity distribution we want recall the relations and boundary conditions that govern the flow in radial impellers and impact on the velocity distribution. This discussion refers to Figure 3.3.

- The average velocity $\widetilde{W}$ is lower at the hub than at the shroud because of the meridional curvature. It decreases from inlet to outlet along the shroud end wall, while it may increase along the hub.
- The SS to PS loading increases from the inlet section until close to the trailing edge. This is a consequence of the increasing inter-blade distance with radius and the increasing influence of the Coriolis force in the radial part eventually attenuated by backward curvature (Equation (3.13)). The zero velocity difference at the trailing edge is in agreement with the Kutta conditions. It is related to an increasing flow angle β_{2fl} resulting from the backward curvature and slip.
- Near the inlet the loading is larger at the shroud than at the hub because:
  - the inlet relative angle is larger at the shroud than at the hub and the flow turning is over a shorter distance (smaller curvature radius $\mathfrak{R}_b$) to reach the outlet direction
  - the inlet radius is larger at the shroud than at the hub and the blade to blade velocity gradient thus applies over a much larger distance (pitch).
- No axial gradient of the velocity from hub to shroud is observed at the exit. This confirms that the spanwise velocity gradient goes to zero when $\gamma = 90.$ and $\mathfrak{R}_n$ goes to infinity (Equation (3.3)). However, this situation may change at off-design operation when a change in mass flow perturbs the equilibrium of forces expressed by Equation (3.1).
- Equations (3.3) and (3.13) can be used in an inverse manner to design an impeller with a prescribed velocity distribution at the hub and shroud as described in Chapter 5.

The first attempts to optimize centrifugal impeller geometries were reported by Johnsen and Ginsburg (1953) and Schnell (1965) considering standardized geometries (Figure 3.10). Blades are generated by straight radial lines at constant x crossing the axis of rotation and a parabolic, elliptic or circular generating line at constant radius R_S. This design procedure is incomplete because it does not provide any information about the optimal meridional contour and has only a restricted design space ($\beta_2 \approx 0$).

The differences in experimental performance presented by Kramer et al. (1960) (Figure 3.11) relate to the discussions about blade curvature and throat section (Figures 2.29 to 2.31). The highest efficiency and pressure ratio are obtained for the impeller with a parabolic blade shape. The largest choking mass flow is observed in the elliptic blade shape. The moderate leading edge curvature of the latter increases the throat section more than the parabolic blades, without excessive acceleration of the fluid on the suction side. The velocity is rather uniform in the throat section and nearly maximum flux occurs over the whole section. The larger curvature of the circular blade results in a larger throat section, but the choking mass flow is smaller than with the parabolic blade. The circular blade does not profit from the larger throat section because of the larger suction to pressure side velocity gradient resulting from the larger curvature. The extra flow deceleration downstream of the larger suction side velocity acceleration initiates stall and is responsible for the zero operating range at the highest Mach numbers. The impeller with the circular blades has the best performance at low Mach number. The impeller with the parabolic blade performs best at high inlet Mach number.

Figure 3.11 also illustrates the impact of the meridional contour on the performance. The excessive diffusion in the impeller with larger exit width (dashed lines) results in a lower pressure rise and efficiency because of premature flow separation.

The velocity distributions on the hub, mean, and shroud surfaces of similar impellers have been calculated by Bhinder and Ingham (1974). The results confirm that parabolar inducers

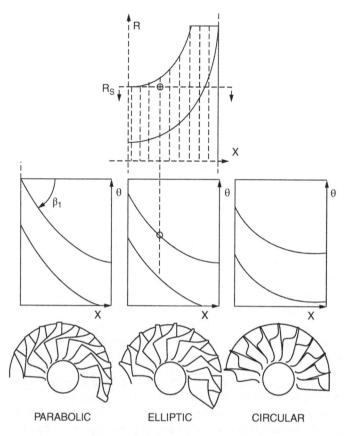

Figure 3.10 Definition of parabolic, elliptic, and circular inducer blades.

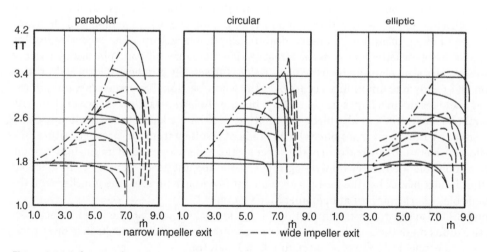

Figure 3.11 Influence of compressor geometry on operating range (from Kramer et al. 1960).

Figure 3.12 Velocity distribution in a radial impeller with parabolic, elliptic, and circular inducer blades (from Bhinder and Ingham 1974).

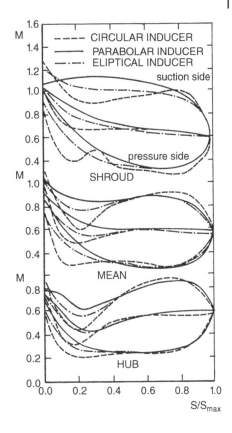

have a lower loading at the inlet and a smoother velocity variation (Figure 3.12). The small increase in the throat section of the parabolic blade in combination with the progressively increasing blade loading results in a nearly constant velocity on the first part of the suction side. The large increase of the throat section with the circular blade results in a rapid deceleration on the suction side followed by a reacceleration. It also gives rise to a steep deceleration on the pressure side. The rapid deceleration on the suction side is unfavorable in terms of stall and surge margin. A deceleration followed by a reacceleration is unfavorable in terms of performance.

The definition of the optimum velocity distribution in centrifugal impellers is still a controversial and complex matter. It should not only be based on criteria for minimum losses but also satisfy the targets in terms of pressure ratio and mass flow range, and respect the relations linking the blade to blade loading and meridional velocity. The critical area is the shroud suction side. As the hub pressure side boundary layers are much less sensitive to separation, a stronger velocity deceleration is allowed. In terms of minimization of the losses, there is a theoretical advantage in not postponing the flow deceleration towards the outlet area but decreasing the velocity as soon as possible (without separation) and continuing the flow at low velocity with lower friction losses towards the trailing edge (Huo 1975). This is restricted by the blade loading requiring a larger suction than pressure side velocity. Decelerations should also be smooth and any acceleration should be avoided (if possible) because it will require a larger subsequent deceleration.

The velocity distributions shown in Figure 3.13 illustrate the impact of the different blade loading distributions on the hub and shroud velocity, and how a modification of the shroud velocity influences the hub velocity.

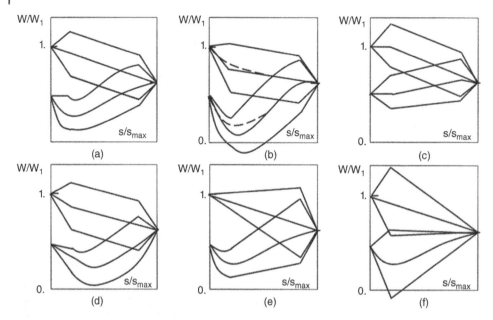

Figure 3.13 Velocity distributions at shroud and hub corresponding to different impeller blade loadings and mean velocity variations.

Figure 3.13a shows a linear variation of the mean velocity at the shroud from inlet to outlet and a constant loading, i.e. a constant difference between the suction and pressure side velocity with a linear increase/decrease near the leading/trailing edge. This results in an undesirable acceleration at the suction side leading edge. The corresponding hub velocity shows a rapid decrease after the leading edge and a reacceleration towards the exit to reach the same velocity as at the shroud.

In order to avoid the shroud suction side acceleration, one could more rapidly reduce the shroud mean velocity by increasing the distance from hub to shroud while keeping the loading unchanged (Figure 3.13b). However, this decrease in the meridional velocity results in a negative velocity at the hub pressure side and the extra blockage, due to recirculation, influences the other velocities, as indicated by the dashed lines.

In Figure 3.13c the mean velocity at the shroud remains constant in the inlet region and decreases more rapidly towards the exit. In combination with the constant blade loading this results in a continuously accelerating velocity at the hub. However, the larger overshoot of the velocity near the shroud leading edge provokes a larger downstream deceleration.

Figures 3.13e and 3.13f show velocity distributions with a linear variation of the mean velocity, such as Figure 3.13a, but with different loading distributions. The first one has an ideal velocity distribution in the front part but an unrealistically large deceleration near the trailing edge to satisfy the Kutta conditions.

The high loading at the inlet in combination with the slow decrease in the average velocity results in a velocity increase on the suction side (Figure 3.13f). At high inlet Mach numbers this is not only unfavorable from the point of view of choking but also in terms of range. A much larger deceleration is needed to reach the given trailing edge velocity and stall is more likely to occur. Friction losses are proportional to W^2 and one should not postpone the deceleration towards the trailing edge. The large blade loading in the front part will also result in larger secondary flows. Hence the velocity distributions shown in Figures 3.13e and 3.13f are far from optimal and must be rejected.

A possible "optimal" compromise is shown in Figure 3.13d. The moderate loading near the leading edge in combination with a limited decrease in the average velocity results in a nearly constant velocity in the first part of the shroud suction side. The small meridional velocity decrease at the shroud leading edge has a favorable impact on the hub velocity by shifting the location of minimum velocity further downstream, reducing the rate of flow deceleration.

Advanced optimization methods for radial impellers use numerical procedures to find the geometry that provides the pressure ratio at the required mass flow with maximum efficiency, taking into account the mechanical and geometrical constraints as well as other flow considerations. They are discussed in more detail in Chapter 5.

3.2 3D Impeller Flow

The real 3D inviscid flow is more complex than a simple addition of the flows on the blade to blade and hub to shroud surfaces. This section explains how this interaction can be used to further modify the flow, followed by a qualitative description of the different secondary flow components and the algebraic relations that govern them. This is illustrated by results from the numerical analysis of the flow in an impeller with large values of $\beta_{2,bl}$.

3.2.1 3D Inviscid Flow

Quasi 3D codes calculate the blade to blade flow on several 2D streamsurfaces between the hub and the shroud, and provide only an approximate picture of the 3D flow. By definition this approach does not allow any velocity perpendicular to axisymmetric streamsurfaces. However Wu (1952) and Vavra (1974) demonstrated that even for inviscid (Beltrami) flows the streamsurfaces cannot be axisymmetric except for very special geometries such as:

- impellers with an infinite number of blades
- impellers with a pure radial hub and shroud wall (two parallel disks) and constant blade shape perpendicular to it, with axially uniform inlet flow.

All other impellers have non-axisymmetric streamsurfaces. The 3D character of the real flow is illustrated by the full 3D inviscid and incompressible flow calculated by Ellis and Stanitz (1952) in an impeller with straight radial blades. The 3D flow structure has been visualized by integrating the velocity along the walls (Figure 3.14).

A streamwise vorticity is observed in a direction opposite to the impeller rotation resulting in a streamline shift to the hub on the pressure side and to the shroud on the suction side. This is the consequence of a different balance between the fluid and pressure forces acting on the flow on both sides.

Ellis and Stanitz (1952) further concluded that the modulus of the velocity and the blade pressure distribution of a quasi 3D calculation are not very different from those of a fully 3D inviscid solution. However, we will see later that the normal velocities and pressure gradients play an important role in viscous flows.

Baljé (1981) has presented an analytical model applied to a simplified geometry that allows an explanation of the non-axisymmetry of the streamsurfaces. Equation (3.1) expresses the inviscid balance of forces in the meridional plane. Contrary to an axial impeller, it is possible to use the meridional curvature of an axial-radial impeller to compensate the centrifugal force due to the peripheral velocity in order to achieve a zero pressure gradient from the hub and shroud.

Imposing $\partial P/\partial n = 0$ in Equation (3.1) results in the following relation

$$\frac{W_m^2}{\Re_n} = \frac{V_u^2}{R}\cos\gamma \tag{3.21}$$

defining the radius of curvature necessary for a zero pressure gradient from hub to shroud.

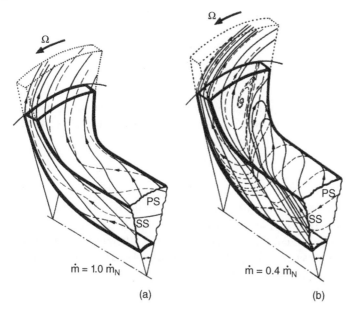

Figure 3.14 Wall surface streamlines for 3D inviscid flow at (a) 100% and (b) 40% mass flow (from Ellis and Stanitz 1952).

Satisfying this relation at all positions between the impeller inlet and outlet for the mean velocity between the pressure and suction sides defines a meridional contour. However, W and hence also W_m increase towards the suction side and decrease towards the pressure side. Assuming the same non-zero flow angle β_{fl} the absolute tangential velocity V_u will be larger than average on the pressure side and smaller at the suction side. Hence satisfying Equation (3.21) for the flow near the pressure side requires a smaller meridional curvature radius. Inversely, if that equation must be satisfied near the suction side, the curvature radius should be larger and the impeller should be longer in the axial direction. It is therefore not possible to define a meridional contour that cancels the spanwise pressure gradient over the complete pitch.

This is illustrated in Figure 3.15, where the influence of the number of blades is also shown.

Impellers with a large number of blades have a smaller difference between the pressure and suction side velocities. Hence, the radii of curvature defined by Equation (3.21) are less different between the two sides.

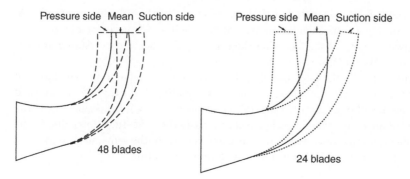

Figure 3.15 Influence of blade number on zero pressure gradient design at the pressure and suction sides (from Baljé 1970).

Impellers with a low number of blades have very different velocities on the suction and pressure sides and hence a larger difference between the meridional shapes for equal pressure at the hub and shroud.

Forcing the pressure side flow to follow the meridional flow path corresponding to zero pressure difference of the pitchwise averaged flow creates a spanwise pressure gradient that is opposite to the one at the suction side. These pressure differences occur in inviscid as well as viscous flows and are at the origin of the non-axisymmetry of the flow similar to that illustrated in Figure 3.14.

W_m increases with increasing volume flow whereas the peripheral velocity component V_u decreases. The balance between the forces on the fluid (Equation (3.21)) is perturbed. A zero pressure gradient would require a meridional contour with a larger meridional curvature radius $\Re_n$ similar to the one on the suction side in Figure 3.15. For an unchanged meridional contour the fluid will have a tendency to move towards the hub side, resulting in an increase in the pressure on that side.

The balance of forces is inverted when decreasing the volume flow. The peripheral velocity component V_u increases whereas the meridional one W_m decreases and more fluid will exit near the shroud.

This is illustrated by the experimental results of Ellis (1964) in a low speed compressor (Figure 3.16). At nominal flow the impeller outlet flow is concentrated on the hub side, where the highest total pressure is also measured. A gradual shift of the fluid towards the shroud is observed when decreasing the volume flow. The zone of local return flow near the shroud disappears and a zone of return flow appears near the hub side. As will be discussed in Section 4.1.3 this spanwise non-uniformity of the incoming flow may have a large impact on the diffuser stability.

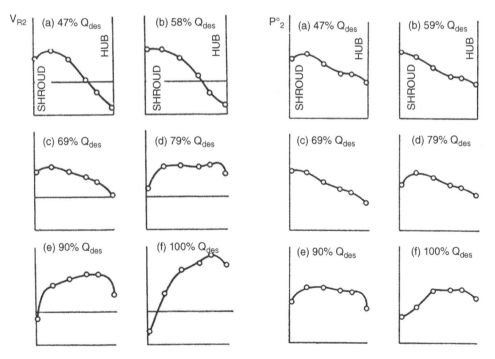

Figure 3.16 Spanwise variation of the radial velocity (left) and total pressure (right) at the impeller outlet at different volume flows (from Ellis 1964).

The other 3D inviscid effects, such as tip leakage flow and blade lean, are discussed in the following sections.

3.2.2 Boundary Layers

Radial compressors differ from axial ones by the presence of Coriolis forces in the blade to blade plane and the curvature of the meridional contour. Before describing the impact of the viscosity on the velocity distribution in radial impellers we will first discuss the impact of curvature and Coriolis on the boundary layers. We present here:

- a qualitative explanation about how a normal to wall pressure gradient, resulting from the blade curvature or the rotation of the channel, influences the state of the boundary layer
- a discussion of the impact this may have on the flow.

The boundary layers along the walls are submitted to the same streamwise pressure gradients as the adjacent free stream. However, the velocity of the fluid inside a boundary layer is much smaller than in the inviscid part of the flow. Separation will occur when the sum of the kinetic energy available in the boundary layer and that added by entrainment and turbulent mixing is dissipated before the pressure rise imposed by the free stream is achieved. This is more likely to happen in compressors and pumps, where the flow is decelerating because of strong adverse pressure gradients, than in turbines where the flow is mainly accelerating.

In a laminar boundary layer, the particles move along parallel paths. The exchange of momentum is very limited because it only occurs on the level of the molecular interaction between direct neighbors. Hence it takes time before the dissipated energy is restored.

In a turbulent boundary layer, the particles move randomly and even perpendicular to the wall: the exchanges of momentum are very intense because of collisions between particles with different velocities. Dissipated energy is more rapidly restored by the mixing of low and high velocity fluid particles and the boundary layer can better overcome adverse pressure gradients.

Laminar boundary layers create less losses than turbulent ones because of the low momentum exchange. On the other hand, they receive less energy from the core flow and are more likely to separate than turbulent ones. This means that it is preferable to have turbulent boundary layers in compressors and pumps in order to avoid flow separation and the associated losses and instabilities as well as the possible decrease in energy input. The flow in turbines is mostly accelerating and the risk of flow separation is much smaller. Laminar boundary layers are preferred in order to maximize efficiency.

It has been shown by Johnston (1974) that a pressure gradient perpendicular to the wall influences the state of the boundary layer. In a centrifugal impeller, this gradient can be due to either Coriolis forces (because of the radial velocity component) or wall curvature (of the blades and meridional walls).

When the flow is in equilibrium, the following relation will be satisfied in every point of the boundary layer:

$$F_P = \frac{1}{\rho} \frac{\partial P}{\partial n} = \frac{w^2}{\mathfrak{R}_C} = F_{curv} \quad \text{(curvature)} \tag{3.22}$$

or

$$F_P = \frac{1}{\rho} \frac{\partial P}{\partial n} = 2\Omega w_R = F_{Cor} \quad \text{(Coriolis)} \tag{3.23}$$

where n is the normal to the wall, and w is the relative velocity inside the boundary layer and is smaller than the free stream velocity W.

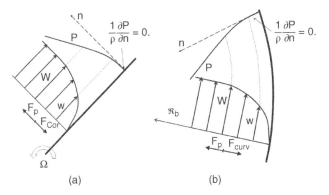

Figure 3.17 Velocity and pressure variation perpendicular to (a) a driving or (b) concave surface.

Consider the pressure variation normal to the blade pressure side (Figure 3.17a) or concave hub wall (Figure 3.17b). The local pressure gradient $\frac{\partial P}{\partial n}$ is a function of the local velocity and increases with increasing distance from the wall. If a fluid particle moves from its original position in the main flow (high velocity) towards the wall, the large dynamic force (F_{Cor} or F_{curv}) will no longer be compensated for by the smaller local pressure gradient and this particle will not return to its original position. It will move further to the wall until the excess of kinetic energy has been dissipated by collision with particles of low kinetic energy. Inversely, if a low energy particle of the boundary layer moves away from the wall, the lower Coriolis and centrifugal forces corresponding to its small velocity will not be able to compensate the stronger pressure gradient and the particle will be pushed further away from its original position. The boundary layer will be continuously fed by high velocity particles coming from the free stream, replacing the particles with low velocity that move to the free stream. The result is a strong exchange of energy between the main flow and the boundary layer. Under these conditions the boundary layer becomes more turbulent and is less likely to separate in the presence of an adverse pressure gradient. It is said that the boundary layer is destabilized.

The opposite behavior is observed at the suction side or along convex walls (shroud) where the pressure gradient increases with the distance from the wall (Figure 3.18). In these cases, a low energy particle that moves away from the wall will face a stronger pressure gradient that is not in balance with the small dynamic force. This particle will therefore be pushed back to its

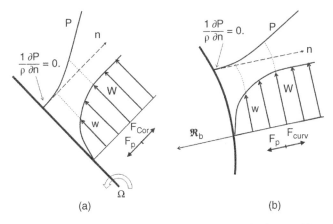

Figure 3.18 Velocity and pressure variation perpendicular to a (a) suction or (b) convex surface.

original position, and the exchanges between the free stream flow and the boundary layer will be very limited. The boundary layer is stabilized, resulting in a lower turbulence, and is more likely to separate under an adverse pressure gradient.

This qualitative explanation demonstrates that, in centrifugal impellers, the boundary layers will be more turbulent along the blade pressure side and the hub endwall. Conversely, they will be more laminar or have lower turbulence along the blade suction side and the shroud endwall. Hence the risk of flow separation is highest in the shroud/suction side corner, an area where the largest velocity deceleration takes place. Special attention should therefore be given to the velocity distribution in this area.

This behavior of the flow in rotating channels has been studied experimentally by Di Sante et al. (2010) and the impact of curvature and Coriolis forces can be accounted for by advanced CFD tools (Van den Braembussche et al. 2010).

3.2.3 Secondary Flows

Secondary flows are defined as the difference between the full 3D inviscid flow and the real viscous one occurring in impellers and diffusers. The main consequence is a redistribution of the low energy fluid by the streamwise vorticity with only a minor influence on the level of the inviscid core velocity and pressure.

The equations describing the rate of increase in the streamwise vorticity ω_s along a relative streamline have been derived by Smith (1957) but the most practical form is the one of Hawthorne (1974), relating the growth of vorticity to the gradients of the rotary stagnation pressure:

$$\frac{\partial}{\partial s}\left[\frac{\omega_s}{W}\right] = \frac{2}{\rho W^2}\left[\frac{1}{\mathfrak{R}_b}\frac{\partial P_{Ro}^o}{\partial n} + \frac{1}{\mathfrak{R}_n}\frac{\partial P_{Ro}^o}{\partial b} + \frac{\Omega}{W}\frac{\partial P_{Ro}^o}{\partial x}\right] \tag{3.24}$$

The first two RHS terms express the generation of vorticity due to the flow turning in the blade to blade surface and the meridional plane, respectively. The last RHS term has its origin in the Coriolis forces and thus will occur only in the radial part of the impeller. The first and third RHS terms generate the so-called passage (PV) and Coriolis (CV) vortex. Both vortices drive low energy fluid from the pressure to the suction surface along the hub and shroud endwall. The second RHS term creates the so-called blade surface vortex (BV). This flow is driven by the meridional curvature and will generate vorticity along the blade surfaces transporting fluid from hub to shroud. The different vortices are schematically illustrated in Figure 3.19.

As explained in Section 1.4 the rothalpy $Ro = h + \frac{W^2}{2} - \frac{R^2\Omega^2}{2}$ is constant along a streamline for both viscous and inviscid flows. It is also constant perpendicular to the streamlines if it

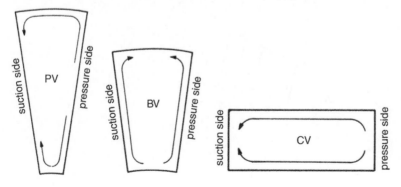

Figure 3.19 Definition of the individual vortices.

is uniform at the impeller inlet. Replacing h by P/ρ allows the rotary stagnation pressure for incompressible flows to be defined:

$$P_{Ro}^o = P + \frac{\rho W^2}{2} - \frac{\rho R^2 \Omega^2}{2} \tag{3.25}$$

The latter will also be constant in the inviscid core of the flow in rotating impellers but not in the viscous layers. Considering that the static pressure is nearly constant over the boundary layer thickness and neglecting the changes in $\Omega^2 R^2$ over the boundary layer thickness, we can conclude from Equation (3.25) that there will be large variations of P_{Ro}^o because W^2 is changing from zero near the wall to the free stream velocity at the boundary layer edge.

Based on previous assumptions we can express Equation (3.24) in terms of relative velocities instead of rotary stagnation pressure by substituting Equation (3.25) in (3.24):

$$\frac{\partial}{\partial s} \left[\frac{\omega_s}{W} \right] = \frac{2}{W} \left[\frac{1}{\mathfrak{R}_b} \frac{\partial W}{\partial n} + \frac{1}{\mathfrak{R}_n} \frac{\partial W}{\partial b} + \frac{\Omega}{W} \frac{\partial W}{\partial x} \right] \tag{3.26}$$

The following comments can be made about the three terms of this expression:

- The passage vortex (PV) (Figure 3.19), characterized by the first RHS term, is due to the turning of the flow in the blade to blade plane. It will usually be stronger in the first half of the impeller channel because of the blade curvature in the axial part of the impeller (smaller $\mathfrak{R}_b$). The curvature decreases and can even change sign further downstream in the blade passage to compensate for the increasing amount of blade loading resulting from the increase in the radius R. The passage vortex starts at the leading edge because the hub and shroud boundary layers have already developed upstream of the impeller. Moreover, the relative velocity W is higher near the shroud than at the hub, and the boundary layer is thicker. Hence the gradient $\partial W / \partial n$ extends over a larger area and more vorticity is thus expected near the shroud than at the hub.

- The blade surface vortex (BV) (Figure 3.19) is generated by the meridional curvature $\mathfrak{R}_n$ (second RHS term) and will develop in the axial-to-radial elbow of the channel. It will vanish progressively towards the radial exit. As the radius of curvature is smaller near the shroud, the intensity of vorticity will increase towards this wall. $W_{SS} > W_{PS}$ near the inlet, hence the gradient $\partial W / \partial b$ is larger in the suction side boundary layer where larger blade surface vortices are expected than on the pressure side.

- The last term of the RHS originates from the Coriolis forces and will be effective in the radial part where an axial gradient of the velocity exists in the hub and shroud wall boundary layers. The corresponding vortex (CV) will contribute to the passage vortex (Figure 3.19c).

Hirsch et al. (1996) have proposed an approximated integration of these three terms in order to evaluate the intensity of the generated vorticity at the exit of the impeller. By doing a similar integration up to a cross section between inlet and outlet the following expressions are obtained.

The growth of the passage vortex (PV) generated by blade curvature is approximated by

$$\Delta \left[\omega_s \right]_{PV} = 2 \left[\frac{W}{\delta_{H,S}} \right] \frac{\Delta s}{\mathfrak{R}_b} \tag{3.27}$$

$\Delta s / \mathfrak{R}_b$ is equal to $\Delta \beta$ in the axial part and to $\Delta \beta - \Delta \theta$ in the radial part. This vorticity increases with blade curvature and is zero for straight blades. $\delta_{H,S}$ is the boundary layer thickness along the hub or shroud. The growth of this vortex is largest in the axial part of the impeller and near the shroud where the velocity deceleration is largest and the boundary layer thicker.

The growth of the blade surface vortex (BV) can be approximated by

$$\Delta[\omega_s]_{BV} = 2\left[\frac{W}{\delta_{SS,PS}}\right]\Delta\gamma \tag{3.28}$$

where $\Delta\gamma$ is the total turning angle of the meridional contour between inlet and the cross section. It is equal to $\Delta m/\Re_n$ and $90°$ at the exit of an axial-to-radial impeller. δ_{SS} and δ_{PS} are the boundary layer thicknesses on the suction and pressure sides, respectively.

The contribution of the Coriolis force to the passage vortex can be approximated by

$$\Delta[\omega_s]_{CV} = \frac{2\Omega\Delta m}{\delta_{H,S}}\sin\tilde{\gamma} \tag{3.29}$$

This vortex grows faster in the radial part of the impeller. $\tilde{\gamma}$ is the average over the distance Δm between the inlet and the cross section.

The vorticity due to Coriolis (CV_S, CV_H) and the passage vorticity (PV_S, PV_H) have a similar effect on the flow. As shown in Figure 3.20 they transport low energy fluid from the pressure to the suction side along the hub and shroud. The blade surface vorticity (BV_{SS}, BV_{PS}) transports fluid from the hub to the shroud along the blade sides. It can be observed that the vorticity is counter rotating in the hub–pressure side corner and in the shroud–suction side corner. Low energy fluid is taken away from the first location and brought together in the latter. The vorticity is co-rotating in the hub–suction side and shroud–pressure side corners. Low energy fluid transported to the corner is removed from it. The overall effect is a transport of low energy fluid from the hub–pressure side corner towards the shroud–suction side corner. This explains why the velocity in the "separated" flow zone, often observed near the shroud suction side corner, has a positive through flow component.

There is no unambiguous way to distinguish between the secondary and main velocity components of the 3D flow field. In a pragmatic approach, the primary flow is approximated by the velocity component in the direction of the streamwise oriented mesh and the velocity components in the crosswise grid surfaces are considered as secondary flows. The intersection of the surfaces on which the secondary flows are shown with the suction and pressure side of the shrouded impeller are indicated in bold and numbered I to IV on the meridional view in Figure 3.21.

The endwall passage vorticity (PV) generated by the blade curvature is already present in cross section I (Figure 3.22). It is much stronger near the shroud because of the higher relative velocity, the thicker inlet boundary layer, and the presence of some prerotation in this shrouded impeller. The latter pushes the low energy fluid towards the suction side, as illustrated in Figure 3.24b.

The blade surface vorticity (BV) due to the meridional curvature is also present near the suction side in section I, which according to Figure 3.21 is located further downstream in the

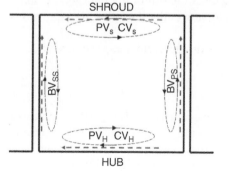

Figure 3.20 Combined vortices in a passage.

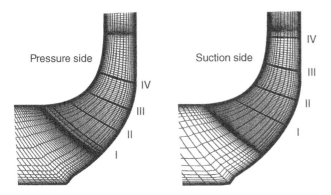

Figure 3.21 Definition of cross sections in the meridional plane.

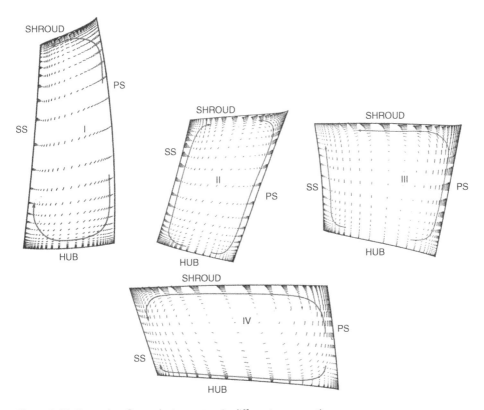

Figure 3.22 Secondary flow velocity vectors in different cross sections.

impeller, where the flow is already subject to the meridional curvature. The blade surface vorticity along the pressure side is very small but gets stronger near the shroud, where the relative velocity is higher.

The blade surface vorticity is further developed on the suction and pressure sides in cross section II, but is still stronger near the suction side. It is superposed to the passage vorticity at the shroud and hub. The combination of both transports low energy fluid to the shroud/suction side corner.

The blade surface vorticity along the suction side has nearly vanished between cross sections II and III because it is no longer fed by the meridional curvature on this side. On the contrary, the vorticity is still very strong along the pressure side because it is located more upstream in the meridional channel. The passage vorticity is now mainly fed by the loading due to the Coriolis force vorticity (*CV*), which is opposite to the local blade curvature effect in this backward leaning impeller. It is very weak near the hub because the local boundary layer is very thin as most of the boundary layer fluid has already been evacuated by the blade vortex on the pressure and suction side. It is much stronger near the shroud because the loading is higher, and the boundary layer is thicker and continuously fed with low energy fluid by the pressure side blade vorticity.

The predicted flow structure is not clearly visible on cross section IV. The passage vorticity along the shroud dominates all the other secondary flow components in such a way that it extends towards the hub endwall, where the velocity in the boundary layer has now reversed and goes from the suction side towards the pressure side. This discrepancy with the theoretical model is due to the fact that even the inviscid flow is not aligned with the grid, as assumed when preparing this figure, but has a large velocity component towards the pressure side because of slip. The predicted flow structure reappears when compensating for the slip by adding a velocity component that is equal but opposite to the one outside the boundary layers near the hub wall. In doing this an even stronger vorticity is obtained at the shroud and the predicted secondary flow near the hub wall appears. The latter is anyway weak because the boundary layer is very thin on that wall. The blade vorticity is still present near the pressure side of cross section IV because this part of the cross section is located further upstream of the radial exit and thus is still influenced by the meridional curvature.

The area with the largest danger of flow separation is the shroud suction side corner where low energy fluid is brought by the secondary flows. In terms of optimum design, geometries that avoid the accumulation of low energy fluid in one area should be sought because this locally weakens the resistance to positive pressure gradients.

3.2.3.1 Shrouded–unshrouded

Figure 3.23 shows a schematic representation of the streamwise vortices in the outlet section of a shrouded and an unshrouded impeller. The flow in the shrouded impeller is in line with the description of secondary flow in the previous section.

The flow in an unshrouded impeller is influenced by the tip leakage flow in the direction opposite to the passage vortex. It blows the low energy fluid away from the shroud suction side corner and modifies the jet/wake flow pattern at the impeller exit. Part of it rolls up and gives rise to the tip leakage vortex. The rest of the leakage flow is entrained by shear on the shroud wall with a velocity equal to the peripheral velocity of the wall, in a direction opposite to the local passage vorticity.

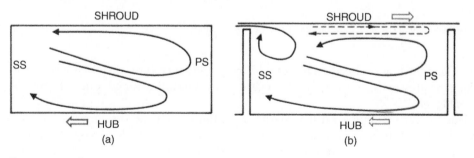

Figure 3.23 Difference between secondary flows in (a) a shrouded and (b) an unshrouded impeller.

The main consequences of the tip leakage flow are:

- a reduction of the pressure rise across the impeller because the tip clearance gap reduces the effective width of the channel on which the blade forces are acting
- an increase in the meridional velocity required to maintain the same overall mass flow in the compressor further reduces the energy input in backward leaning impellers
- even return flow ($V_m < 0$) is likely to occur through this gap and depending on the amount of backward lean it may create recirculating flows ($V_m < 0$) in the impeller, requiring extra energy input
- the roll-up of the clearance flow in the tip vortex creates extra blockage and losses; this vortex starts at the leading edge and has an important impact on the impeller flow stability (Yamada et al. 2011; Hazby and Xu 2009)
- the shear force on the shroud wall is in the opposite direction to the impeller and creates a fluid movement towards the pressure side, resulting in extra vorticity losses and a further decrease in the pressure rise because of increasing slip.

The shroud passage vorticity increases with the inlet boundary layer and can be controlled by a careful design of the inlet duct. In the absence of prerotation, the absolute flow on the shroud wall of an unshrouded impeller has a small axial velocity and zero peripheral velocity. The corresponding relative velocity reinforces a movement of the fluid from the suction to the pressure side, i.e. a movement that is opposite to the shroud passage vortex (Figure 3.24a) and has a favorable effect on the secondary flows.

The shear forces on a rotating shroud induce a tangential velocity component in the direction of rotation, i.e. pointing towards the suction side and reinforcing the secondary flow in the impeller (Figure 3.24b). This prerotation may be enhanced by the leakage flow upstream of the leading edge of a shrouded impeller. The swirl velocity of the recirculating flow can even be larger than the shroud velocity because it is defined by the conservation of tangential momentum from impeller exit to the inlet shroud and by the friction forces in the shroud cavity. Important here is the width and orientation of the gap in a shrouded impeller (Figure 3.25) (Mischo et al. 2009). A too small gap perpendicular to the shroud wall may create a jet perturbing the flow and thickening the shroud inlet boundary layer. Reducing the radial component by widening and inclining the gap in the streamwise direction favors the attachment of the leakage flow on the shroud wall by the Coanda effect.

Unshrouded impellers are preferred because they are easier to manufacture and provide high performance if the clearance can be controlled. Shrouded impellers are mostly used in multistage compressors where thermal dilatation of the shaft prevents an accurate control of the clearance in all stages. The leakage flow is then limited by the axial seals on the shroud, which are insensitive to axial displacements (Figure 3.25). Shrouded impellers are also used to better control the leakage flow in low *NS* compressors.

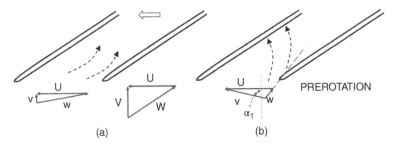

Figure 3.24 Effect of prerotation on secondary flows near the shroud.

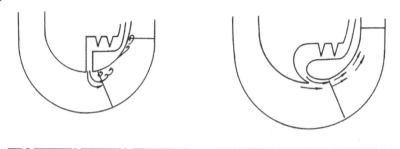

Figure 3.25 Leakage flow near the shroud.

3.2.4 Full 3D Geometries

Blade lean (Figure 3.26) is commonly used in axial compressors and turbines to modify the spanwise pressure gradient in the same way as a change in the meridional curvature. The main consequences are a change in blade loading and a spanwise redistribution of the flow influencing the outlet velocity distribution by modified secondary flows.

Lean is defined here as the angle between the leading edge stacking line and the meridional plane (containing the axis of rotation). This definition is different from the one commonly used in axial compressors and turbines where the lean angle is measured in a plane perpendicular to the blade chord. A circumferential shift of the blade sections defined by a stacking line is more convenient for centrifugal impellers because the blade chord is not well defined. Near the trailing edge the angle between the blade and the meridional plane is commonly called rake. Changing the lean modifies the trailing edge rake because they are related by the wrap angle (Equation (5.13)). Lean also modifies the stress distribution in the blade.

The lean angle can be constant (straight stacking line) or can vary along the blade height (Sugimura et al. 2012; Hazby et al. 2017; Hehn et al. 2017). Conventionally, the lean is positive when the angle between the suction side and the hub wall is obtuse ($>90°$). Positive lean results in more forward rake at the trailing edge.

The influence of lean on the pressure gradients in a centrifugal impeller is illustrated in Figures 3.27 and 3.28. The meridional curvature and centrifugal forces create a hub to shroud pressure gradient defined by Equation (3.1). It is schematically shown in Figure 3.27a, where

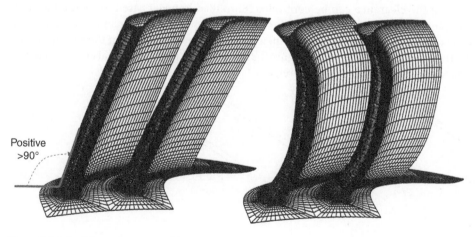

Positive
$>90°$

Figure 3.26 Straight and compound lean.

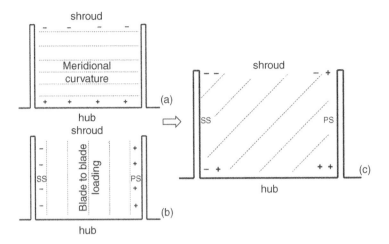

Figure 3.27 Pressure distribution in a crosswise plane with zero lean.

the blade to blade pressure gradient is assumed to be zero. Regions with higher pressure are indicated with a + sign, regions with lower pressure have a − sign. The blade to blade loading in the circumferential direction increases the pressure on the pressure side and lowers it on the suction side. This is shown in Figure 3.27b, where the hub to shroud pressure gradient is assumed to be zero. Combining the two pressure gradients provides a pressure field over the cross section, as shown in Figure 3.27c. The highest pressure occurs in the hub pressure side corner and the lowest one in the shroud suction side corner. The slope of the iso-pressure lines depends on the relative strength of the meridional and blade to blade pressure gradients.

Repeating the same exercise on a cross section with positive lean results in the pressure distributions shown in Figure 3.28. The hub to shroud pressure gradient, being a function of the pitchwise averaged velocity and the meridional curvature, is unchanged (Figure 3.28a). Shifting the individual blade sections in the circumferential direction does not, in case of zero blade loading, alter the pressure at the hub and shroud because the blades do not exert any force on the flow. The suction to pressure side pressure variation is a function of the blade curvature and Coriolis forces (Equation (3.13)). As there is no mechanism to create a pressure gradient from

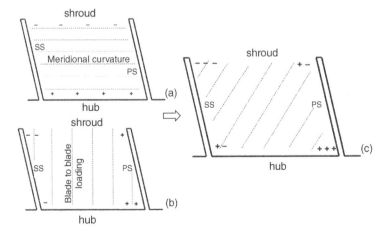

Figure 3.28 Influence of the blade lean on the pressure distribution in a crosswise plane.

hub to shroud, iso-pressure lines must remain perpendicular to both walls (Figure 3.28b). As a consequence the average pressure at the hub is larger than at the shroud. This is reflected in the combined pressure field shown in Figure 3.28c. The increase in the static pressure at the hub and the decrease at the shroud is similar to what would result from a decrease in the meridional curvature radius $\Re_n$. Hence lean is an additional degree of freedom in the design of an impeller.

The change in the spanwise pressure gradient has an impact on the meridional velocity distribution and hence also on the blade loading. This has been experienced in a parametric study comparing the flow in radial impellers with leans corresponding to $0°$ and $45°$ rake. Figure 3.29 shows the change in the reduced static pressure ($P_{Ro} = P - \frac{\rho U^2}{2}$). An increase in the average pressure at the hub and a decrease at the shroud can be observed. Flow changes start at the leading edge and act up to the trailing edge. The hub to shroud pressure gradient resulting from lean may be relatively small but applies along the complete blade length.

Lean allows the velocity distribution on the impeller vanes to be changed without changing the $\beta(s)$ or meridional contour. Although the blade forces are limited to the impeller, they have also an impact on the flow just upstream and downstream of the impeller. Lean may result in a non-negligible change in the radial velocity distribution and absolute flow angle at the impeller exit (Figure 3.30).

It turns out that lean and rake modify the position of the low energy zone (wake) and the total pressure distribution at the impeller exit. The change in diffuser inlet velocity, shown in Figure 3.30, impacts on the flow in a vaneless diffuser, as illustrated in Figure 3.31. This phenomenon has been studied by Rebernick (1972) and Ellis (1964) and will be further elaborated in Section 4.1.3. Everitt and Spakovszky (2013) claim that the spanwise non-uniformity defines the type of diffuser rotating stall (spike or modal) (Section 8.5).

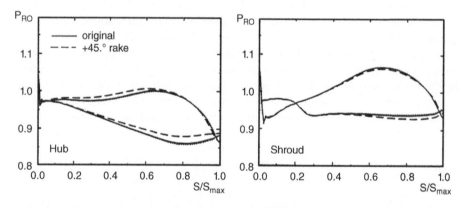

Figure 3.29 Blade to blade pressure variation at the hub and shroud for an impeller with and without lean.

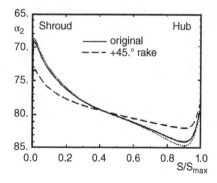

Figure 3.30 Diffuser inlet flow angle with and without lean.

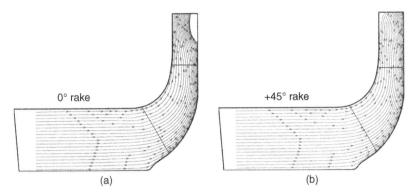

Figure 3.31 Streamlines in the meridional plane of a radial impeller and diffuser at (a) 0° rake and (b) 45° rake.

The centrifugal forces increase the tension on the side of the sharp angle and increase the compression on the side of the obtuse angle when introducing lean to an uncambered rotor blade of an axial turbomachine. The way lean influences the stress concentrations in a centrifugal impeller is much more complex because of the meridional curvature and the double camber of the blades. Only a detailed finite element stress analysis can provide an accurate stress distribution.

Optimization studies by Verstraete et al. (2008) demonstrated that for a particular impeller geometry a 10° negative leading edge lean (opposite to the impeller rotation) results in lower stresses and an improved efficiency. The impact of lean on the compressor efficiency has also been studied by Oh et al. (2011b) and Xu and Amano (2012). The CFD results of the latter show that a 4° negative lean results in the highest peak efficiency, while 4° positive lean results in a wider operating range. Optimum lean is still a subject of research in pumps, compressors, and turbines. Optimization tools to find the optimum combination of design parameters are valuable to make optimal use of this degree of freedom.

Changing the sign of the lean from hub to shroud results in blades with compound lean. The consequence is an increase in the average pressure at the hub and shroud with a decrease at midspan (Figure 3.32). Inside the blade passage the mass flow converges towards the midspan, resulting in a local increase in blade loading. The unloading of the blades at the hub and shroud sections has a favorable impact on the development of secondary flows. The spanwise pressure gradient on the blade surfaces also favors the migration of the low energy fluid towards the midspan and hence favorably influences the flow structure at the trailing edge.

It has been shown by Poulain and Janssens (1980) that it is possible to design backward leaning impellers with radial fibered blades (blades that are generated by radii). Combining positive rake with values of $\gamma < 90°$ allows the generation of backward leaning blades with radial sections at the different axial positions as show by the points B and B′ on Figure 3.33.

Figure 3.32 Influence of compound lean on the spanwise pressure gradients.

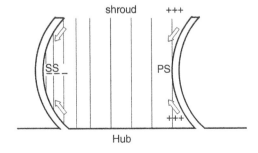

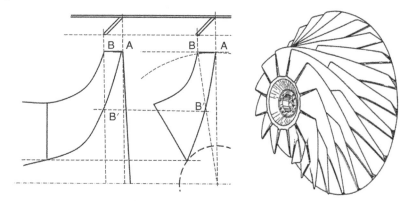

Figure 3.33 Backward leaning impeller with radial fibered blades (from Poulain and Janssens 1980).

It is expected that the stresses in these radial fibered blades will be closer to the ones in pure radial compressor blades $\gamma < 90°$.

3.3 Performance Predictions

Performance prediction methods for centrifugal compressors are very different from the ones for axial compressors because a large part of the energy input results from deflecting the flow from the axial to the radial direction, i.e. by increasing the peripheral velocity between the inlet and outlet $(U_2^2 - U_1^2)/2$. This large amount of additional energy input by the centrifugal forces does not create additional friction losses and can be considered as isentropic. It decreases the relative importance of the diffusion losses which play a major role in axial compressor performance predictions. Flow separation no longer prevents high performance or stable operation and centrifugal compressors can operate with separated flow, as illustrated by the measurements of Eckardt (1976) (Figure 3.34). Contrary to axial compressors, a net positive outflow is observed in the separated flow zone, filled with low energy fluid by the secondary flows (Section 3.2.3).

The following sections present two types of performance prediction models for radial impellers, the diffusion model and the two-zone model.

3.3.1 Flow in Divergent Channels

Before discussing the viscous flow in radial impellers, we first recall some basic ideas about the flow in diverging channels. The flow at the exit of a divergent channel can be defined in two different ways depending on the geometry.

The velocity at the exit of channels with a small divergence angle ($2\theta < 9°$) (Figure 3.35a) can be calculated by applying continuity between the inlet and outlet section, corrected for the boundary layer blockage ϵ_2:

$$\dot{m} = \rho_2 \, W_2^i \, A_2 \, (1 - \epsilon_2) = \rho_2 \, W_2^i \, A_2^i = \rho_1 \, W_1 \, A_1 \tag{3.30}$$

The static pressure is assumed to be uniform over the outlet section and is defined by the isentropic velocity in the central part where the total pressure is equal to the inlet total pressure. Hence

$$P_2 = P_1^o - \frac{\rho \, W_2^{i,2}}{2} \tag{3.31}$$

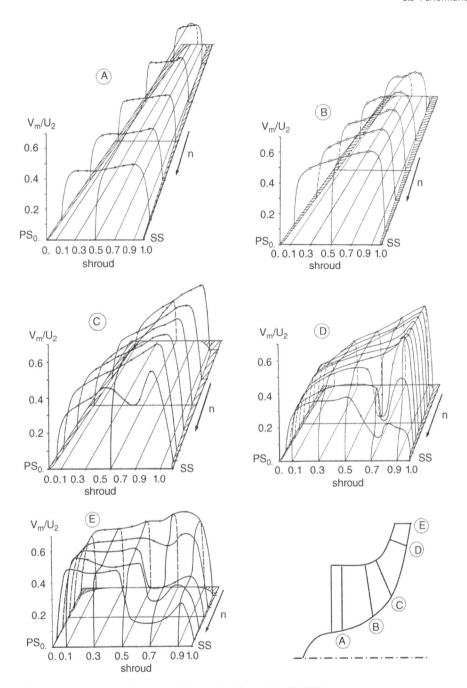

Figure 3.34 Separated flow in radial impellers (from Eckardt 1976).

and the outlet static pressure is a function of the blockage factor ϵ_2, which in turn depends on the Reynolds number and divergence angle.

When increasing the divergence angle above a limit, separation will occur before the end of the channel outlet. It is known that once the flow is separated no further diffusion takes place. Hence $W_2^i = W_{SEP}$ and the static pressure remains constant from the separation point to the

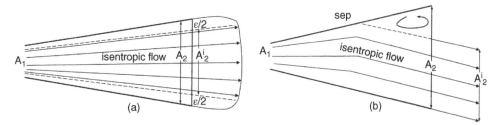

Figure 3.35 Flow in channels with (a) a small and (b) a large divergence angle.

exit of the channel. The outlet flow conditions are no longer a function of the blockage but of the diffusion that took place upstream of the separation point (Figure 3.35b). Downstream of the separation point appears a zone, commonly called a jet, with almost isentropic flow, and a death water zone, commonly called a wake, filled with low energy fluid without net mass flow. The extension of the latter depends on the velocity in the isentropic jet W_j, assumed to be equal to the velocity at separation point W_{SEP}:

$$W_j \approx W_{SEP} = \frac{W_1}{DR} \tag{3.32}$$

The diffusion ratio DR depends on the channel geometry and inlet flow distortion. The exit static pressure is defined by the isentropic jet velocity

$$P_2 = P_1^o - \frac{\rho\, W_{2,j}^2}{2} \tag{3.33}$$

In straight channels the pressure in the wake is assumed to be equal to that in the jet. The jet velocity being known, the size of the wake (ϵ_2) can be defined from continuity assuming zero net outflow in the wake:

$$\dot{m} = \rho_2\, W_{2,j}\, A_2\, (1 - \epsilon_2) = \rho_2 \frac{W_1}{DR} A_2\, (1 - \epsilon_2) \tag{3.34}$$

3.3.2 Impeller Diffusion Model

A common way to predict the impeller performance is by correlating the losses to the overall diffusion (Lieblein et al. 1953). The flow model for centrifugal compressors, described here, is due to Vavra (1970) and is an extension of the diffusion model by considering also the change in peripheral velocity. The average flow characteristics at the inlet and outlet are related by constant rothalpy along a streamline:

$$\frac{Ro_1}{C_p} = T_1 + \frac{\widetilde{W}_1^2}{2C_p} - \frac{\widetilde{U}_1^2}{2C_p} = T_2 + \frac{\widetilde{W}_2^2}{2C_p} - \frac{U_2^2}{2C_p} = \frac{Ro_2}{C_p} \tag{3.35}$$

The model distinguishes between the energy input by centrifugal forces and by internal diffusion:

$$T_2 - T_1 = \frac{U_2^2 - \widetilde{U}_1^2}{2C_p} + \frac{\widetilde{W}_1^2 - \widetilde{W}_2^2}{2C_p} = (T_u - T_1) + (T_2 - T_u) \tag{3.36}$$

The first term depends only on the radius change and is independent of the flow in the impeller. It is assumed that the corresponding pressure rise ($P_u - P_1$) occurs without losses

Figure 3.36 T–S diagram of radial compressor (diffusion model).

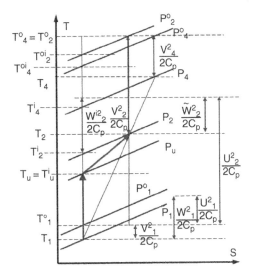

and can therefore be calculated by isentropic flow relations (Figure 3.36):

$$\frac{P_u}{P_1} = \left(\frac{T_u}{T_1}\right)^{\frac{\kappa}{\kappa-1}} = \left(1 + \frac{U_2^2 - \tilde{U}_1^2}{2C_p T_1}\right)^{\frac{\kappa}{\kappa-1}} \tag{3.37}$$

The second part of the temperature rise is due to the deceleration of the relative flow in the impeller blade channels:

$$T_2 - T_u = \frac{\widetilde{W}_1^2 - \widetilde{W}_2^2}{2C_p} \tag{3.38}$$

and results in an entropy increase from S_1 to S_2 (Figure 3.36). The isentropic outlet temperature T_2^i is defined by

$$T_2^i - T_u = \frac{\widetilde{W}_1^2 - W_2^{i,2}}{2C_p} \tag{3.39}$$

where W_2^i is the isentropic velocity at the outlet and, according to Section 3.3.1, a function of the diffusion ratio DR or blockage ϵ_2. The efficiency of this flow deceleration from P_u to P_2

$$\eta_W = \frac{\Delta T^i}{\Delta T} = \frac{T_2^i - T_u}{T_2 - T_u} = \frac{\widetilde{W}_1^2 - W_2^{i,2}}{\widetilde{W}_1^2 - \widetilde{W}_2^2} \tag{3.40}$$

is called wheel diffusion efficiency. It is lower than in axial compressors because of the complexity of the flow channel where curvature in the axial and radial plane limits the overall diffusion and increases the losses.

The corresponding static pressure rise is then defined by the isentropic relation

$$\frac{P_2}{P_u} = \left(\frac{T_2^i}{T_u}\right)^{\frac{\kappa}{\kappa-1}} \tag{3.41}$$

The main problem in using this model for the impeller outlet flow calculation is the estimation of η_W as a function of the flow and impeller geometry. In analogy with the diffusion factor for axial compressors, Vavra (1970) has tried to correlate η_W to the velocity deceleration $\widetilde{W}_2/W_{1,S}$

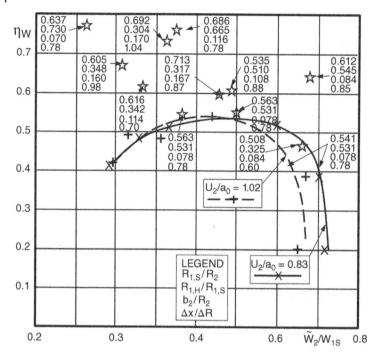

Figure 3.37 Variation of wheel diffusion efficiency with divergence (from Vavra 1970).

(Figure 3.37) where $\widetilde{W}_2$ is the average velocity at the impeller outlet. The stars in Figure 3.37 represent the experimental values of η_W for a number of centrifugal impellers.

The following trends are observed:

- the wheel diffusion efficiency decreases for $\widetilde{W}_2/W_{1,S}$ values larger than 0.5 except for one geometry
- at lower values of $\widetilde{W}_2/W_{1,S}$ the wheel efficiency varies between 0.5 and 0.78 without a clear relation to the geometrical parameters listed in the legend next to each point.

The four values are $R_{1,S}/R_2$, $R_{1H}/R_{1,S}$, b_2/R_2, and $\Delta x/(R_2 - \tilde{R}_1)$, where Δx is the axial length of the impeller. No correlation between these geometrical parameters and η_W could be derived.

Substituting Equations (3.38) and (3.39) into Equation (3.40) gives the following expression for the wheel diffusion efficiency:

$$\eta_W = \frac{\widetilde{W}_1^2 - W_2^{i,2}}{\widetilde{W}_1^2 - \widetilde{W}_2^2} = \frac{\widetilde{W}_1^2 - \widetilde{W}_2^2 - (W_2^{i,2} - \widetilde{W}_2^2)}{\widetilde{W}_1^2 - \widetilde{W}_2^2} = 1 - \frac{W_2^{i,2} - \widetilde{W}_2^2}{\widetilde{W}_1^2 - \widetilde{W}_2^2} \tag{3.42}$$

The last part of this equation can be interpreted as

$$\eta_W = 1 - \frac{\Delta P_{fr}^o/\rho}{\widetilde{W}_1^2 - \widetilde{W}_2^2} = 1 - \omega_{fr} \tag{3.43}$$

The pressure losses increase because of increasing friction when decreasing the outlet section (increasing values of $\widetilde{W}_2/W_{1,S}$ on Figure 3.37) whereas the difference between the inlet and outlet average relative velocity ($\widetilde{W}_1^2 - \widetilde{W}_2^2$) decreases. This explains the decrease in wheel diffusion efficiency for increasing values of $\widetilde{W}_2/W_{1,S}$.

The flow is likely to separate in the impeller for large values of the outlet width (small values of $\widetilde{W}_2/W_{1,S}$). The difference between W_2^i and $\widetilde{W}_2$ (the enumerator in the first RHS term of Equation (3.42)) will not change very much with increasing exit section because it is limited by the maximum diffusion ratio DR ($W_2^i = W_{1,S}/DR$):

$$\eta_W = \frac{\widetilde{W}_1^2 - (W_{1,S}/DR)^2}{\widetilde{W}_1^2 - \widetilde{W}_2^2} \tag{3.44}$$

The denominator increases with impeller exit width, hence the wheel diffusion efficiency depends on the achieved DR. The efficiency at low values of $\widetilde{W}_2/W_{1,S}$ (wider outlet sections) depends on the quality of the diffusing devise. The further downstream separation can be postponed by increasing DR, the smaller the value of $\widetilde{W}_{1,S}/DR$ and the higher will be the maximum wheel diffusion efficiency.

The lines with symbols (+ and x) on Figure 3.37 are experimental data corresponding to the operating lines of a centrifugal impeller with radial ending blades at two different RPM. Changing the mass flow in a given impeller (fixed outlet width) results in a change in divergence of the streamlines between inlet and outlet (Figure 3.38). The phenomena that occur are the same as when changing the divergence angle of a diffuser. At low mass flow (large inlet flow angle) the distance between two upstream streamlines is small and increases very much towards the exit. Flow separation is likely to occur because of low values of $\widetilde{W}_2/W_{1,S}$.

At high mass flow the distance between the two stagnation streamlines at the inlet is large and results in a smaller diffusion towards the exit (Figure 3.38b). The latter is limited by the channel geometry and by the blockage at the exit. It results in large values of $\widetilde{W}_2/W_{1,S}$.

The point of maximum efficiency is likely to be at the limit between the two zones, i.e. where maximum diffusion occurs without separation. Decreasing the mass flow below this value will not increase the diffusion. Only the separated flow zone will become larger (ϵ_2 increases). Increasing the mass flow above the optimal value decreases the diffusion $W_2^2 - W_1^2$ and hence the energy input.

In conclusion, the wheel diffusion efficiency η_W depends on many geometrical parameters. Some of them are shown on Figure 3.37 but no reliable correlation could be found. Some prediction models try to estimate the value of η_W by detailing and adding up the different loss sources in the impeller (Herbert 1980; Oh et al. 1997; Aungier 2000; Swain 2005).

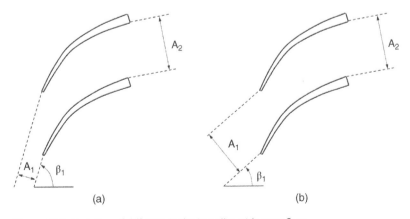

Figure 3.38 Variation of diffusion in the impeller with mass flow.

The following section explains a performance prediction method that is directly related to the different types of flow (unseparated or separated) and depends on only three parameters that have a more direct physical meaning.

3.3.3 Two-zone Flow Model

The inviscid flow calculations predict a large suction side and low pressure side outlet velocity, as shown on the right of Figure 3.39. Flow separation and secondary flows, however, result in a complete rearrangement of the impeller outlet velocity distribution, as shown on the left of Figure 3.39. The unseparated flow has a higher relative velocity and remains attached to the pressure side in a jet-like flow pattern. The low energy fluid is transported to the suction side by secondary flows where it mixes with the tip leakage flows. Although this region of low relative velocity appears already in the impeller it is commonly called wake.

This wake/jet flow model, used to calculate the real outlet flow conditions, is schematically shown in Figure 3.40. It is a representation of the velocity distributions measured by Eckardt (1976) at different cross sections between the impeller inlet and outlet (Figure 3.34). Such a flow model was first presented by Carrard (1923) and then further elaborated by Dean et al. (1972). The flow in the impeller is divided into two parts that are treated separately:

- the inducer, between the inlet and the separation point (SEP)
- the impeller flow between the separation point and the outlet.

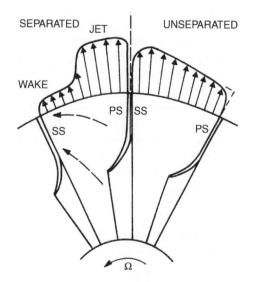

Figure 3.39 Impeller outlet circumferential velocity variation for unseparated and separated flows.

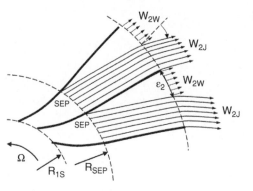

Figure 3.40 Jet and wake flow calculation model.

The latter is further divided into:

- a jet flow (SEP → 2j)
- a wake flow (SEP → 2w).

This division of the rotor flow into subflows is an important feature in the present approach because some real flow phenomena can more easily be accounted for in this way.

One of the most important parameters determining the rotor losses is the wake width ε_2, which is a function of the velocity at the separation point. If the inducer is well designed, an important deceleration can be achieved before separation occurs. In that case the separation point will occur at a lower velocity further downwards to the impeller outlet and the wake will be smaller. In more divergent impeller channels and in case only low diffusion is achieved, separation will occur closer to the inlet and the wake will be much wider. In the present model the impeller losses are mainly due to the wake flow.

The critical area for flow separation is the shroud because of the convex curvature and largest flow deceleration. The relative velocity at the separation point is therefore related to the shroud leading edge velocity by the diffusion ratio DR:

$$DR = \frac{W_{1,S}}{W_{SEP}} \tag{3.45}$$

The larger diffusion realized in the impeller, the smaller will be the velocity at the separation point and hence the smaller the inviscid velocity at the exit.

Limits of maximum diffusion have been given by Dean (1972) for different diffusing devices (Figure 3.41). Only moderate values of DR are observed in centrifugal impellers. The complex 3D nature of the flow results in important velocity gradients which, together with the Coriolis and curvature effects in the boundary layers, limit the maximum achievable diffusion. A larger DR can be achieved in axial compressor blades because there is only one curvature in the blade

Figure 3.41 Diffusion ratio for various compression systems (from Dean 1972).

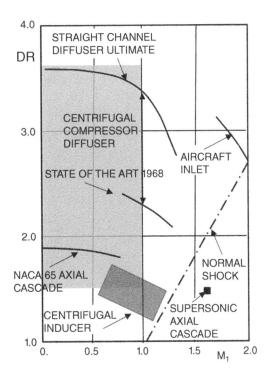

to blade plane. Straight diffusers (no curvature at all) allow the highest *DR*. However, the predicted ultimate value on this graph is too optimistic and has not yet been reached with simple geometries.

The flow deceleration by a normal shock depends on the upstream Mach number and can be very large. However, the increased blockage by shock boundary layer interaction limits the real pressure rise, as shown by the experimental data for a supersonic axial cascade (Breugelmans 1972).

According to Dean (1972), the efficiency of a compressor with degree of reaction $r \approx 0.5$ relates to *DR* and CP_D (pressure recovery in the downstream diffuser), as shown in Figure 3.42. For given CP_D, the efficiency increases with *DR*. However, the increase becomes rather small for values of $DR > 1.4$. It turns out to be more interesting to pay attention to the diffuser pressure rise than to a further increase in the impeller diffusion.

Critical values of the radial impeller relative velocity ratio have also been studied by Dallenbach (1961). He predicts maximum values of *DR* between 1.48 and 1.6 depending on the velocity gradient, which is in agreement with the data in Figure 3.41.

The expected value for *DR* is an input to the analysis program and the performance prediction will be accurate only if the assumed *DR* is achieved. The *DR* is mainly a function of the rotor geometry, which means that the value used in this 1D analysis method must still be confirmed by a detailed analysis of the 3D design.

The following assumptions are proposed by Dean (1972) to calculate the flow in the impeller:

- The Mach number remains constant in the jet between the separation point and the impeller exit.
- The jet flow is considered as isentropic although some friction and clearance losses may be introduced.
- The mass flow in the wake is not zero but contains all the low energy fluid resulting from friction and leakage losses.

The static temperature at any position can be calculated from the condition of constant rothalpy in the impeller:

$$C_p \left(T - T_1 \right) = \frac{W_1^2 - W^2}{2} + \frac{U^2 - U_1^2}{2} \tag{3.46}$$

Hence T_{SEP} can be calculated as a function of *DR* and U_{SEP}:

$$T_{SEP} = T_{1,S} + \frac{W_{1,S}^2}{2C_p} \left(1 - \frac{1}{DR^2} \right) + \frac{U_{SEP}^2 - U_{1,S}^2}{2C_p} \tag{3.47}$$

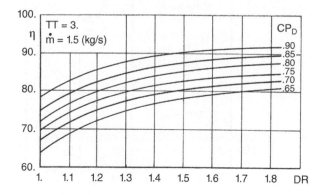

Figure 3.42 Influence of *DR* and CP_D on compressor efficiency (from Dean 1972).

Figure 3.43 *T–S* diagram of a centrifugal compressor (separated flow model).

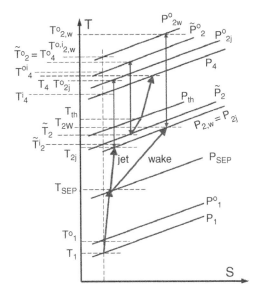

The entropy increase from the inlet to the separation point, shown in Figure 3.43, results from an estimation of the incidence and, in the case of transonic inlet flow, also from the shock losses (Rodgers 1962):

$$\Delta h_{inc} = \left[0.833 \left(\frac{i - i_{opt}}{C^{te} \Delta i} \right)^2 + 0.1667 \frac{i - i_{opt}}{C^{te} \Delta i} \right] \frac{W_{1,S}^2}{2} \tag{3.48}$$

where $C^{te} = 2.5$ for incidences larger than optimum and $C^{te} = 2$ for incidences smaller than optimum. i_{opt} is defined in Figure 2.26. Δi is defined by

$$\Delta i = 2.5 + 0.15(12.5 - 0.1 \, \tilde{\beta}_1) \frac{(\tilde{M}_1 - 1.2)^2}{2}$$

The following additional enthalpy loss occurs when the shroud inlet Mach number exceeds 1:

$$\Delta h_{M>1} = 0.1(M_{1,S}^2 - 1) \frac{W_{1,S}^2}{2} \tag{3.49}$$

Assuming a constant Mach number in the jet, defined by the separation point Mach number, $M_{2j} = M_{SEP}$, the jet velocity is defined by

$$W_{2j} = W_{SEP} \frac{T_{2j}}{T_{SEP}}$$

which results in

$$T_{2j} = T_{1,S} + \frac{W_{1,S}^2}{2C_p} \left(1 - \frac{1}{DR^2} \frac{T_{2j}}{T_{SEP}} \right) + \frac{U_2^2 - U_{1,S}^2}{2C_p} \tag{3.50}$$

Analyzing the impact of the radial position of the separation point on performance (Equation (3.50)) indicates that an error on R_{SEP} has only a small influence on T_{2j}. The separation point is therefore assumed to be located halfway between the shroud inlet and outlet.

Assuming an isentropic jet means that further losses can only be due to the wake. It is possible to account more explicitly for friction losses (dependent on geometry, Reynolds number, and roughness) and clearance losses (function of clearance over outlet width δ_{cl}/b_2) (Figure 1.26). Their impact may be particularly important in low specific speed impellers and allows a distinction between shrouded and unshrouded impellers. It is assumed that the losses affect the outlet static pressure by the same amount as the total pressure.

The friction losses are estimated by means of the hydraulic diameter at inlet (DH_1) and outlet (DH_2) of the impeller, and the hydraulic length LH of the impeller blades.

The hydraulic diameter at the inlet is approximated by

$$DH_1 = 4 \frac{\pi \frac{(R_{S1}^2 - R_{H1}^2)}{Z_r} \cos \beta_1}{2\pi \frac{(R_{S1} + R_{H1})}{Z_r} \cos \beta_1 + 2(R_{S1} - R_{H1})} \tag{3.51}$$

and at the outlet by

$$DH_2 = 4 \frac{\frac{2\pi R_2 b_2}{Z_r}(1 - \epsilon_2) \cos \beta_{2bl}}{2(1 - \epsilon_2)\frac{2\pi R_2}{Z_r} \cos \beta_{2bl} + 2b_2} \tag{3.52}$$

The blade hydraulic length is approximated by

$$LH \approx \frac{1}{2}\left(x_2 - x_1 + R_2 - \frac{(R_{H1} + R_{S1})}{2} \right) \frac{\pi}{2} \frac{1}{\cos \frac{(\tilde{\beta}_1 + \beta_{2bl})}{2}} \tag{3.53}$$

and the friction losses are defined by

$$\Delta h_{fr} = 4C_f LH \frac{W_j^2}{2}\left(\frac{1}{DH_1} + \frac{1}{DH_2} \right) \tag{3.54}$$

The wall friction coefficient C_f is a function of Reynolds number (Re) and relative wall roughness k_s/DH_2. It is defined by the implicit formula of Colebrook (1939) (Equation (1.16)).

The clearance losses can be approximated by

$$\Delta h_{cl} = 2.43 \frac{\delta_{cl}}{b_2}\left(1 - \frac{R_{1,S}^2}{R_2^2} \right) U_2^2 \tag{3.55}$$

The jet static pressure and isentropic temperature are then defined by

$$T_{2,j}^i = T_{2,j} - \frac{1}{C_p}\left(\Delta h_{inc} + \Delta h_{M>1} + \Delta h_{fr} + \Delta h_{cl} \right) \tag{3.56}$$

$$P_{2,j} = P_{1,S}\left(\frac{T_{2j}^i}{T_{1,S}} \right)^{\frac{\kappa}{\kappa-1}} \tag{3.57}$$

The mass flow in the jet is defined by

$$\dot{m}_j = 2\pi R_2 b_2 k_{b2} W_{2j} \rho_{2j} \cos \beta_{2j}(1 - \epsilon_2) \tag{3.58}$$

where $k_{b2} = \frac{Z_r \delta_{th}}{2\pi R_2} \frac{1}{\cos \beta_{2bl}}$ stands for the blockage by the blades and $\rho_{2j} = \frac{P_{2j}}{R_G T_{2j}}$.

The non-zero mass flow in the wake can be characterized either as a fraction λ of the total mass flow

$$\lambda = \frac{\dot{m}_w}{\dot{m}} \tag{3.59}$$

or by the parameter v, defining the ratio of the wake over jet relative velocity:

$$v = \frac{W_{2w}}{W_{2j}} \tag{3.60}$$

By fixing v, instead of λ, the mass flow in the wake will depend also on the extend of the wake and hence on the width of the exit section. Obviously v is not a constant, but correlates to the geometry, secondary flow, clearance, wake width ϵ_2, etc. However, at the present state of the design process this relation is unknown. Based on the experimental results of Eckardt (1976) and Fowler (1966), this value is estimated at $0.2 < v < 0.3$.

Another important feature of this model is that the same relative outlet angle β_2 is used for the jet and the wake:

$$\beta_{2j} = \beta_{2w} = \beta_2 \tag{3.61}$$

The wake outlet static temperature is calculated from the conservation of rothalpy, applied separately along a jet and a wake streamline. Starting at the same inlet flow conditions and because U_2^2 is equal for jet and wake the temperature difference between the jet and wake is function only of the local relative velocities:

$$C_P \left(T_{2w} - T_{2j} \right) = \frac{W_{2j}^2 - W_{2w}^2}{2} = W_{2j}^2 \left(1 - v^2 \right) \tag{3.62}$$

This indicates that the wake static temperature is considerably higher than the one in the jet.

The zero loading condition at the impeller exit (equal pressure on suction and pressure side at the trailing edge) and the condition of equal pressure on both sides of the border between the jet and the wake justify the assumption that the average static pressure is the same in the jet and the wake (Figure 3.44):

$$\widetilde{P}_{2w} = \widetilde{P}_{2j} \tag{3.63}$$

A more detailed calculation of the circumferential pressure distribution would require more information about the streamline curvature in the jet and wake at the impeller exit. As shown on the T–S diagram (Figure 3.43) the same static pressure for the jet and wake results in a much larger entropy in the wake, reflecting the much larger losses.

Equations (3.62) and (3.63) allow the calculation of the density in the wake

$$\rho_{2w} = \frac{P_{2w}}{R_G \, T_{2w}} \tag{3.64}$$

and hence the mass flow

$$\dot{m}_w = \lambda \dot{m} = 2\pi R_2 b_2 k_{b2} \cos \beta_{2w} \rho_{2w} \epsilon_2 W_{2w} \tag{3.65}$$

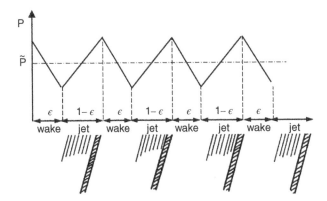

Figure 3.44 Circumferential variation of the impeller exit static pressure.

The value of ϵ_2 can now be defined by comparing the sum of the jet and wake mass flow to the total mass flow:

$$\dot{m} = 2\pi R_2 b_2 k_{b2} \cos\beta_2 \left[(1-\epsilon_2)\rho_{2j} W_{2j} + \epsilon_2\rho_{2w}W_{2w}\right] \tag{3.66}$$

The W_2^i in the calculation of the wheel diffusion efficiency (Equation (3.42)) is equivalent to the isentropic jet velocity W_{2j} at $v = 0$. Substituting the W_{2j} corresponding to different values of DR provides the wheel diffusion efficiency as a function of $\widetilde{W}_2/W_{1,S}$, i.e. for different values of the impeller exit width or for different mass flows in a given impeller (Figure 3.38). The wheel diffusion efficiencies for $DR = 1.4$ and 1.6 are shown in Figure 3.45. Hence the diffusion of the relative velocity in the impeller is the main parameter defining η_W.

At low values of $\widetilde{W}_2/W_{1,S}$ the flow is separated and the performance depends on the diffusion realized prior to separation. Higher values of the diffusion ratio DR means a lower velocity at separation and a lower value for W_{2j}, hence smaller wakes and a higher wheel diffusion efficiency. At increasing values of $\widetilde{W}_2/W_{1,S}$ (a narrower impeller or a given impeller operating at larger mass flow) the extend of the separation zone decreases and the wheel diffusion efficiency increases until the flow is no longer separated. This is the point of maximum impeller efficiency because of maximum diffusion and minimum wake width. A further decrease in the exit width will result in a larger relative velocity with increasing friction and hence a decreasing efficiency.

The diffusion and wake/jet model are in some way equivalent. The scatter in the experimental wheel diffusion efficiencies can be correlated with the change in DR that has been realized. The latter has the advantage that the performance is more directly related to the impeller geometry and to the change in operating point.

The diffusion is very limited at high mass flow (Figure 3.38b) and separation may not occur upstream of the trailing edge. As explained in Section 3.3.1, the outlet velocity can then be defined by the blockage model. A minimum wake width $\epsilon_{2,min}$ in the jet and wake model must be defined to account for the unseparated boundary layers developing along the impeller walls. The boundary layer blockage inside the impeller has been studied by Pampreen (1981). The largest

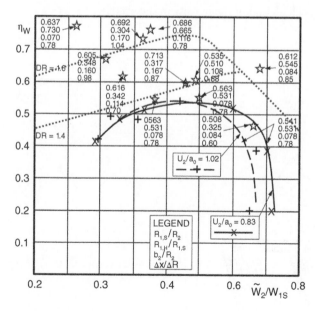

Figure 3.45 Variation of the wheel diffusion efficiency based on *DR* versus experimental wheel diffusion efficiency.

blockage occurred at low mass flow, which corresponds to the operation with large separation zones. He also observed that when increasing the mass flow the blockage does not decrease below 20% or 30%, which must be accounted for by imposing a minimum wake width $\epsilon_{2,min}$.

When Equation (3.66) predicts a wake width $\epsilon_2 < \epsilon_{2,min}$, the jet velocity must be increased until $\epsilon_2 = \epsilon_{2,min}$. Hence less diffusion takes place in the impeller and the efficiency will decrease, as illustrated in Figure 3.45. The change in efficiency is no longer dependent on the DR but on the value of ϵ_{min}. The value of $\widetilde{W}_2/W_{1,S}$, separating the two regimes, depends on DR and on $\epsilon_{2,min}$. For large values of DR the flow decelerates to smaller velocities and maximum efficiency is obtained in impellers with wider exit sections (smaller $\widetilde{W}_2/W_{1,S}$ values). For lower values of DR a smaller exit width is required to avoid separation and a lower η_{max} occurs at a larger value of $\widetilde{W}_2/W_{1,S}$.

The difference between the impeller outlet flow direction and blade angle is quantified by the slip factor σ or work reduction factor μ discussed in more detail in Section 3.4. Making a first guess of β_2 allows the velocity triangles of jet and wake to be defined (Figure 3.46):

$$V_{u2j} = -W_{2j} \sin \beta_2 + U_2$$
$$V_{m2j} = W_{2j} \cos \beta_2$$
$$V_{u2w} = -W_{2w} \sin \beta_2 + U_2$$
$$V_{m2j} = W_{2w} \cos \beta_2$$

Except for forward curved blades ($\beta_2 < 0$), this results in a larger absolute velocity in the wake than in the jet.

The outlet total temperature is a measure for the energy input:

$$T^\circ_{2j} = T_{2j} + \frac{V^2_{2,j}}{2C_P} \quad \ll \quad T^\circ_{2w} = T_{2w} + \frac{V^2_{2,w}}{2C_P} \tag{3.67}$$

It is higher in the wake than in the jet because both the static temperature and the absolute velocity are higher in the wake.

The jet total pressure at the impeller outlet can be defined by the isentropic relation

$$P^o_{2,j} = P_{2,j} \left(\frac{T^o_{2,j}}{T_{2,j}} \right)^{\frac{\kappa}{\kappa-1}} \tag{3.68}$$

In the same way one can also calculate the wake total pressure $P^o_{2,w}$ at the impeller outlet.

3.3.4 Calculation of Average Flow Conditions

The average impeller outlet (diffuser inlet) flow conditions can be calculated by assuming a sudden mixing at the impeller exit resulting in an instantaneous uniformization of the flow

Figure 3.46 Jet and wake velocity triangles.

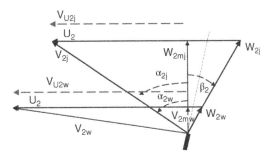

(Johnston and Dean 1966). The following equations are used to calculate the losses and mixed out flow conditions. They express the conservation of the jet and wake radial momentum:

$$\dot{m}_j V_{R2j} + \dot{m}_w V_{R2w} = \dot{m}\widetilde{V}_{R2} + (\widetilde{P}_2 - P_{2j})b_2 2\pi R_2 \tag{3.69}$$

the jet and wake tangential momentum

$$\dot{m}_j V_{u2j} + \dot{m}_w V_{u2w} = \dot{m}\widetilde{V}_{u2} \tag{3.70}$$

the conservation of total energy

$$\dot{m}_j C_p T^o_{2j} + \dot{m}_w C_P T^o_{2w} = \dot{m}C_P \left(\widetilde{T}^o_2 - \Delta T_{diskfriction} \right) \tag{3.71}$$

and mass flow

$$2\pi b_2 R_2 V_{R2} \rho_2 = \dot{m} \tag{3.72}$$

The temperature rise $\Delta T_{diskfriction}$ is a non-adiabatic heat addition due to friction in the side cavities, quantified later. The average total temperature $\widetilde{T}^o_2$ and average tangential velocity $\widetilde{V}_{u2}$ are readily obtained from Equations (3.71) and (3.70). Making a first guess of $\tilde{\rho}_2$ allows a first estimation of $\widetilde{V}_{R2}$ from Equation (3.72), so that $\widetilde{P}_2$ can be obtained from Equation (3.69).

The average static temperature downstream of the impeller $\widetilde{T}_2$ can then be defined by

$$\widetilde{T}_2 = \widetilde{T}^o_2 - \frac{\widetilde{V}^2_{u2} + \widetilde{V}^2_{m2}}{2C_P} \tag{3.73}$$

which allows an update of the mean density $\tilde{\rho}_2 = \widetilde{P}_2/(R_G \widetilde{T}_2)$. Previous procedure is repeated until convergence, i.e. until the calculated value of $\tilde{\rho}_2$ does not change anymore.

$\widetilde{V}_{u2}$ and $\widetilde{V}_{R2}$ constitute the diffuser inlet flow conditions. The isentropic diffuser inlet temperature is defined by

$$\widetilde{T}^i_2 = T^o_1 \left(\frac{\widetilde{P}_2}{P^o_1} \right)^{\frac{\kappa-1}{\kappa}}$$

and the impeller total to static isentropic efficiency is

$$\eta_{r,T-S} = \frac{\widetilde{T}^i_2 - T^o_1}{\widetilde{T}^o_2 - T^o_1} = T^o_1 \frac{(\widetilde{P}_2/P^o_1)^{\frac{\kappa-1}{\kappa}} - 1}{\widetilde{T}^o_2 - T^o_1} \tag{3.74}$$

3.3.5 Influence of the Wake/Jet Velocity Ratio *v* on Impeller Performance

Except for the incidence, friction, clearance, and shock losses, the compressor performance depends on *DR*, *v* or *λ* and ϵ_{min}. Defining relevant values for those parameters is of utmost importance for an accurate performance prediction.

From the continuity (Equation (3.66)) and neglecting the difference in density, we can derive

$$\epsilon_2 \cong \frac{1-C}{1-v} \tag{3.75}$$

where

$$C = \frac{\dot{m}}{2\pi R_2 b_2 \cos \beta_2 \rho_{2j} W_{2j}} \cong \frac{\widetilde{W}_2}{W_{2j}} \tag{3.76}$$

C is a value indicating the part of the outlet section filled by the jet if there is no mass flow in the wake ($v = 0$). For a fixed impeller geometry it decreases with decreasing mass flow. At constant mass flow it decreases with increasing impeller outlet area.

Combining Equations (3.59), (3.65), and (3.66) we find the following expression for λ:

$$\lambda = \frac{\dot{m}_w}{\dot{m}} \approx \frac{v}{1-v}\frac{1-C}{C} \tag{3.77}$$

The influence of the wake/jet velocity ratio on the impeller efficiency is evaluated from the following expressions for static to static efficiency. An isentropic jet flow corresponds to

$$\eta_{j,S-S} \approx 1 \tag{3.78}$$

The static to static wake efficiency η_w is different from the velocity diffusion efficiency η_W and defined by

$$\eta_{w,S-S} = \frac{T_{2j} - T_1}{T_{2w} - T_1} \tag{3.79}$$

and according to Equation (3.62) can also be expressed as

$$\eta_{w,S-S} = T_1 \frac{(P_{2w}/P_1)^{\frac{\kappa-1}{\kappa}} - 1}{T_{2j} + (1-v^2)\frac{W_j^2}{2C_p} - T_1} \tag{3.80}$$

The impeller static to static efficiency, based on the average outlet flow conditions (Figure 3.43), is

$$\tilde{\eta}_{r,S-S} = \frac{\tilde{T}_2^i - T_1}{\tilde{T}_2 - T_1} \quad \text{or} \quad \tilde{\eta}_{r,S-S} = T_1 \frac{(\tilde{P}_2/P_1)^{\frac{\kappa-1}{\kappa}} - 1}{\tilde{T}_2 - T_1} \tag{3.81}$$

The wake and impeller efficiency are functions of the impeller inlet and outlet flow conditions. The last ones depend on the peripheral speed, the diffusion ratio DR, blade angles and impeller outlet width (characterized by C), and the wake/jet velocity ratio v.

Figure 3.47 shows the results of some calculations that have been made for an impeller with 40° backward leaning blades operating in air at 400 m/s peripheral speed. It illustrates the variation of the impeller static efficiency with v in a given geometry ($C = $ constant). Higher v values result in more efficient wakes but less efficient impellers because of the corresponding increase in the mass flow in the wake.

The wake efficiency increases from 83.6% (at $v = 0$) to 87.0% (at $v = 0.5$) and is independent of C. The low efficiency in the wake, at low values of v, results in a higher static temperature than in the jet (Equation (3.62)). However, this does not contribute to the pressure rise because the pressure is not obtained by diffusion but is set equal to the pressure in the jet. The impact of the low efficiency of the wake at lower values of v on the rotor performance is smaller than at large values of v because of the lower mass flow involved.

At $v = 0$ there is no mass flow in the wake so the corresponding losses are entirely due to the sudden expansion mixing losses of the isentropic jet at the impeller exit.

The points where $v = C$ correspond to a wake covering the entire outlet section with no jet flow. This corresponds to maximum wake losses (at the given v value) and defines the minimum impeller efficiency. Higher C values result in an increasing overall efficiency because more flow passes in the highly efficient jet.

There is some discussion about what condition, $v = C^{te}$ or $\lambda = C^{te}$, provides the most realistic prediction of the flow. Doubling the impeller exit width (decreasing C from 0.75 to 0.50) at constant value of v results in a decrease in the rotor efficiency by 5% to 7% points, which seems

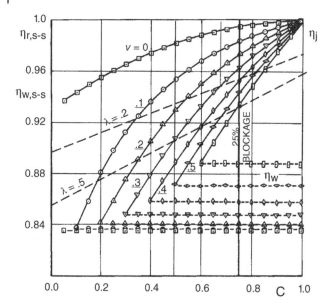

Figure 3.47 Variation of rotor and wake efficiency with C and v.

reasonable for such a large change in the geometry (Figure 3.47). Assuming unchanged mass flow in the wake ($\lambda = C^{te}$) while doubling the impeller exit width results in a decrease in the rotor efficiency of less than 3% (Figure 3.47). Based on comparisons with experimental results it is estimated that the efficiency drop at constant v is more realistic than at constant λ.

When the impeller outlet width is reduced (increased C) at constant value of v, the wake width ϵ_2 will become smaller until it reaches a minimum value $\epsilon_{2,min}$, defined by boundary layer blockage. Although there is no flow separation at C values above this limit (lower values of b_2/R_2) there will be losses by mixing the jet with the boundary layers.

A further decrease in the impeller exit width gives rise to less flow deceleration in the impeller and hence a decrease in η_r. Lower values of b_2 result also in narrow cross sections for the jet and therefore lower outlet hydraulic diameters (DH_{2j}). We can conclude from Equation (3.54) that friction losses will increase considerably. Equation (3.55) shows that clearance losses also increase with decreasing impeller width.

The variation of the different impeller losses (separation, clearance, and friction) with exit width is shown in Figure 3.48 in a qualitative way. It makes clear that there exists an optimum value for b_2/R_2 where the total impeller losses are a minimum. The optimum impeller outlet width, based on maximum overall compressor efficiency, may be different from this value because the diffuser and volute losses also depend on b_2/R_2.

3.4 Slip Factor

The impeller outlet dimensions are primarily defined by the required total enthalpy rise $\Delta H = U_2 V_{u2} - U_1 V_{u1}$ where V_{u2} is function of the outlet flow angle β_2, the relative velocity W_2, and peripheral velocity U_2, as shown by the velocity triangle in Figure 1.33. U_2 is defined by the known outlet radius and RPM, so that the problem reduces to the prediction of W_2 and $\beta_{2,fl}$.

Due to the rotationality of the flow in the radial part of the impeller with a limited number of blades, the relative flow will not be tangent to the blade at the rotor exit. The parameters

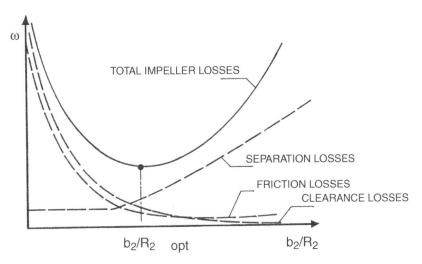

Figure 3.48 Separation, clearance, and friction losses as a function of the impeller outlet width b_2.

commonly used to characterize this effect are the slip factor σ or the work reduction factor μ (Section 1.5.2).

Both definitions are the same for radial ending blades ($\beta_{2,bl} = 0$) or $\dot{m} = 0$ because in those cases $V_{u2}^{\infty} = U_2$:

$$\mu_{\dot{m}=0} = 1 - \frac{\Delta V_u}{V_{u2}^{\infty}(\dot{m} = 0)} = 1 - \frac{\Delta V_u}{U_2} = \sigma_{\dot{m}=0}$$

These factors express the reduction in energy input due to the passage vortex and are therefore very important in any performance prediction. They characterize the amount of energy that has not been added to the fluid but have no direct influence on efficiency because they are not the consequence of losses. Slip factors are normally defined by expressions derived from theoretical considerations and corrected by experimental observations.

Extensive studies on slip and work reduction factors have been made by Stiefel (1965) and Wiesner (1967). The later compared different slip factor correlations with experimental data and concluded that the expression of Busemann (1928) was the most generally applicable one. This correlation is derived from analytical considerations for incompressible, frictionless flow in an impeller with pure radial blades, at zero mass flow. The results are given in graphs in Figure 3.49 and define the slip factor as a function of the inlet over outlet radius ratio $R_{1,S}/R_2$, blade number Z_r, and blade angle $\beta_{2,bl}$. Different graphs are shown for different values of $\beta_{2,bl}$.

Wiesner (1967) confirmed the good agreement of the Busemann correlation with experimental data and proposed the following approximated expressions

$$\sigma = 1 - \frac{\sqrt{\cos \beta_{2bl}}}{Z_r^{0.70}} \tag{3.82}$$

applicable up to a limited blade solidity (sol_{lim}) defined by

$$sol_{lim} = \frac{R_{1,S}}{R_2} \approx \frac{1}{e^{\left(\frac{8.6 \cos \beta_{2bl}}{Z_r}\right)}} \tag{3.83}$$

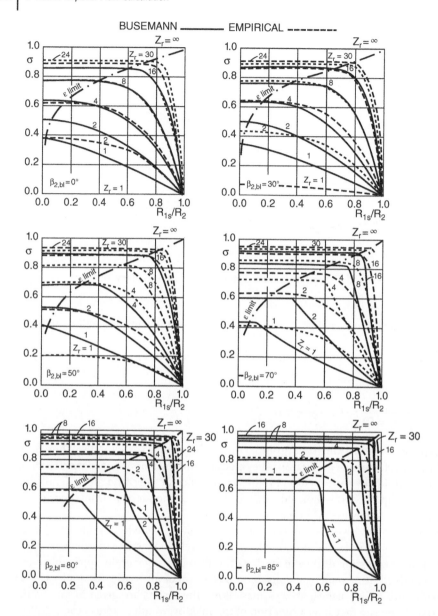

Figure 3.49 Variation of the slip factor according to Busemann and Wiesner (from Wiesner 1967).

The flow is not well guided by the blades inside impellers with a radius ratio $R_{1,S}/R_2$ in excess of this limit and the following relation should be used:

$$\sigma = \left(1 - \frac{\sqrt{\cos\beta_{2bl}}}{Z^{0.70}}\right)\left[1 - \left(\frac{R_{1,S}/R_2 - sol_{lim}}{1 - sol_{lim}}\right)^3\right] \tag{3.84}$$

Equation (1.88) relates the work reduction factor to the blade trailing edge angle and the number of impeller blades. It has been shown by Traupel (1962) that the work reduction factor is not a constant but changes with flow coefficient ϕ_2 (Figure 3.50).

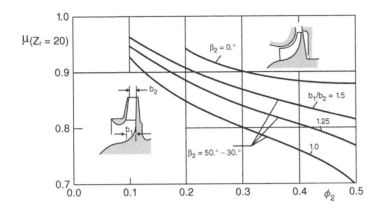

Figure 3.50 Variation of the slip factor with flow coefficient (from Traupel 1962).

A systematic experimental study by Stiefel (1965) indicated that, besides the blade number Z_r, radius ratio R/R_2, blade angle β_{2bl}, and flow coefficient ϕ_2, the work reduction factor is also influenced by:

- the clearance: larger clearance results in smaller work reduction factors by the following amount

$$\mu = \mu_{\left(\frac{\delta_{cl}}{b_2}=0.005\right)} - \left(\frac{\delta_{cl}}{b_2} - 0.005\right) 2.5 \tag{3.85}$$

- the inclination of the meridional streamlines at impeller outlet: an increase in the work reduction factor is observed at a mixed outlet ($\gamma_2 < 90°$)

$$\mu = 1 - (1 - \mu_{(\gamma_2=90°)}) \sin \gamma_2 \tag{3.86}$$

- the diffuser geometry: small differences in the work reduction factor are measured when an impeller is operated with a vaned or vaneless diffusers
- the efficiency: lower slip factors are observed at higher efficiency but the limited amount of experimental data did not allow this to be quantified.

The influence of flow separation on the work reduction factor has been evaluated by Eckert and Schnell (1961). Reducing the passage width to the jet (Figure 3.51), the same vorticity is applied to a smaller section and the work reduction factor changes as follows:

$$\mu_{SEP} = \frac{\mu}{1 - \epsilon_2(1 - \mu)} \tag{3.87}$$

Figure 3.51 Influence of flow separation on the work reduction factor (from Eckert and Schnell 1961).

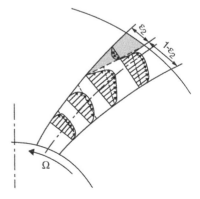

where μ is the work reduction factor for non-separated flows defined by the previous correlations.

Another way to account for flow separation and mass flow changes is to assume that the separated flow is tangent to the blades (zero slip), as proposed by Japikse (1985). The separated flow zone is increasing for decreasing mass flow and as a consequence the work reduction factor is increasing because more mass flow is leaving the impeller with a work reduction factor equal to 1.

The slip factor may further change with geometry (in particular the shape of the trailing edge), test conditions or the running time due to dust deposit.

The following expression for the work reduction factor

$$\mu = \frac{\widetilde{V}_{u2}}{\widetilde{V}_{u2\infty}} = \frac{U_2 - \widetilde{V}_{m2} \tan \beta_{2fl}}{U_2 - \widetilde{V}_{m2} \tan \beta_{2bl}} \tag{3.88}$$

allows the calculation of the outlet relative flow angle as a function of the average outlet meridional velocity component. The last one is calculated by Equations (3.69) to (3.72) and depends on β_{2fl} so that an iterative procedure is required.

Even if the uncertainty on the predicted slip factors is within 5%, it constitutes a rather large error because it is in the first place meant to estimate the amount of energy that has not been added $(1 - \sigma)$ and not the amount that has been added. More accurate predictions may be obtained by comparing slip factors within families of similar impellers, a common practice in industry.

3.5   Disk Friction

The fluid in the cavity between the back side of the rotor disk and the stationary wall of the casing is on one side rotating with the rotor disk, but is without swirl on the fixed wall side (Figure 3.52). The swirling fluid is centrifuged outwards and pushes the other fluid downwards along the stationary casing. The large swirl component generated on the rotating wall is then dissipated by friction on the stationary wall. Once the fluid has reached the bottom of the cavity it is aspirated by the rotating wall and the tangential velocity increases again with the radius when centrifuged outward. This fluid rotation produces intensive whirl and a continuous energy dissipation.

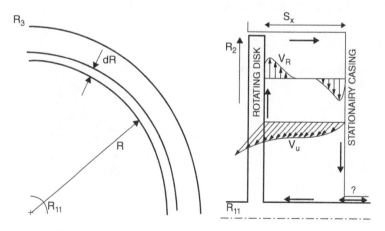

Figure 3.52 Flow inside the disk cavity.

The main effects are:

- extra torque due to the friction on the rotating wall
- heat production when dissipating the swirl velocity, which is conducted through the rotor disk to the impeller fluid or dissipated through the stationary walls.

The extra torque due to this friction is given by

$$M_o = \int_{R_{11}}^{R_2} 2\pi R^2 \tau_w dR \tag{3.89}$$

where τ_w is the wall shear stress. It is proportional to the square of the velocity and function of a friction coefficient:

$$\tau_w = C_f \frac{\rho U^2}{2} \tag{3.90}$$

hence

$$M_o = \int_{R_{11}}^{R_2} C_f 2\pi R^2 \frac{\rho \Omega^2 R^2}{2} \; dR = C_f \pi \frac{\rho \; R_2^5 \Omega^2}{5} \tag{3.91}$$

assuming that R_{11}^5 is negligible compared to R_2^5.

It is common to replace the friction coefficient by a torque or momentum coefficient defined by

$$C_m = \frac{4\pi}{5} C_f \approx 2.5 C_f \tag{3.92}$$

which results in the following expression for the energy dissipation:

$$Pw_{diskfriction} = M_o \Omega = \frac{1}{4} C_m \rho R^5 \Omega^3 \tag{3.93}$$

In the case of a shrouded impeller the disk friction takes place on both sides of the impeller.

A detailed experimental study by Daily and Nece (1960) has shown that the momentum coefficient C_m depends on the type of flow which in term depends on the distance S_x between the rotating and fixed wall (Figure 3.53). Four flow regimes can be observed:

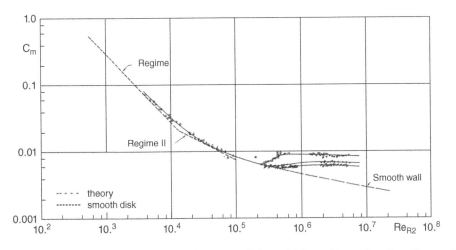

Figure 3.53 Variation of the disk cavity friction coefficient with Reynolds number (from Nece and Daily 1960).

I Two interacting laminar boundary layers.
II Laminar boundary layers separated from each other.
III Two interacting turbulent boundary layers.
IV Turbulent boundary layers separated from each other.

Their existence depends on the axial gap S_x between the two walls (Figure 3.52) and the Reynolds number based on the disk radius R_2 and peripheral velocity U_2

$$Re = \frac{\Omega R_2^2}{\mu} \rho \qquad (3.94)$$

The momentum coefficient for turbulent boundary layers as proposed by Nece and Daily (1960) also accounts for the surface roughness of the disk (Figure 3.53). The empirical equations that best approximate the experiments of Daily and Nece (1960) are summarized in Table 3.1.

According to Vavra (1970) there is an optimum gap with minimum friction losses, corresponding to the limit between regime IV and the others (Figure 3.54). Too small a ratio of S_x/R_2 results in a more rapid increase than too large a gap and should be avoided.

Table 3.1 Disk cavity friction coefficient as a function of Reynolds number and S_x

Regime	I	II	III	IV
C_m	$\dfrac{2\pi}{\frac{S_x}{R_2} Re}$	$\dfrac{3.7\left(\frac{S_x}{R_2}\right)^{0.1}}{Re^{0.5}}$		$\left(\dfrac{1}{3.8\log\left(\frac{R_2}{k_s}\right) - 2.4\left(\frac{S_x}{R_2}\right)^{0.25}}\right)^2$
$\frac{S_x}{R_2}$	$Re_{min} \Rightarrow Re_{max}$	$Re_{min} \Rightarrow Re_{max}$	$Re_{min} \Rightarrow Re_{max}$	$Re_{min} \Rightarrow Re_{max}$
0.01	$\Rightarrow 10^5$	Impossible	$10^5 \Rightarrow 10^9$	$10^9 \Rightarrow \infty$
0.02	$\Rightarrow 3 \times 10^4$	$3 \times 10^4 \Rightarrow 10^5$	$10^5 \Rightarrow 2 \times 10^7$	$2 \times 10^7 \Rightarrow \infty$
0.05	$\Rightarrow 4 \times 10^3$	$4 \times 10^3 \Rightarrow 3 \times 10^5$	Impossible	$3 \times 10^5 \Rightarrow \infty$
0.20	Impossible	$\Rightarrow 3 \times 10^5$	Impossible	$3 \times 10^5 \Rightarrow \infty$

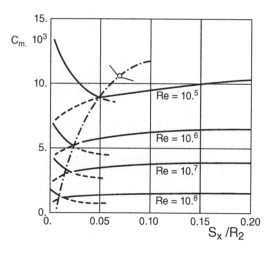

Figure 3.54 Influence of the cavity geometry on disk friction losses (from Vavra 1970).

In an adiabatic system, the disk friction energy is transmitted to the fluid and results in a temperature rise at the rotor discharge:

$$\Delta T_{diskfriction} = \frac{Pw_{diskfriction}}{\dot{m}C_p} \tag{3.95}$$

Adding energy to the small quantity of fluid in the cavity results in a temperature rise which in some extreme cases (very high pressure ratio) may have a detrimental effect on the mechanical resistance of the disk. The heat may be evacuated by leakage flow through the shaft seal or in the case of a shrouded impeller at the shroud seal. However, this leakage influences the energy dissipation. Numerical predictions by Chew and Vaughan (1988) have shown that the value of C_m can be twice the value shown in Figure 3.54 if there is some leakage flow in the shaft seal. In multistage compressors the leakage flow can enter the cavity from the seal side because the pressure at the exit of the return channel is higher then the one at the impeller exit.

4

The Diffuser

The amount of kinetic energy available at the diffuser inlet depends on the degree of reaction. It easily amounts to 50% of the total energy added by the impeller (Figure 1.37) and an efficient transformation of this energy into pressure is an important design issue.

The decrease in the kinetic energy from $V_2^2/2$ to $V_4^2/2$ (Figure 3.36) results in an increase in the total to static efficiency from

$$\eta_{TS} = \frac{T_2^i - T_1^o}{T_4^{o,i} - T_1^o} \qquad \text{to} \qquad \eta_{TS} = \frac{T_4^i - T_1^o}{T_4^{o,i} - T_1^o} \tag{4.1}$$

The losses in the diffuser, however, result in a decrease in the total to total efficiency from

$$\eta_{TT} = \frac{T_2^{o,i} - T_1^o}{T_4^o - T_1^o} \qquad \text{to} \qquad \eta_{TT} = \frac{T_4^{o,i} - T_1^o}{T_4^o - T_1^o} \tag{4.2}$$

Assuming parallel iso-pressure lines in the T–S diagram the difference between η_{TT} and η_{TS} is

$$\eta_{TT} \approx \eta_{TS} + \frac{V_4^2/(2C_p)}{(T_4^o - T_1^o)} \tag{4.3}$$

The difference becomes almost negligible when V_4^2 is small ($M_4 < 0.1$). In the case where the compressor exit is connected to a reservoir with a large cross section, it is of interest to reduce the exit velocity V_4 to a minimum because the corresponding increase in the exit static pressure reduces the energy dissipation in the reservoir. There is no reason to reduce V_4 below the velocity that is needed to transport the fluid in a pipe or the inlet section of the next stage of a multistage compressor.

The performance of a diffuser is normally not characterized by the efficiency but by the pressure coefficient CP_{2-4} relating the amount of kinetic energy that has been transformed into static pressure (Figure 3.36) to the inlet kinetic energy:

$$CP_{2-4} = \frac{\widetilde{P}_4 - \widetilde{P}_2}{\widetilde{P}_2^o - \widetilde{P}_2} \tag{4.4}$$

The impact of the diffuser static pressure rise on the stage efficiency is shown in Figure 3.42. Depending on the degree of reaction an increase in CP of 0.1 results in an increase in the total to static efficiency up to 5 points.

The diffuser pressure rise coefficient can also be written as

$$CP_{2-4} = 1 - \frac{(\widetilde{P}_4^o - \widetilde{P}_4)}{\widetilde{P}_2^o - \widetilde{P}_2} - \frac{(\widetilde{P}_2^o - \widetilde{P}_4^o)}{\widetilde{P}_2^o - \widetilde{P}_2} \tag{4.5}$$

Design and Analysis of Centrifugal Compressors, First Edition. René Van den Braembussche.
© 2019, The American Society of Mechanical Engineers (ASME), 2 Park Avenue, New York, NY, 10016, USA (www.asme.org).
Published 2019 by John Wiley & Sons Ltd.

For isentropic flows ($P_4^o = P_2^o$) and assuming that the square of the average velocity equals the average of the velocity square we obtain the following expression for the isentropic pressure rise coefficient

$$CP_{2-4}^i = 1 - \frac{(\widetilde{P}_2^o - \widetilde{P}_4)}{\widetilde{P}_2^o - \widetilde{P}_2} \approx 1 - \frac{\rho_4 \widetilde{V}_4^2}{\rho_2 \widetilde{V}_2^2} \tag{4.6}$$

as a function of the inlet and outlet kinetic energy. Neglecting the change in density when applying continuity provides the following relation between $CP_{2-4,inc}^i$ and the inlet and outlet cross section area:

$$CP_{2-4,inc}^i \approx 1 - \frac{\widetilde{V}_4^2}{\widetilde{V}_2^2} = 1 - \frac{A_2^2}{A_4^2} = 1 - \frac{1}{AR^2} \tag{4.7}$$

A_2 and A_4 are the cross sections perpendicular to the velocity vectors $\vec{V}_2$ and $\vec{V}_4$.

After substitution of Equations (4.7) and (4.6) in Equation (4.5) we obtain the following approximate relations for $CP_{2-4,inc}$:

$$CP_{2-4,inc} \approx CP_{2-4,inc}^i - \omega_{2-4} \approx 1 - \frac{\widetilde{V}_4^2}{\widetilde{V}_2^2} - \omega_{2-4} = 1 - \frac{A_2^2}{A_4^2} - \omega_{2-4} \tag{4.8}$$

This equation illustrates how the static pressure rise coefficient depends on the isentropic value $CP_{2-4,inc}^i$ and total pressure loss coefficient ω_{2-4}. It also shows that for a given geometry ($A_2/A_4 = C^{te}$), the sum of the pressure rise coefficient and the total pressure loss coefficient is constant.

The effect of compressibility on the static pressure rise in the divergent channel of a vaned diffuser can be illustrated by considering CP^i. The diffuser outlet (4) to throat (th) velocity ratio for compressible flows is defined by continuity:

$$\frac{\widetilde{V}_4}{\widetilde{V}_{th}} = \frac{\rho_{th} A_{th}}{\rho_4 A_4} \tag{4.9}$$

After substitution of this velocity ratio in Equation (4.6) we obtain:

$$CP^i = 1 - \frac{\rho_{th} A_{th}^2}{\rho_4 A_4^2} = 1 - \frac{\rho_{th}}{\rho_4} \frac{1}{AR} \tag{4.10}$$

As the density ρ_4 is larger than ρ_{th} for compressible flows, the value of CP^i will be larger for compressible than for incompressible flows (Equation (4.6)). The variation of the isentropic pressure coefficient for different values of the AR and throat Mach number has been calculated by Runstadler and Dean (1969). The results, shown in Figure 4.1, indicate an increase in CP^i with throat Mach number. The effect is much larger at low values of AR because the flow in long diffusers with large AR decelerates much more and becomes nearly incompressible in the downstream part of the diffuser.

In a vaneless diffuser the static pressure rise is accomplished by a reduction with radius of the radial and swirl velocity as defined by continuity ($\rho R b V_m = C^{te}$) and the conservation of the angular momentum ($R V_u = C^{te}$). Depending on the inlet angle α_2, this may be along long streamlines with high friction losses resulting in a lower pressure rise than the isentropic one.

The streamlines in vaneless diffusers make an angle of $90° - \alpha$ with the nearly concentric isobars (Figure 4.2a). The whole pressure gradient is applied to the radial velocity component, which has already suffered a diffusion in the impeller and has a small kinetic energy. As a consequence the flow in the boundary layers is deflected to a more tangential direction than the main flow.

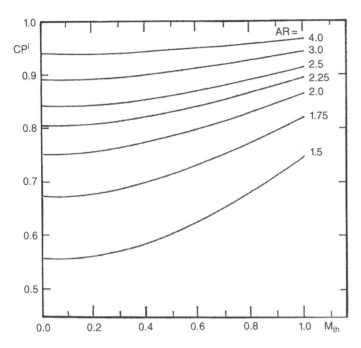

Figure 4.1 Influence of the inlet Mach number on the isentropic static pressure rise in a vaned diffuser (from Runstadler and Dean 1969).

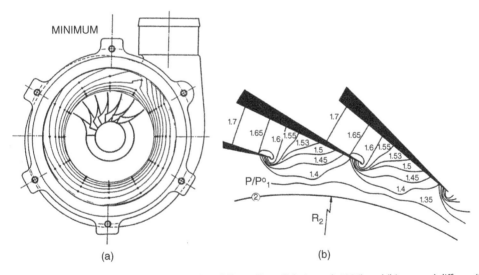

Figure 4.2 Iso-pressure lines in (a) a vaneless diffuser (from Sideris et al. 1986) and (b) a vaned diffuser (from Krain 1981).

One way to improve the performance is by a larger reduction of the tangential velocity and a shortening of the flow path by means of vanes turning the flow more radially. The radial velocity component being defined by continuity remains unchanged except for the effect of increased density and blade blockage or by varying the diffuser width. Diffuser vanes have a favorable effect on the 3D boundary layers because they create a pressure gradient which is better aligned with the main velocity vector (Figure 4.2b).

Vaneless diffusers are better suited for off-design operation because they allow a large variation of absolute inlet flow angles α_2. In the case of vaned diffusers, a too large positive incidence may result in flow separation on the suction side and give rise to rotating stall and surge, while a too large negative incidence leads to choking. Vaneless diffusers can only choke in the very improbable case that the radial component of the Mach number exceeds unity.

The vaned diffuser is separated from the impeller by a short vaneless diffuser where the flow is different from the one in a complete vaneless diffuser by the upstream influence of the diffuser vanes. The description in the next section of the flow in vaneless diffusers is to a large extent also applicable to the vaneless part of the vaned diffuser.

4.1 Vaneless Diffusers

The relative flow leaving the impeller is highly non-uniform in tangential and axial direction because of the boundary layers developing along the walls and the secondary flow and flow separation in the impeller. The circumferential non-uniformity of the relative velocity results in an unsteady absolute flow with strong variations in velocity magnitude and direction (Figure 4.3b) favoring the circumferential uniformization of the flow. The axial non-uniformity of the relative flow is responsible for the skewness of the absolute flow at the diffuser inlet (Figure 4.3a). This non-uniformity is enhanced by the radial pressure gradient in the diffuser and can cause return flow in the side wall boundary layers triggering diffuser stall.

Full 3D unsteady Navier–Stokes calculations are needed to calculate this flow. There are no simple models to do this with sufficient accuracy and acceptable cost. Depending on the application and the required degree of accuracy, the problem may be simplified.

- Assume uniform flow at the diffuser inlet, after an instantaneous mixing, similar to the one described in Section (3.3.4), and assume it remains like that further downstream in the diffuser. Such a flow is axisymmetric and uniform in the spanwise direction. The flow properties change only with radius.
- A more sophisticated approach is one where only the flow uniformization in the spanwise direction is instantaneous. The mixing of the unsteady, circumferentially non-uniform flow in the vaneless diffuser is then calculated by an approximating 2D method.
- A third approach results from the assumption that the wake/jet mixing is instantaneous and that there is only a spanwise non-uniformity of the incoming flow. The flow is axisymmetric with 3D boundary layers near the walls. The flow can be calculated by a steady 3D Navier–Stokes solver. An approximate analytical method is presented because it provides a more physical insight into the flow.

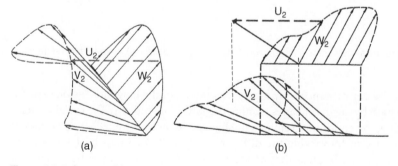

(a) (b)

Figure 4.3 Influence of the (a) axial and (b) circumferential distortion on the diffuser inlet flow.

4.1.1   One-dimensional Calculation

The **isentropic uniform flow** in a vaneless diffuser is easily predicted by imposing the conservation of mass

$$2\pi Rb\rho V_R = 2\pi R_2 b_2 \rho_2 V_{R,2} \tag{4.11}$$

and angular momentum

$$RV_u = R_2 V_{u,2} \tag{4.12}$$

Combining both equations it turns out that the flow angle is constant for incompressible flows in parallel wall diffusers.

$$\tan \alpha = V_u/V_R = V_{u,2}/V_{R,2} = \tan \alpha_2 \tag{4.13}$$

As illustrated in Figure 3.7b, a constant α does not result in a straight streamline except for $\alpha_2 = 0$.

According to the Bernoulli equation, the static pressure for isentropic incompressible flow increases with decreasing velocity square. Hence

$$P^i - P_2 = \frac{\rho V_2^2}{2} - \frac{\rho V^{i2}}{2} = \frac{\rho V_2^2}{2}\left(1 - \frac{R_2^2}{R^2}\right) \tag{4.14}$$

and the static pressure rise coefficient for isentropic incompressible flow in a parallel vaneless diffuser

$$CP^i_{(2-4)} = \left(1 - \frac{R_2^2}{R_4^2}\right) \tag{4.15}$$

depends only on the square of the diffuser outlet over inlet radius ratio.

As shown in Figure 4.4 most of the pressure rise takes place near the diffuser inlet. 75% of the inlet kinetic energy has already been recuperated at radius ratio 2. Doubling the diffuser length to a radius ratio of 3 results in an increase in CP^i of less than 0.14.

The impact of compressibility or a variation of the diffuser width b is easily evaluated by combining Equations (4.11) and (4.12). This results in the following expression for the inviscid

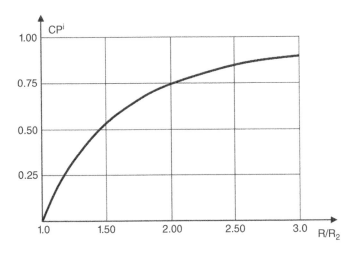

Figure 4.4 Static pressure recovery in vaneless diffusers with 2D inviscid flow.

flow angle:

$$\tan \alpha = \tan \alpha_2 \frac{b\rho}{b_2\rho_2} \tag{4.16}$$

Compressibility results in a more tangential flow while convergence turns the flow outwards, as illustrated in Figure 4.5.

One-dimensional methods to calculate the **non-isentropic compressible flow** in vaneless diffusers with constant or variable width have been presented by Stanitz (1952) and Traupel (1962). The flow is considered uniform in both the circumferential and spanwise directions (Figure 4.6). Viscous effects are distributed over the whole flow and accounted for by distributed body forces. The tangential and meridional component of the friction forces are calculated by means of a friction coefficient C_f defined in Equation (1.16) as a function of Reynolds number and wall roughness.

The system of equations to be solved is:

(i) the continuity equation, which after differentiation with respect to the meridional direction results in

$$V_m bR \frac{d\rho}{dm} + \rho bR \frac{dV_m}{dm} + \rho V_m \frac{d(bR)}{dm} = 0 \tag{4.17}$$

(ii) the momentum equation in the tangential direction, expressing the balance between shear and inertia forces:

$$C_f \frac{\rho V^2}{2} \frac{V_u}{V} \frac{2}{b} + \rho \frac{V_m}{R} \frac{dRV_u}{dm} = 0 \tag{4.18}$$

(iii) the momentum equation in the meridional direction, expressing the equilibrium between the friction force, centrifugal force, impulse, and pressure gradient:

$$C_f \frac{\rho V^2}{2} \frac{V_m}{V} \frac{2}{b} - \rho \frac{V_u^2}{R} \sin \gamma + \rho V_m \frac{dV_m}{dm} - \frac{dP}{dm} = 0 \tag{4.19}$$

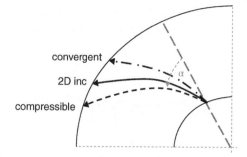

Figure 4.5 Effect of compressibility and convergence in vaneless diffusers.

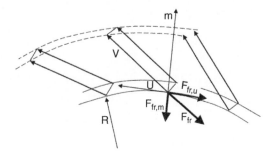

Figure 4.6 Two-dimensional viscous flow model in vaneless diffusers.

The geometric relations between the velocity components, the equation of state

$$\frac{1}{P}\frac{dP}{dm} = \frac{1}{\rho}\frac{d\rho}{dm} + \frac{1}{T}\frac{dT}{dm} \qquad (4.20)$$

and conservation of total energy

$$C_p T^o = C_p T + \frac{V^2}{2} = C^{te} \qquad (4.21)$$

allow the calculation of the spanwise average flow parameters along the streamlines by a stepwise integration in the meridional direction. These equations are valid for radial ($\gamma = 90°$) and mixed flow diffusers ($\gamma \neq 90°$).

Traupel (1962) proposed replacing the radial momentum equation by the energy dissipation equation obtained by integrating the friction force along a streamline:

$$\frac{dP}{\rho} = -d\frac{V^2}{2} - \frac{C_d \rho V^3}{\rho_2 V_2 b_2 R_2} R dR \qquad (4.22)$$

The following relation between C_d and C_f is proposed by Schmalfuss (1972):

$$C_d = C_f + 0.0015 \qquad (4.23)$$

Comparisons by Van den Braembussche et al. (1987) have shown that in most cases the difference between the results of these approaches is negligible.

The main outcomes of such a calculation are the radial variation of the static pressure and the absolute flow angle (Figure 4.7). Friction losses result in a reduction of the static pressure rise, proportional to the total pressure losses. This can even lead to a decreasing static pressure near the outlet if the diffuser is too long.

The decrease in the flow angle for isentropic flow in Figure 4.7 results from a smaller decrease in the radial velocity component due to the convergence of the diffuser ($b_4 < b_2$). The tangential component of the friction force results in a more rapid reduction in the tangential velocity than for inviscid flow. The corresponding decrease in the flow angle is further enhanced by boundary layer blockage and a slower increase in the density because of the smaller static pressure rise. Hence the streamlines for viscous flow are more radial and shorter.

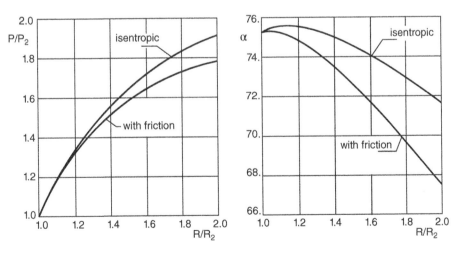

Figure 4.7 Isentropic and non-isentropic static pressure rise and flow angle in a convergent vaneless diffuser (from Stanitz 1952).

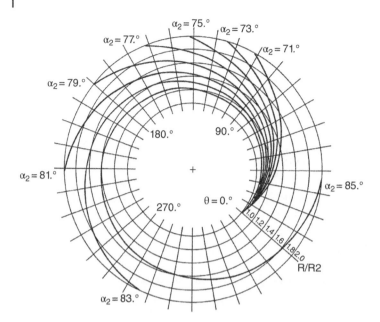

Figure 4.8 Average flow streamlines in a spanwise convergent diffuser for different values of α_2 (from Stanitz 1952).

The diffuser inlet flow angle is closely related to the diffuser width. More tangential flow (higher values of α_2) results in longer streamlines swirling around the diffuser (Figure 4.8). The flow in a vaneless diffuser with radius ratio 2 and 85° inlet flow angle will travel circumferentially over 400° before leaving the diffuser.

When evaluating the impact of the diffuser width on performance, it is necessary to account for the change in inlet conditions resulting from a proportional change in the impeller exit width. Figure 4.9 shows the variation of the diffuser static pressure and total pressure loss coefficient with mass flow at different values of the impeller exit/diffuser inlet width. The change in the diffuser inlet conditions ($V_{u,2}$ and $V_{R,2}$) with mass flow is predicted by the two-zone flow model described in Section 3.3.3 for an impeller with 40° backward lean and $\pi_{T-T} \approx 1.4$.

The full lines are for varying mass flow along a performance curve (constant RPM and given diffuser width). They suggest a considerable performance improvement when widening the diffuser. The decrease in the loss coefficient when widening the diffuser at constant α_2 is due to a decrease in the friction losses proportional to L/DH ($DH = 2b$) but is also because the losses are distributed over a larger mass flow.

Friction has only a limited impact on the streamline shape when changing the diffuser width at constant inlet flow angle (Figure 4.7). Hence the inlet over outlet cross sections of the streamtubes do not change very much and CP^i will also be almost unchanged. According to Equation (4.8) the increase in pressure rise coefficient is then a consequence of the decrease in loss coefficient.

When decreasing the mass flow in narrow diffusers, the nearly constant or slightly increasing static pressure rise changes into a rapidly decreasing one because of the increasing friction losses due to the longer flow path at increasing value of α_2. This rapid decrease is not observed with wider diffusers because they cannot operate at such small mass flows because diffuser rotating stall already occurs at lower values of α_2.

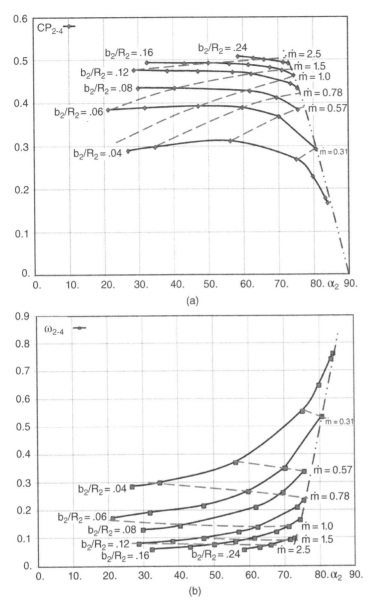

Figure 4.9 Influence of diffuser width on (a) the static pressure rise and (b) the total pressure loss coefficient in a parallel vaneless diffuser.

The full lines allow an estimation of the stable operating range of a vaneless diffuser as a function of b_2/R_2. Optimizing a compressor outlet/diffuser inlet width turns out to be a compromise between a narrower diffuser, stable up to smaller mass flows but with higher losses, and a wider one with lower losses but already stalling at a larger mass flow.

The dashed lines are for constant mass flow and indicate a considerable increase in the pressure rise coefficient and decrease in the loss coefficient when widening the diffuser (and impeller). Widening the impeller exit width of a 40° backward leaning impeller results in a decrease in the radial velocity and an increase in the tangential velocity at the diffuser inlet. The increase in pressure rise by increasing the diffuser width goes with an increase in the diffuser inlet angle. Hence the decrease in the loss coefficient is not the result of a large decrease in the friction losses because the corresponding increase in α_2 gives rise to a longer flow path and only a small change in L/DH. The decrease in the loss coefficient along a line of constant mass flow results mostly from non-dimensionalizing the losses by a larger inlet kinetic energy because of the larger tangential velocity at the outlet of a wider backward leaned impeller.

As explained in Section 8.2, the diffuser inlet angle is a critical factor for vaneless diffuser stability. Flows entering a vaneless diffuser with a flow angle α_2 exceeding a critical value α_{2c} give rise to vaneless diffuser rotating stall. Hence widening the vaneless diffuser is limited by stability to the critical inlet flow angles, shown by the dash double dot line in Figure 4.9. Designing too close to this limit will limit the possible reduction of the mass flow and hence reduce the operating range of a compressor. It turns out that that optimal value of b_2 is a compromise between maximizing CP_{2-4} and maximizing the range between design mass flow and the minimum mass flow for stability.

Decreasing the mass flow at constant diffuser width results in an increasing inlet flow angle α and, according to Figure 4.9, a reduction in the diffuser static pressure rise coefficient. Hence the pressure rise versus mass flow curve of vaneless diffusers has a positive slope ($dCP_{2-4}/d\dot{m} > 0$) which, as will be discussed in Section 8.6.1, has a destabilizing effect on centrifugal compressors.

4.1.2   Circumferential Distortion

The pitchwise variation of the flow at the exit of the impeller is schematically represented in Figure 4.10. The low velocity part is due to the blade boundary layers and other low energy fluid accumulated by the secondary flows in the wake near the suction side. The high velocity part represents the inviscid jet flow. Because of impeller rotation, this non-uniformity results in an unsteady absolute flow at the diffuser inlet. The unsteady change in magnitude and direction of the jet and wake velocity vectors are at the origin of an intense energy transfer between jet and wake, which, together with the friction forces, results in a rapid uniformization of the flow, commonly called wake/jet mixing. Uniform flow is obtained after a much shorter distance than in a steady parallel flow jet and wake mixing process. This wake/jet mixing is also responsible for the large total pressure loss observed near the inlet of vaneless diffusers.

The calculation of the impeller outlet average flow conditions, discussed in Section 3.3.3, assumes an instantaneous mixing of the impeller outflow. Although this model does not describe the real mechanism of energy exchange and increased friction that occurs at the impeller exit, it provides satisfactory results in terms of static pressure rise and losses. Inoue (1978) explains why the simple sudden expansion model predicts losses in the rotating case very well and can be accepted for performance predictions.

The following mixing model, first proposed by Dean and Senoo (1960) and extended to compressible flows by Frigne et al. (1979), provides some insight into the mixing mechanisms and also predicts the extent over which the uniformization takes place. The approach is similar

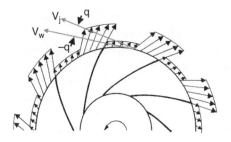

Figure 4.10 Schematic presentation of the jet and wake flow pattern at the diffuser inlet.

to the 1D calculation of the uniform flow in vaneless diffusers described in the previous section. However, conservation of mass and angular and radial momentum are applied separately to the jet and wake, taking into account the forces $\pm q$ acting on the interface between jet and wake. The method is based on the following assumptions (Figure 4.10):

- The same relative flow angle β for jet and wake.
- Uniform jet and wake relative velocity distributions at the impeller outlet.
- The boundary layer blockage and friction along the side walls are uniformly distributed over the diffuser width.
- The blade blockage at the impeller discharge is negligible.
- The static pressures are identical for jet and wake, as assumed when predicting the impeller outlet flow.

Figure 4.11 shows the results of a calculation performed for an impeller discharge wake width $\epsilon_2 = 0.71$ and $v = 0.075$, corresponding to a wake mass flow of 17.5%. It can be seen on Figure 4.11a,b that in the first part of the mixing process ($R/R_2 < 1.06$) the total temperature and pressure of the wake are increasing at the cost of a decrease in the temperature and stagnation pressure in the jet. The jet and wake continue to exist separately with an energy transfer between them. This energy transfer and the losses due to the friction on the walls and on the jet–wake interface are responsible for the decrease in total pressure and temperature in the jet. The corresponding pressure and temperature rise in the wake is much larger than

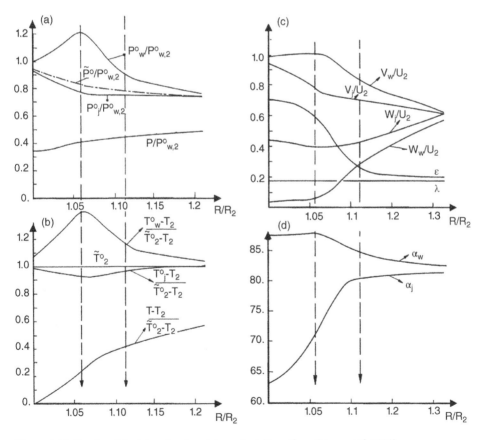

Figure 4.11 Results of the jet and wake mixing calculations (from Frigne et al. 1979).

the drop in the jet because it contains only 17.5% of the mass flow. The wake width ϵ remains nearly constant with a slight decrease in the relative velocity in the jet and an increase in the wake (Figure 4.11c). The jet absolute flow angle increases rapidly because of the force q acting on the interface and because of the rapid decrease in the jet radial velocity needed to catch up with the strong radial pressure gradient due to the centrifugal forces resulting from the large V_u in the wake (high absolute velocity at very large α).

For $R/R_2 > 1.06$ a breakdown of the wake flow takes place and ϵ reduces very fast to its final value defined by its part of the mass flow $\lambda = 0.175$. The direction of the energy transfer is reversed and the total temperature T_j^o increases. It can be seen in Figure 4.11d that the jet flow becomes more tangential until it is parallel to the wake. The total pressure of the jet is not increasing during this process because the energy, transferred from the wake to the jet, is completely dissipated by the friction on the diffuser walls.

At $R/R_2 = 1.12$, the wake width is already reduced to 1.1 times its final value. The mixing process now progresses slowly to a complete uniformization. The calculations are stopped when the difference between the relative velocities is less than 5%.

Figure 4.12 shows the total pressure variation along the relative streamlines in jet and wake as derived from the measurements by Senoo and Ishida (1975). This confirms the predicted increase in total pressure in the wake up to $R/R_2 = 1.1$ and a decrease in the jet followed by a zone of nearly constant total pressure. These authors have also shown that this energy exchange relates to the circumferential static pressure gradient.

The high total pressure losses observed at the diffuser inlet are well predicted by this method, as shown in Figure 4.13 where experimental data and theoretical predictions are compared. They are normally called wake/jet mixing losses, suggesting that they are due to the shear stresses between jet and wake. The shear friction coefficient C_M (≈ 0.094), acting on the surface between jet and wake, is about 20 times larger than a typical wall friction coefficient C_f (≈ 0.005). However, the first one applies to a much smaller surface and a smaller velocity difference than the wall friction coefficient. The corresponding forces are therefore much smaller:

$$C_f V^2 2\pi R\Delta R \gg C_M (W_j - W_w)^2 Z_r b\Delta R \tag{4.24}$$

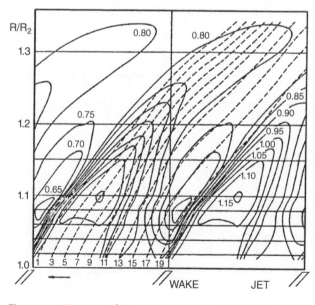

Figure 4.12 Variation of the total pressure in jet and wake (from Senoo and Ishida 1975).

Figure 4.13 Influence of jet and wake mixing on the static pressure rise and losses in a vaneless diffuser (from Dean and Senoo 1960).

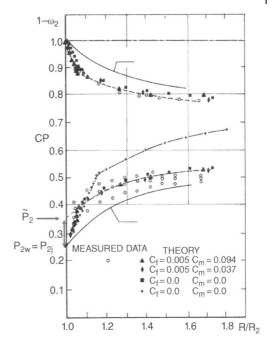

It can be seen in Figure 4.12 that the total pressure gradient between jet and wake does not change much in the streamwise direction. It is therefore believed that the shear forces along this border are not the main contribution to the uniformization of the flow. The mixing process is not by an exchange of mass between jet and wake (turbulent mixing), but rather by a transfer of energy. The extra losses at the diffuser inlet are mostly due to the increased friction of the flow on the side walls resulting from the non-uniformity of the flow. Inoue (1978) shows that the shortening of the mixing process by energy exchange has a favorable effect on these extra friction losses due to the more rapid uniformization of the flow.

4.1.3 Three-dimensional Flow Calculation

An instantaneous mixing of the jet and wake at the exit of the impeller results in a circumferentially uniform relative velocity and hence a steady absolute flow in the diffuser. The spanwise non-uniformity of the relative flow results in a skewed 3D velocity profile at the diffuser inlet (Figure 4.3). This 3D character of the flow is further enforced downstream in the vaneless diffuser by the radial pressure gradient making an angle α with the mean velocity vector.

The problem can be solved with a steady 3D Navier–Stokes solver, providing the inlet conditions are known. The following summarizes an analytical approach, which is no longer up to date but helps understanding of the flow mechanisms that govern this kind of flow and has been used to define the vaneless diffuser stability limit (Jansen 1964b; Senoo and Nishi 1977a,b). Assuming that the axial velocity component is negligible, the problem reduces to satisfying the radial and circumferential component of the Navier–Stokes equation. In order to facilitate the solution, the flow is subdivided into three regions (Figure 4.14):

- the viscous region on the shroud wall corresponding to the 3D boundary layer
- the viscous layer close to the hub wall corresponding to the 3D boundary layer

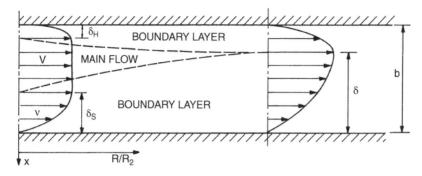

Figure 4.14 Model for a 3D viscous flow calculation in a vaneless diffuser (from Senoo et al. 1977).

- an inviscid flow zone ($\tau_R = \tau_u = 0$) characterized by a linear variation of the tangential and radial velocity, between the two boundary layers. This zone reduces to one point when the two boundary layers have merged together in a fully developed flow.

An integral approach is used to solve the equations. It consists of an analytical integration of the following equations with respect to x:

- the radial and tangential momentum equations for the boundary layers on each side of the diffuser
- the radial and tangential momentum integral equations for the inviscid flow between δ_H and $b - \delta_S$
- the continuity equation

followed by a numerical integration with respect to R of the resulting system of ordinary differential equations.

To close the problem, we must define the boundary layer velocity profile and the wall shear stresses τ_{wu} and τ_{wR}.

The 3D boundary layers are characterized by the thickness δ and the angle ε between the limiting wall streamline and the main flow streamline (Figure 4.15).

The velocity profile is represented by the superposition of the velocity component in the main flow direction

$$\frac{v_s}{V} = \left(\frac{x}{\delta}\right)^{1/n} \tag{4.25}$$

and the crossflow component

$$\frac{v_n}{V} = \tan\varepsilon\left(1 - \frac{x}{\delta}\right)^m \frac{v_s}{V} \tag{4.26}$$

Figure 4.15 Definition of the 3D boundary layer.

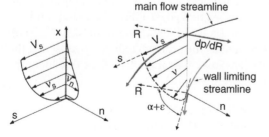

where x is the distance from the wall, and m and n are empirically defined. Typical values of m are between 2 and 3, while n is a function of the local Reynolds number:

$$n = 2.667 Re_\delta^{1/8} \qquad Re_\delta = \frac{V\delta}{\nu} \tag{4.27}$$

The wall shear stress is in the direction of the limiting wall streamline and makes an angle ε with the main flow direction. The shear forces at the wall (τ_ε) and in the main flow direction (τ_s) are

$$\tau_\varepsilon = C_{fw}\frac{\rho V^2}{2} \qquad\qquad \tau_s = C_{fv}\frac{\rho V^2}{2} \tag{4.28}$$

The wall friction coefficient C_{fw} relates to the main flow component (C_{fv}) by

$$C_{fw} = C_{fv} / \cos \varepsilon \tag{4.29}$$

The 2D friction coefficient can be approximated by

$$C_{fv} = K_f Re_\delta^{1/4} \tag{4.30}$$

The value of $K_f = 0.051$ proposed by Senoo et al. (1977) is higher than the one proposed by Johnston (1960) for 2D flows to account for the high intensity of the turbulence at the impeller exit.

The calculations predict the evolution from thin boundary layers at the diffuser inlet with very tangential flow into a fully developed flow downstream. The most interesting results are (Figure 4.16):

- the free stream flow angles α_S and α_H at the edge of the boundary layers
- the flow direction of the wall streamline ($\alpha_H + \varepsilon_H$ and $\alpha_S + \varepsilon_S$).

The latter have an important influence on diffuser stability and its variation with radius can be explained as follows. The tangential component of the absolute velocity is very large in the

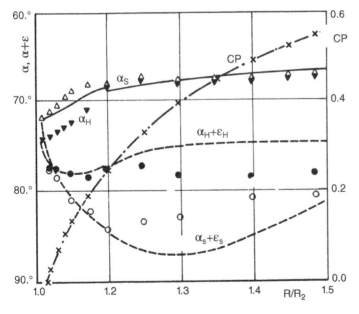

Figure 4.16 Results of a 3D diffuser flow calculation and comparison with experimental results (from Senoo et al. 1977).

boundary layer at the diffuser inlet, and the corresponding centrifugal forces help to overcome the radial pressure gradient. However, this component reduces very quickly because of friction on the walls, and the radial pressure gradient turns the flow in the boundary layers backward, which is reflected by the rapid increase in $\alpha + \varepsilon$. The blockage due to the boundary layer growth gives rise to an increase in the radial velocity in the core flow. The growing difference between the freestream α and $\alpha + \varepsilon$ near the walls increases the shear stresses between the core flow and the boundary layer and helps the fluid near the walls to overcome the radial pressure gradient. The result is a decrease in $\alpha_H + \varepsilon_H$ starting at $R/R_2 \approx 1.15$ at the hub and at $R/R_2 \approx 1.2$ at the shroud.

Senoo (1984) has used this integral boundary layer method to evaluate the influence of the inlet flow distortion on the static pressure rise and total pressure losses. The inlet distortion is mainly on the radial velocity component and is expressed by:

$$B_f = \frac{\int_0^b \rho V_R^2 ds / \int_0^b \rho V_R ds}{\int_0^b \rho V_R ds / \int_0^b \rho ds} \tag{4.31}$$

The variation of the static pressure rise coefficient CP_{2-4} and total pressure losses ω_{2-4} as a function of B_f are plotted in Figure 4.17. The important deterioration of diffuser performances with increasing inlet distortion is confirmed by the experimental data indicated by circles.

The results shown in Figure 4.16 are for an asymmetric inlet flow. The impact of a spanwise variation of the energy input and velocity at the impeller outlet on the flow in the vaneless diffuser has been studied by Ellis (1964) and Rebernik (1972). They have shown that viscosity is not the only origin of flow separation in the diffuser but that the latter may also be due to the tangential vorticity resulting from the spanwise non-uniformity of the flow at the diffuser inlet.

Assuming steady absolute frictionless flow, the equation of motion (1.53) reduces to

$$\vec{V} \vee (\nabla \vee \vec{V}) = \nabla \left(\frac{P}{\rho} + \frac{V^2}{2} \right) = \frac{\nabla P^o}{\rho} \tag{4.32}$$

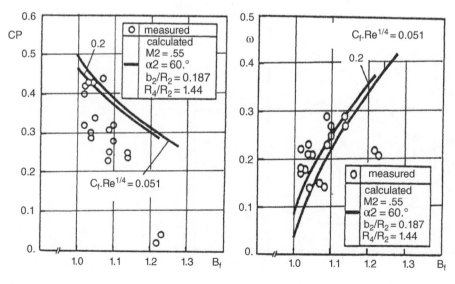

Figure 4.17 Variation of static pressure rise and losses with flow distortion at the inlet of a vaneless diffuser (from Senoo 1984).

Written in polar coordinates (R, x, u) the vector product $\vec{V} \vee (\nabla \vee \vec{V})$ is

$$\vec{i}_x \frac{V_R}{R} \left(\frac{\partial V_R}{\partial \theta} - \frac{\partial R V_u}{\partial R} \right) + \vec{i}_R \frac{V_u}{R} \left(\frac{\partial R V_u}{\partial R} - \frac{\partial R V_R}{\partial \theta} \right)$$

$$+ \vec{i}_u \left(\frac{V_u}{R} \left(\frac{\partial R V_u}{\partial x} - \frac{\partial V_x}{\partial \theta} \right) - V_R \frac{\partial V_x}{\partial R} + V_R \frac{\partial V_R}{\partial x} + \frac{V_R^2}{\mathfrak{R}_n} \right) \tag{4.33}$$

For inviscid $\left(\frac{1}{\rho} \frac{\partial P^o}{\partial R} = 0 \right)$ and circumferentially uniform flow $\left(\frac{\partial V_x}{\partial \theta} = 0, \frac{1}{\rho} \frac{\partial P^o}{\partial \theta} = 0 \right)$ at the inlet of a radial diffuser with straight walls $(\mathfrak{R}_n = \infty)$ the $\vec{i}_u$ component of Equation (4.32) is

$$\frac{\partial V_R}{\partial x} - \frac{\partial V_x}{\partial R} = \frac{1}{V_R} \left(\frac{1}{\rho} \frac{\partial P^o}{\partial x} - \frac{V_u}{R} \frac{\partial R V_u}{\partial x} \right) \tag{4.34}$$

This equation indicates the existence of a tangential vorticity whose sign depends on the relative strength of the spanwise total pressure gradient and energy input. Neglecting $\frac{\partial V_x}{\partial R}$, which must anyway be very small, we can conclude that the radial velocity component in a vaneless diffuser will increase towards the side of the largest P^o except if $\frac{1}{\rho} \frac{\partial P^o}{\partial x}$ is smaller than the absolute value of $\frac{V_u}{R} \frac{\partial R V_u}{\partial x}$. However, the latter decreases rapidly in the diffuser because both V_u and $1/R$ decrease with radius, whereas the total pressure gradient remains almost unchanged in wide diffusers. This means that the total pressure gradient will dominate the flow further downstream in the diffuser by increasing the radial velocity and hence concentrating the flow on the side of the highest total pressure.

Ellis (1964) has demonstrated that this concentration of the flow on one side can lead to local return flow on the opposite side. The streamlines shown in Figure 4.18 are obtained with an inviscid calculation of the axisymmetric flow in a diffuser with a spanwise variation of the total pressure and uniform axial vorticity at the inlet. The flow separation predicted on the hub side has been experimentally confirmed. These results are typical for a wide diffuser where the vorticity of the free stream is more important than the shear stresses near the walls.

Different flow patterns in a vaneless diffuser downstream of a centrifugal impeller (Figure 4.19) have been measured by Rebernik (1972), including flows with separation on one or both sides of the diffuser. Inviscid flow calculations starting from the measured inlet conditions at 110% design mass flow predict the measured shift of the flow towards the shroud side with separation on the hub side, at the location where the pressure gradient starts dominating the angular momentum. Although hot wire traces indicate rotating stall at 63% design mass flow, the same calculations confirm the measured concentration of the streamlines

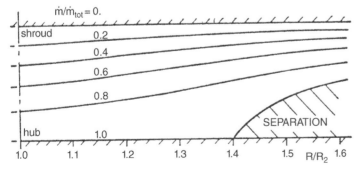

Figure 4.18 Streamlines in a vaneless diffuser calculated for an inviscid flow with a spanwise variation of the inlet P^o (from Ellis 1964).

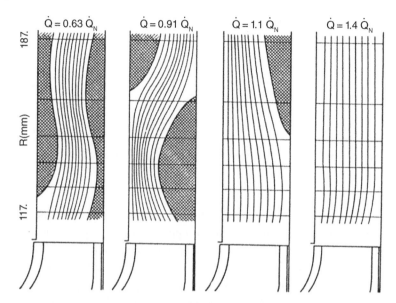

Figure 4.19 Separated flows in vaneless diffusers (from Rebernik 1972).

towards the center of the diffuser. The shift of separation from one side to the other at 91% design mass flow could not be reproduced by the inviscid method but detailed measurements reveal an inversion of the spanwise gradient of the P^o once separation occurs. This change in pressure gradient continues downstream of the separation point until flow reversal takes place on the other side. It is still unclear whether this switch from one side to the other is the consequence of a smaller energy dissipation in the separated flow zone or due to a spanwise energy transfer in the separated flow zone.

The main consequence of this alternating separation is a non-linear variation of the losses with mass flow resulting in a local decrease in diffuser efficiency, as observed both by Ellis (1964) and Rebernik (1972). Such a dip in the diffuser performance is reflected in the overall performance curve by a change in the slope of the pressure rise versus flow rate (Figure 4.20) which, as will be explained in Section 8.6.1, has a destabilizing effect.

The RHS of Equation (4.34) depends on the hub to shroud variation of the impeller work input and losses, and is influenced by the secondary flow defining the spanwise position of the wake region near the impeller trailing edge. One way to modify it was experimentally demonstrated by Ellis (1964) in a series of tests in which the spanwise gradient of the work input was modified by a cut back of the outlet radius at the shroud side, as shown in Figure 4.21. It could be verified that the corresponding change in work input near the shroud has a rather small effect on the spanwise distribution of the radial velocity at the diffuser inlet but resulted in an inversion of the radial velocity distribution at the diffuser exit. Return flow at the diffuser exit shifted from the hub side towards the shroud side.

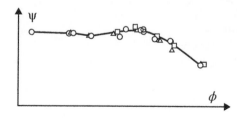

Figure 4.20 Effect of nonlinear diffuser losses on compressor pressure rise (from Ellis 1964).

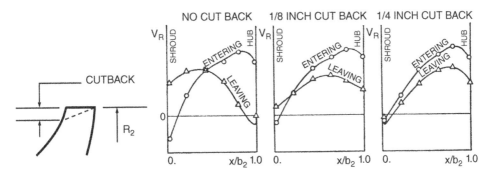

Figure 4.21 Effect of trailing edge cutback on the flow in a vaneless diffuser (from Ellis 1964).

4.2 Vaned Diffusers

Two main types of vaned diffusers can be distinguished. The first one is composed of curved vanes, similar to those used in axial compressors (Figure 4.22a,b) where the diffusion is achieved by turning the flow more radially than the undisturbed logarithmic flow path. The flow between two successive vanes is not symmetric but varies between the pressure and a suction side. The latter are not related to the convexity or concavity of the surface but to the way the flow approaches the vane leading edge and the pressure gradient resulting from the flow deviation by the blades.

The other type of diffuser consist of straight divergent channels (Figure 4.22c,d). They are shaped by straight plates or triangular vanes that are fixed between the diffuser walls. They are therefore called channel diffusers or vaned island diffusers. The deceleration of the flow is controlled by the divergence of the channel. A large diffusion can be achieved because of the symmetry of the flow in these straight channels. An additional increase in the static pressure may occur at the exit of the vaned island diffuser because of the large trailing edge thickness, but at the cost of sudden expansion losses. The latter may be avoided by curving the channel tangentially, which in turn results in a decrease in the pressure rise in the channel (Sagi and Johnston 1967).

A special type of channel diffuser is the pipe diffuser shown in Figure 4.22e,f. Each channel is composed of two conical pipes in the plane of the diffuser, a short converging one upstream of the throat section and a longer diverging one downstream.

A large variety of intermediate diffuser geometries exist. The difference between the two types is less in the geometry than in the way the performance is predicted, i.e. by predicting the flow around the vanes or by predicting the flow in the channels. Curved vane diffusers are commonly used at lower pressure ratios ($\pi < 3$), whereas channel and pipe diffusers are better suited for higher pressure ratios ($\pi > 3$) and inlet Mach numbers.

4.2.1 Curved Vane Diffusers

The undisturbed streamline of an incompressible inviscid flow in a vaneless diffuser is a logarithmic spiral of constant flow angle α. Turning the flow more radially by means of vanes is similar to turning the flow more axially in an axial blade row. It results in a decrease in the tangential momentum, hence in an additional decrease in the tangential velocity (Figure 4.23). The change in the radial velocity between the inlet and outlet makes it more difficult than in an axial cascade to imagine what will be the resulting velocity and pressure distribution on a given vane geometry.

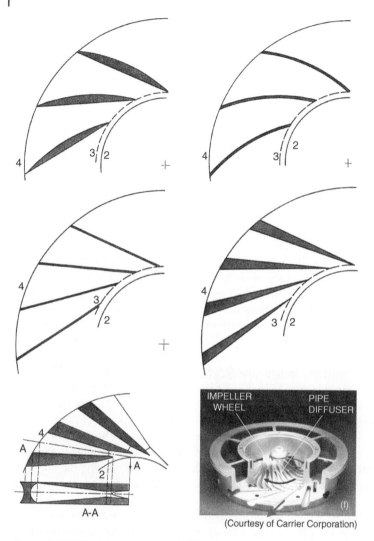

Figure 4.22 Different types of vaned diffusers.

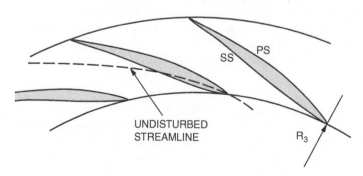

Figure 4.23 Influence of a diffuser vane on the flow in a radial diffuser.

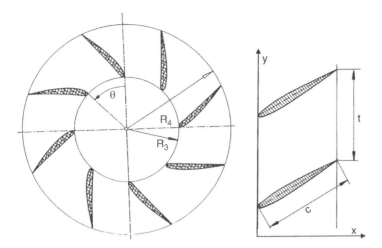

Figure 4.24 Transformation of a vaned diffuser into a straight cascade.

Singularity methods are well suited to analyze 2D flows in curved vane diffusers (Imbach 1964; Kannemans 1977). This inviscid method is quite fast and simple, but is restricted to 2D potential flows with uniform inlet flow. Conformal transformation of the geometry is another 2D method relating the flow to experimental or theoretical results obtained in straight cascades (Pampreen 1972; Yoshinaga et al. 1980).

The geometry transformation is very simple for the 2D incompressible flow (Figure 4.24):

$$x = \ell n R \qquad y = \theta \tag{4.35}$$

The velocity components in the linear cascade and diffuser are related by

$$V_x = R V_R \qquad V_y = R V_u \tag{4.36}$$

The transformation for compressible flow is more complex because it requires a correction as a function of the Mach number. Furthermore, the amount of useful experimental data obtained in straight cascades is limited because of the higher stagger angles used in radial diffusers. The method is further limited to potential flows, hence not applicable to transonic flows, unless a large vaneless region is present.

These approximate methods have anyway lost a lot of interest since accurate solvers and more powerful computers have become available. They are useful only to make a first design before starting the detailed Navier–Stokes analyses.

Multi-row or tandem diffusers aim for a larger pressure rise. Placing rows of vanes one after the other allows a new boundary layer to be started on each blade row and hence a larger flow deceleration can be achieved before it is limited by separation (Pampreen 1972; Startsev et al. 2015; Oh et al. 2011a). Pampreen used the conformal transformation technique to design a radial diffuser with three successive blade rows in tandem configuration (Figure 4.25). The purpose of the first blade row is to turn the flow to a direction which is optimal for the second and third ones, in which the diffusion is achieved. Only a small increase in compressor efficiency was measured but there was an important increase in the stall free operating range. An important design parameter is the circumferential position of the downstream blades relative the upstream ones. The leading edge of the downstream blade should not be in the wake of the upstream one because that would weaken the boundary on the downstream blade (Oh et al. 2011a). One of the major drawbacks of this type of diffuser is the large number of vanes that are required and the large radial extension.

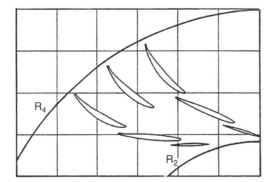

Figure 4.25 Use of tandem vanes in a radial diffuser (from Pampreen 1972).

Vaned diffusers have the potential to be more efficient than vaneless diffusers but have a smaller operating range because of choking in the diffuser throat section, limiting the maximum mass flow, and leading edge stall at high incidence, limiting the minimum mass flow. Senoo et al. (1983) and Pampreen (1989) have shown that these disadvantages can be avoided and that a larger flow range and high efficiency can be obtained with low solidity diffusers (LSDs). The blades of LSDs do not overlap so there is no clear geometrical throat and the choking mass flow may be as large as with a vaneless diffuser (Figure 4.26).

One of the consequences of the small radial velocity in combination with the large radial pressure gradient resulting from the large tangential velocity at low mass flow is the inward flow along the side walls. The increase in stability with LSDs results from the prevention of the development of large zones of return flow near the walls (Figure 4.27). The increase in efficiency is the consequence of turning the flow more radially at midspan, which increases the pressure rise while shortening the streamline length.

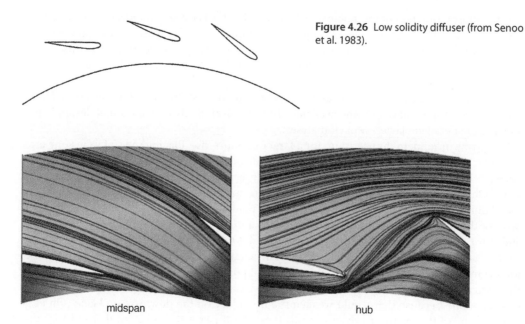

Figure 4.26 Low solidity diffuser (from Senoo et al. 1983).

midspan

hub

Figure 4.27 Streamlines in an LSD at stall limit (from Bonaiuti et al. 2002).

4.2.2 Channel Diffusers

Channel diffusers are mainly used in high pressure ratio compressors where the diffuser inlet Mach number is highly subsonic or supersonic. A typical geometry is shown in Figure 4.28, where three different flow regions can be recognized:

- the vaneless space between the rotor exit (R_2) and the diffuser leading edge (R_3)
- the semi-vaneless space between the diffuser leading edge (R_3) and the throat (th) section
- the diffuser channel between the throat (th) and the exit (R_4).

At low exit pressure the throat will be choked ($M = 1$) and the flow in the divergent channel accelerates supersonically until it is decelerated by a normal shock. The low efficiency at low exit pressure is the consequence of the large shock losses and subsequent flow separation (Figure 4.29a). The shock starts moving forward when increasing the back pressure. The Mach number upstream of the shock decreases, as do the shock losses. The shock will stay in the divergent channel as long as the flow in the throat remains sonic. The mass flow and hence also the flow conditions upstream of the throat remain unchanged, which explains the vertical part of the performance curve (Figure 4.29b).

The most interesting part of the constant RPM operating line is when the throat is unchoked and the flow in the divergent channel is fully subsonic. The static pressure increases continuously towards the exit of the divergent channel unless it is limited by flow separation. The flow and pressure in the vaneless and semi-vaneless space as well as the mass flow change with increasing diffuser exit pressure. The impeller and diffuser incidence increases as does the pressure rise between the impeller inlet and the throat section. Maximum pressure rise is when the impeller or diffuser stalls. The optimization of the inlet geometry for varying incidence and Mach number is the most important part of vaned diffuser design to assure maximum efficiency and operating range between surge and choke.

Although the flow in the three parts is strongly interdependent, the flow in each part can normally be predicted separately. The relation between the different parts is expressed in terms of flow conditions on the mutual boundaries.

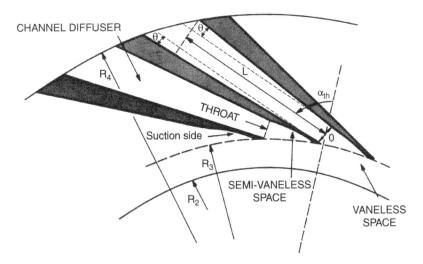

Figure 4.28 Geometry of a vaned island diffuser.

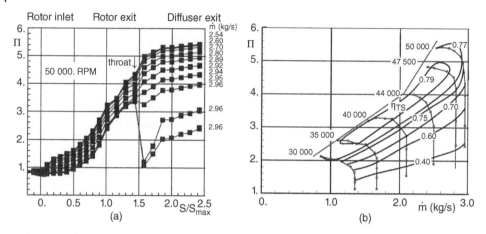

Figure 4.29 (a) Static pressure distribution in a vaned island diffuser at different back pressures and (b) the performance map of a centrifugal compressor with a choked diffuser (from Krain et al. 2007).

4.2.3 The Vaneless and Semi-vaneless Space

The vaneless and semi-vaneless space are functions of the diffuser leading edge to impeller exit radius ratio (R_3/R_2), the number of vanes, and the shape of the suction side between the leading edge and the throat (Figure 4.30). The flow in this area has been experimentally and numerically studied by Kenny (1970), Verdonk (1978), Deniz et al. (2000), Ziegler et al. (2003a,b), Filipenco et al. (2000), Ibaraki et al. (2007), Trébinjac et al. (2008, 2009), and many others. However, no clear design rules have been formulated up to now.

The flow upstream of the throat section is very complex because of the spanwise non-uniformity of the incoming flow, the unsteadiness of the flow in both the absolute (diffuser) and relative (impeller) frame as a consequence of circumferential non-uniformity of the relative flow at the impeller exit, and the upstream influence of the diffuser vanes. As shown by the numerical results of Trébinjac et al. (2009) the Mach number distribution depends on the position of the impeller with respect to the diffuser (Figure 4.31) and is different for the main blades (MB) and splitter blades (SB). As will be discussed in Chapter 7, unsteady calculations require the simultaneous solution of the flow in the complete impeller and diffuser, and are therefore computationally very expensive.

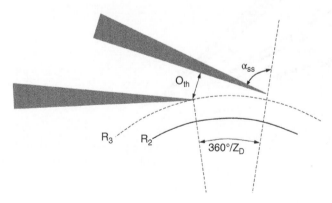

Figure 4.30 Geometry of the vaneless and semi-vaneless space.

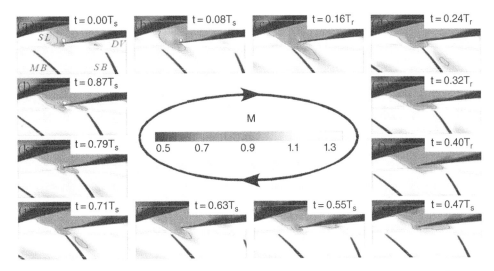

Figure 4.31 Variation of the Mach number in the vaneless and semi-vaneless space with impeller position (from Trébinjac et al. 2009).

The large number of impeller and diffuser blades results in a high frequency unsteadiness with limited spatial extend and impact on the overall performance. Hence the error in performance predictions by steady flow calculations in one impeller blade passage and one diffuser channel, connected by a mixing plane, is rather small and mostly within tolerances. The same instantaneous circumferential mixing of the flow at the impeller/diffuser interface is also used in the mean-line analysis methods presented in Section 3.3.

Four types of losses exist in the vaneless and semi-vaneless space:

- Viscous losses in the sidewall and vane suction side boundary layers. They increase with radius ratio.
- The jet and wake mixing losses, discussed in Section 4.1.2.
- Shock and shock boundary layer interaction losses in the case of transonic flow. They increase with local absolute Mach number and are a function of radius ratio and suction side shape.
- Unsteady flow losses due to the variation of the vane incidence when the flow is not yet fully mixed out at the leading edge.

The time averaged iso-pressure lines in Figure 4.32 show a rapid change from almost circumferential to nearly perpendicular to the velocity at the throat section and further downstream in the diffuser channel. In the 1D performance prediction the wake/jet mixing losses are assumed to be part of the impeller losses and the flow between impeller exit and diffuser leading edge is approximated by means of the vaneless diffuser calculation method described in Section 4.1.1. The upstream influence of the diffuser vanes is thereby neglected.

The throat static pressure and Mach number are calculated from continuity

$$V_{th} = \frac{\dot{m}/Z_D}{A_{th}(1-bl)\rho_{th}} \qquad (4.37)$$

with

$$T_{th} = T_2^0 - \frac{V_{th}^2}{2C_p} \qquad (4.38)$$

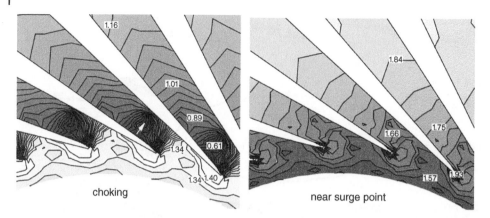

Figure 4.32 Measured pressure distribution (P/P_1^o) in the vaneless and semi-vaneless space at choking and near surge operation (from Kang et al. 2000).

Losses are accounted for by the blockage *bl* and the throat static pressure is defined by the inviscid core flow

$$P_{th} = P_2^o \left(\frac{T_{th}}{T_2^o} \right)^{\frac{\kappa}{\kappa-1}} \tag{4.39}$$

Calculation of the flow conditions in the throat requires an iterative procedure. Starting from a first guess of the throat blockage we can obtain a first approximation of the velocity (Equation (4.37)), static temperature (Equation (4.38)), and pressure (Equation (4.39)), which allows an update of the density in the throat.

Two problems remain to be solved:

- the prediction of the boundary layer blockage (*bl*) in the throat as a function of the inlet flow conditions
- the definition of the limiting value of the static pressure rise, between diffuser leading edge and throat, at which the flow is no longer a function of the boundary layer blockage but separated.

Kenny (1970) has correlated the throat blockage to the static pressure rise between the vane leading edge and the throat (Figure 4.33):

$$CP_{3-th} = \frac{P_{th} - P_3}{P_3^o - P_3} \tag{4.40}$$

This allows updating the throat blockage when calculating the flow conditions in the throat (Equations (4.37) to (4.39)).

Figure 4.33 indicates an increase in the throat blockage with increasing pressure rise. At supersonic leading edge Mach number, a normal shock will occur producing an additional pressure rise but also extra losses and throat blockage by the shock boundary layer interaction.

At maximum mass flow the fluid enters the diffuser with negative incidence and accelerates from the leading edge to the throat, where choking may occur (Figure 4.34c). This acceleration results in a pressure decrease ($CP_{3-th} < 0$), which explains the low throat blockage.

At reduced mass flow, the flow enters the diffuser more tangentially at zero or positive incidence and decelerates from the leading edge to the throat (Figure 4.34a). $CP_{3-th} > 0$ results in a larger throat blockage. The expansion of the flow around the leading edge at positive incidence and the subsequent deceleration towards a throat section, eventually enhanced by a shock, contribute to a further increase in the throat blockage. The local suction side diffusion is much

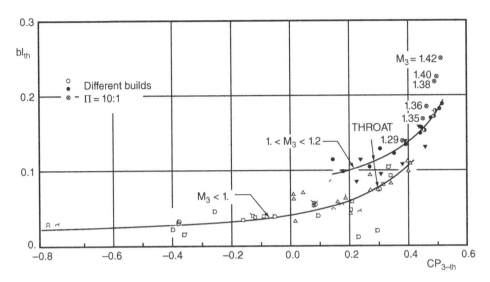

Figure 4.33 Throat blockage versus vane leading edge to throat static pressure rise (from Kenny 1970).

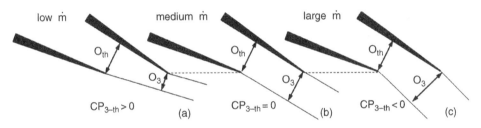

Figure 4.34 (a) Divergent, (b) parallel, and (c) convergent stagnation streamlines at the vaned diffuser inlet.

larger than the one corresponding to the area ratios. Figure 4.35 is based on the same data as Figure 4.33 and shows that the maximum incidence is limited to $+2°$.

Conrad et al. (1980) have studied the dependence of the throat blockage on the diffuser vane suction side angle (α_{SS}). Figure 4.36 shows that larger vane setting angles result in an increased throat blockage as a consequence of the longer flow path of the more tangentially incoming flow.

The minimum mass flow (stall) depends on the maximum achievable pressure rise CP_{3-th} and will be discussed in more detail in Section 8.5. A more explicit accounting of the blockage and losses at off-design operation of subsonic curved vane diffusers is presented by Ribi and Dalbert (2000).

There are no well-established methods that define the optimal geometry of the semi-vaneless space and predict its impact on performance and range. The following is an overview of some relations and findings about the flow characteristics that allow a better understanding of the relation between geometry and performance.

The number of vanes Z_D, the location R_3, and the thickness of the leading edge, the suction side angle α_{SS} and the throat section area O_{th} of a vaned island diffuser are geometrically related. Any change in one parameter has a direct impact on at least another one (Figure 4.30). The relation between throat area and setting angle, shown in Figure 4.37, is for $R_3/R_2 = 1.05$, zero vane leading edge thickness and a straight suction side between the leading edge and the throat. The total throat area increases with decreasing number of vanes. The decrease in the

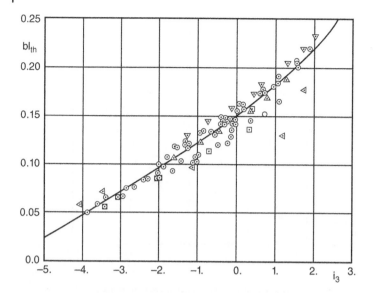

Figure 4.35 Throat blockage versus diffuser incidence angles.

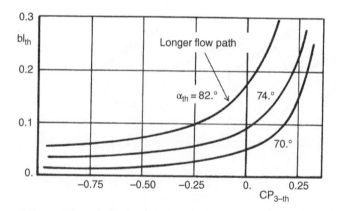

Figure 4.36 Variation of the throat blockage with diffuser vane setting angle (from Conrad et al. 1980).

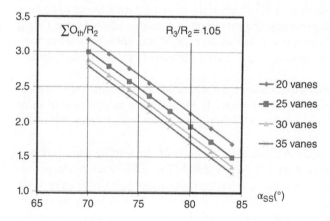

Figure 4.37 Relation between the diffuser vane number, vane setting angle, and throat area for $R_3/R_2 = 1.05$.

number of throat sections is more than compensated for by an increase in the throat area. At a given suction side angle, the choking mass flow, being directly related to the throat area, can be increased by reducing the number of diffuser vanes, but only as far as the maximum mass flow is not limited by impeller choking.

Another way to increase the throat area and hence the choking mass flow is by decreasing the vane setting angle α_{SS}. It is likely that doing this will also shift the stall limit to higher mass flows because of the corresponding increase in the diffuser incidence angle and pressure rise coefficient CP_{3-th}. The stable operating range is between surge and choking. As both may shift by the same amount to larger mass flows, the range will remain unchanged but could be better adapted to the operating range of the impeller. The opposite occurs when increasing the vane setting angle at a fixed number of vanes.

The geometrical relation illustrated in Figure 4.30 can be influenced by changing the suction side shape between the leading edge and the throat. A logarithmic spiral at zero incidence, i.e. a concave suction side as in Figure 4.28, corresponds to a straight suction side in the equivalent linear cascade. Hence it will result in a flow pattern similar to the undisturbed one in a vaneless space. The velocity deceleration up to the throat section will be proportional to $1/R$ and the flow in the throat will be less uniform (lower velocity nearer the suction side than the pressure side). This constitutes a kind of inlet distortion for the downstream diffuser channel.

A straight suction side between the leading edge and the throat corresponds to a convex curved suction side in the corresponding linear cascade. It creates a local flow acceleration along that side and might result in a more uniform flow in the throat by compensating for the flow deceleration at increasing radius described previously.

A further increase in the throat section by an outward curvature of the suction side will result in a further acceleration of the flow along that side (Figure 4.38). At high Mach numbers this will strengthen the normal shock upstream of the throat section and the neighboring diffuser vane. This will thicken the boundary layer and the corresponding increase in the throat blockage could well be larger than the geometrical increase in the throat area. This is similar to what is explained in Section 2.2.3 for the axial inducer. The effective throat area could be smaller than with a straight suction side shape and, as will be discussed later, the increased blockage is also detrimental for the pressure rise in the downstream diverging channel.

When operating axial compressors at transonic and supersonic inlet Mach numbers, the corresponding curvature is normally limited to only a few degrees in order to achieve an optimum compromise between the increase in the suction side Mach number and the increase in the throat section (Figure 2.31). In vaned diffusers this corresponds to an almost straight suction side. The influence of the diffuser vane leading edge shape has been experimentally studied by Bammert et al. (1976, 1983) and Clements and Artt (1989). They recommend a straight vane suction side upstream of the throat.

No clear criterion is available to define the optimal diffuser leading edge radius ratio. A value of R_3/R_2 between 1.05 and 1.12 is generally accepted. The lower values result in a moderate throat blockage but relatively high leading edge Mach number and noise because of the

Figure 4.38 Impact of suction side curvature on the diffuser throat section.

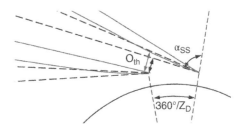

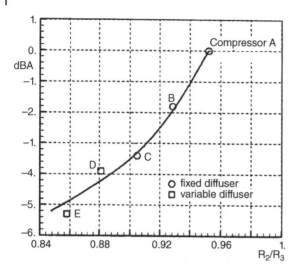

Figure 4.39 Sound level change with impeller–diffuser spacing (from Salvage 1998).

interaction with the wake/jet flow. Larger values of R_3/R_2 result in a lower noise level because of the more steady mixed outflow at the leading edge (Figure 4.39).

Transonic and unsteady flow losses can be avoided by increasing the radius ratio up to values where the flow is subsonic and fully mixed out. At constant vane number and unchanged suction side shape, this results in a proportional increase in the throat section. Lengthening the flow path upstream of the throat increases the friction losses and pressure rise from the impeller exit to the throat. Both have a negative impact on the throat blockage and, as shown in Section 8.5, the latter negatively impacts on the stall limit. The critical inlet flow angle for vaneless diffuser rotating stall also decreases with the radial extend of the vaneless space (Section 8.2). The increased throat blockage reduces the net effect of the geometrical increase in the throat section on choking mass flow. It is therefore uncertain what will be the effect of an increase in R_3/R_2 on the operating range. At a fixed diffuser exit radius, an increase in the semi-vaneless space leaves less room for the diverging channel where the static pressure rise will then be lower.

Ziegler et al. (2003a,b) and Robinson et al. (2012) have studied the impact of the radial extension of the vaneless space. They confirm that smaller radial gaps lead in most cases to a higher diffuser pressure recovery, resulting in a higher compressor efficiency. By changing the radial extent of the vaneless space they also modified the throat section area and, at unchanged outlet radius, also the length of the diffuser channel. Robinson et al. (2012) decided to compensate for the higher velocity at the smaller leading edge radius by a proportional decrease in the throat section by reducing the diffuser width, at unchanged vane setting angle and impeller exit width. Numerical predictions indicated a higher overall static pressure rise but somewhat lower total to total efficiency.

It can also not be excluded that by reducing the distance between the impeller exit and the vane leading edge the impeller energy input is modified due to the increased upstream influence of the vanes. All this explains the difficulty in coming to clear conclusions.

Similar geometrical relations exist for pipe diffusers. However, there are additional restrictions because the diameter of the inlet cone up to the throat is limited by the impeller exit width. The throat area can be reduced by increasing the convergence of the inlet cone. It can be increased by adding pipes, but as a consequence they will intersect at a larger radius, resulting in a larger semi-vaneless space (Bennett et al. 1972). The number of pipes can be lowered by increasing the pipe diameter, requesting side expansion of the diffuser walls to values larger than the impeller exit width, or by using oval pipes instead of circular ones. Rodgers and Sapiro (1972) compared the pressure rise in a pipe and vaned channel diffuser and found a slightly higher pressure rise for the latter and a considerably larger range between surge and choking. Other researchers claim superior performance with pipe diffusers (Kenny 1970), which illustrates the complexity of the optimal vaned diffuser design.

4.2.4 The Diffuser Channel

The flow in straight channel diffusers downstream of the throat has been studied experimentally by Reneau et al. (1967), Runstadler et al. (1969), and many others. They found that the pressure rise coefficient

$$CP_{th-4} = \frac{P_4 - P_{th}}{P^o_{th} - P_{th}} \tag{4.41}$$

depends on

- throat Mach number, M_{th}
- throat blockage, bl
- Reynolds number based on V_{th} and O_{th}
- aspect ratio, $AS = b_{th}/O_{th}$ (Figure 4.40)
- area ratio, $AR = A_4/A_{th}$
- length over width ratio, $LWR = L/O_{th}$
- cross sectional shape (circular or rectangular).

Typical results are shown in Figure 4.41 and allow the definition of the static pressure rise coefficient as a function of previous parameters.

At small values of the divergence angle ($2\theta < 8°$) the deceleration of the fluid is controlled by the geometrical area ratio corrected for the boundary layer blockage. The lines of constant CP are almost parallel to the lines of constant AR. Decreasing the opening angle for the same area ratio results in longer diffusers, hence thicker boundary layers and lower CP values.

For large values of the diffuser opening angle ($2\theta \geq 20°$), separation already occurs near the diffuser inlet at a small value of AR, which results in low values of CP. No diffusion takes place downstream of the separation point and CP is not increasing anymore by making the diffuser longer. Lines of constant CP are now parallel to lines of constant opening angle. The value of CP depends on the pressure at separation point. This is in agreement with the idea of maximum diffusion ratio, already explained at the impeller flow calculation.

Reneau et al. (1967) observed that the locations for maximum CP are independent of the inlet boundary layer blockage. In long diffusers ($L/O_{th} = 15$) maximum diffusion occurs at small opening angles ($2\theta = 6°$) and at larger opening angles ($2\theta = 18°$) in short diffusers ($L/O_{th} = 2$).

The locus of maximum CP may be independent of the throat blockage, the value of CP is not. Runstadler et al. (1969) have made a systematic study of different diffuser geometries with

Figure 4.40 Geometry of a vaned diffuser channel.

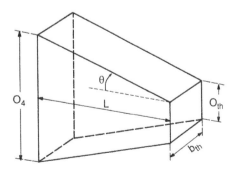

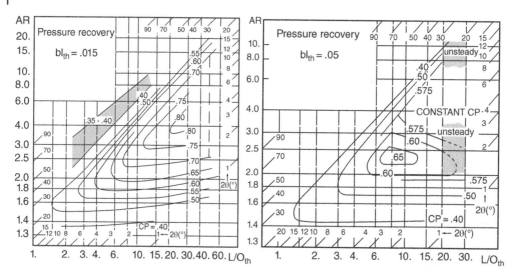

Figure 4.41 Static pressure rise as a function of the channel geometry and inlet blockage (from Reneau et al. 1967).

a large variety of inlet flow conditions. Results have been published on graphs (Figure 4.42) similar to the ones of Reneau et al. (1967). They allow the prediction of the diffuser performance for a large variety of inlet Mach numbers, inlet blockage, Reynolds numbers, and aspect ratios. At large inlet blockage the static pressure rise coefficient increases along the diffuser length until the blockage due to friction becomes larger than the area increase and the boundary layer growth prevents any further deceleration of the flow.

CP increases with increasing values of L/O_{th}. However, L is a geometrical function of the diffuser outlet to throat radius ratio (R_4/R_{th}) and the vane suction side angle α_{SS} (Figure 4.28). L/O_{th} can be increased at constant L by reducing O_{th}, i.e. by increasing the number of diffuser channels which, however, results in an increase in AS_{th}. Results of Runstadler et al. (1969), summarized in Figure 4.43, show that maximum CP occurs at $AS_{th} = 1$. The highest performance is obtained when the ideal AS (maximum DH) occurs at the location of maximum velocity, i.e. in the throat section. The pressure rise coefficient decreases only slowly when increasing the AS above the optimal value because the optimal aspect ratio will occur downstream of the throat due to the divergence of the channel. An increase in the number of vanes may be a possible way to improve the diffuser performance. However, Figure 4.37 indicates that at unchanged α_{SS} an increase in the number of diffuser vanes results in a decrease in the total throat area and hence of choking mass flow.

The maximum achievable CP decreases rapidly for values of $AS_{th} < 1$ because the less than optimal AS decreases further downstream in the diffuser channel. A decrease in L/O_{th} and an eventual decrease in AS below the optimal value are the negative consequences of a decrease in the number of vanes to increase the throat area.

Clements and Artt (1987b) observed that increasing the diffuser divergence angle up to the limit where transient stall occurs increases the stage efficiency but decreases the operating range. This decrease in range was thought to be due to the diffuser channel not being able to tolerate the higher levels of throat blockage without separating. Designing the diffuser for a smaller than optimal divergence angle leaves a margin for stable operation when the throat blockage is increasing at reduced mass flow. Clements and Artt (1987b) correlated maximum throat blockage at surge with the channel divergence angle for diffusers with $L/O_{th} \approx 3.75$ and showed that the value cannot be unique but depends on the diffuser inlet geometry.

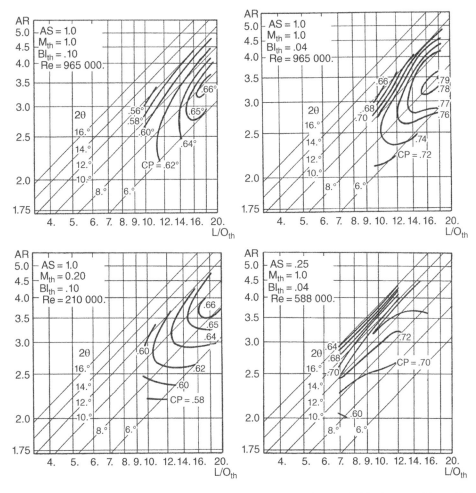

Figure 4.42 Variation of diffuser performance with inlet Mach number, inlet blockage, and Reynolds number (from Runstadler and Dean 1969).

From these observations we can conclude that the design of a channel diffuser is in the definition of the optimum opening angle for a given L/O_{th} and in optimizing the inlet section to minimize the throat blockage.

Extra pressure rise for a given throat section without increasing the diffuser outlet radius is possible by starting with a larger than optimal opening angle after the throat but reducing the diverging later on by installing splitter vanes similar to what is commonly done in impellers (Benichou and Trébinjac 2015). Figure 4.41 shows that short diffuser channels can have a larger opening angle (2θ) but with a limited length and area ratio. Adding splitter vanes reduces the divergence angle to half and diffusion can continue. Care must be taken not to create a second throat limiting the choking mass flow. The split diffuser channels are shorter than the initial ones but the throat section is almost half the original value so that L/O_{th} is larger. The new boundary layer starting at the splitter vane leading edge makes only a small contribution to the blockage in the second throat. The static pressure rise is the sum of the pressure rise in the more divergent inlet section and the one in the less diverging downstream part.

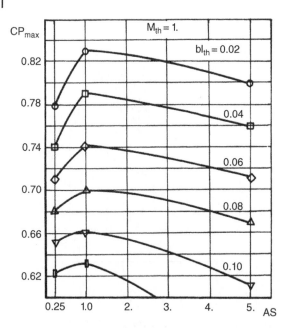

Figure 4.43 Variation of maximum *CP* with aspect ratio and throat blockage (from Runstadler and Dean 1969).

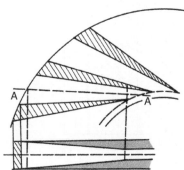

Figure 4.44 Side wall divergence in vaned island diffusers.

Filipenco et al. (2000) claim that "the generally accepted sensitivity of the diffuser static pressure recovery performance to inlet flow distortion and boundary layer blockage can be largely attributed to inappropriate quantification of the average dynamic pressure at diffuser inlet". However, this conclusion is based on the flow conditions at the diffuser leading edge and a throat blockage resulting from flow distortion generated by side wall aspiration or blowing. It is not specified if the blockage was a consequence of a decrease in total pressure due to friction or of a change in flow direction at nearly constant total pressure.

Pipe diffusers have the advantage of lower friction losses because of a large hydraulic diameter and because the optimum aspect ratio is maintained up to the diffuser exit.

Channel diffuser pressure recovery can also be improved by using divergent side walls downstream of the throat such that the optimum AS is conserved throughout the channel (Figure 4.44) (Came and Herbert 1980). However, this complicates an eventual restaggering of the diffuser vanes.

5

Detailed Geometry Design

Accurate CFD techniques are now available and routinely used for the analysis of turbomachinery components. They provide pressure, temperatures, and velocity components in a huge number of points with great accuracy but do not provide information about what geometrical changes are needed to improve the performance or to reach the prescribed targets. The designer has to find it out him- or herself. This explains the growing interest in numerical techniques to accelerate the design process and achieve improved performance.

Advanced design systems must allow full use of all the 3D geometrical features that may improve performance, including lean and sweep. Any limitation of the geometry is acceptable only if it is imposed by mechanical, manufacturing or cost limitations.

The optimal performance can only be guaranteed if all real flow phenomena are taken into account, i.e. if the design is based on accurate flow analyses by means of 3D Navier–Stokes solvers. Any use of inaccurate performance measuring systems may lead to a suboptimal geometry. Simplified methods can only be accepted as a first step of the design and must be complemented by accurate ones.

Design systems must also be sufficiently fast and automated to achieve all these requirements within an acceptable time and cost. Manual modification of the geometry, meshing, and analysis is inefficient. Accurate design systems require large amounts of computer analyses and research focuses on acceleration techniques to limit the effort without compromising on the result.

The system must also provide realistic designs that satisfy the mechanical and geometrical constraints and guarantee the requested lifetime of the device. Satisfying all these objectives requires not only multidisciplinary optimization involving acoustics, stress, and heat transfer analysis but also a compromise between conflicting objectives. High performance must also be guaranteed over a sufficiently large operating range (multipoint optimization and robust design).

The following describes the two basic approaches to the design of compressor components:

- inverse methods defining the geometry corresponding to a prescribed "optimal" pressure or velocity distribution
- optimization techniques defining the geometry that provides the targets such as maximum efficiency, prescribed pressure ratio and mass flow, minimum operating range, etc. while satisfying geometrical and mechanical constraints.

5.1 Inverse Design Methods

The first problem in inverse design is the definition of the velocity or pressure distribution that guarantees the best performance and satisfies the geometrical design constraints. A good understanding of flow mechanisms is needed for this. Boundary layer and loss considerations

Design and Analysis of Centrifugal Compressors, First Edition. René Van den Braembussche.
© 2019, The American Society of Mechanical Engineers (ASME), 2 Park Avenue, New York, NY, 10016, USA (www.asme.org).
Published 2019 by John Wiley & Sons Ltd.

in combination with inverse boundary layer methods have led to what is known as controlled diffusion blades for axial compressors (Papailiou 1971) and optimal pressure distributions for turbine blades (Huo 1975). Finding an optimal distribution may be feasible for 2D flows in axial turbomachines, when secondary flows can be neglected, but it is unclear what an optimal distribution should look like in centrifugal impellers in the presence of strong 3D effects and secondary flows as well as flow separation. Defining a velocity distribution that also performs well at off-design (different incidence for compressors or backpressure for transonic turbines) is an additional challenge.

Furthermore, special attention is required when prescribing the optimal velocity or pressure distribution to satisfy the constraints in terms of blade thickness, camber, lean, etc. Except for incompressible 2D potential flows (Lighthill 1945) there is still no rigorous theory that guarantees the existence of a solution for a given pressure distribution. Some conditions can be formulated that help to control the geometry of axial turbomachines (Van den Braembussche 1994) but there is very little information for radial impellers. Modifying the velocity distribution of an existing geometry instead of defining a completely new one has been shown to be very helpful in this respect.

Another problem of inverse design methods is that the numerical domain is the outcome of the problem and hence is unknown at the start of the design. This complicates the use of the available numerical tools. Older design methods make use of conformal mapping or transpose the problem in the hodograph plane. However, they have a very restricted domain of application (2D, inviscid, and subsonic) and therefore are of no interest for centrifugal compressors.

The main advantage of the inverse design method is the ability to design in detail very specific blade geometries, such as controlled diffusion blades and shock-free transonic blades, that would be very difficult to obtain with any other method.

Two types of inverse design methods are of practical use:

- The first one makes use of an inviscid analytical approach to relate the geometry to the target velocity distribution. These methods provide an approximate answer to the problem and hence are useful only for a first design.
- The second method starts from an existing geometry and uses a physical model in combination with CFD to define the target geometry in an iterative way. The outcome of these methods is as accurate as the numerical technique they are based on (viscous or inviscid, 2D and 3D).

5.1.1 Analytical Inverse Design Methods

The first method is based on the quasi 3D flow model described in Section 3.1. The procedure starts from the prescribed velocity distribution on the suction and pressure side at the hub and shroud (Figure 3.3). The average velocity $\widetilde{W}$ at both locations is used to define the meridional contour and the suction to pressure side velocity difference defines the blade angles at the hub and shroud.

Equation (3.3) can be used to define the meridional curvature radius of axisymmetric streamsurfaces as a function of the average velocity $\widetilde{W}$, the gradient $\frac{\delta\widetilde{W}}{dn}$, and an estimation of the flow angle:

$$\mathfrak{R}_n = \frac{\widetilde{W}\cos^2\beta}{\frac{\delta\widetilde{W}}{\delta n} + \cos\lambda\sin\beta\left(\frac{\widetilde{W}\sin\beta}{R} + 2\Omega\right)} \tag{5.1}$$

The average velocity $\widetilde{W}$ at the hub and shroud can be calculated from the prescribed suction and pressure side velocity (Figure 3.3) and approximated at the intermediate spanwise position by means of a linear interpolation. The circumferentially averaged flow angle β is the one of a first guess or obtained from the blade to blade calculation at previous iteration.

Approximating $\frac{\delta \widetilde{W}}{\delta n}$ by $(\widetilde{W}_S - \widetilde{W}_H)/\Delta n$ together with the velocity $\widetilde{W} = 0.5(\widetilde{W}_H + \widetilde{W}_S)$, Equation (5.1) provides the value of the curvature radius $\mathfrak{R}_n$ of the mean streamsurface between the hub and the shroud. The curvature radius at the shroud is assumed to be $\mathfrak{R}_n - \Delta n/2$ and the coordinates $X_S(m + \Delta m), R_S(m + \Delta m)$ of the shroud can easily be defined by the simple geometrical construction shown in Figure 5.1, where

$$\gamma_S(m + \Delta m) = \gamma_S(m) + \frac{\Delta m_S}{\mathfrak{R}_{n,S}} \tag{5.2}$$

and $\Delta m_S = \Delta s_S \cos \beta_S$.

The corresponding point $X_H(m + \Delta m), R_H(m + \Delta m)$ on the hub contour is defined by the distance Δn between the shroud and hub, and the distance $\Delta m_H = \Delta s_H \cos \beta_H$ (Figure 3.3).

The distance Δn is obtained by applying continuity along the quasi orthogonal:

$$\dot{m} = \int_{n_S}^{n_H} \left(2\pi R - \frac{Z_r \, \delta_{bl}}{\cos \beta} \right) \rho \widetilde{W} \cos \beta \cos(\gamma - \xi) dn \tag{5.3}$$

assuming a linear variation of $\delta_{bl}, \widetilde{W}$, and β between the hub and the shroud.

The average flow direction $\widetilde{\beta}$ is updated by integrating Equation (3.13) along the hub and shroud surfaces as a function of the difference between the required suction and pressure side velocity. The blade angle equals the flow angle except near the trailing edge, where a correction for slip is required. The blade angle is then obtained from inverting Equations (3.8) to (3.12).

Once the blade angle is known, the blade circumferential position θ is easily defined by an integration from leading edge to trailing edge according to Equation (5.11) and is illustrated in Figure 5.12. The blades are further defined by straight lines between the points $R(m), X(m)$, and $\theta(m)$ at the hub and shroud, together with the prescribed blade thickness distributions.

Redesigning an impeller geometry starting from the velocity distribution obtained from the quasi 3D analysis of the existing impeller indicates a good agreement between the original and

Figure 5.1 Construction of shroud and hub contour.

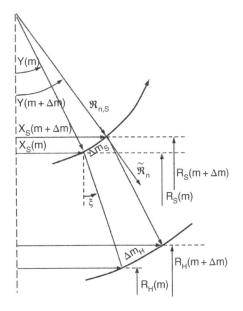

the redesigned one except for some small discrepancies near the trailing edge (Van den Braembussche et al. 1993).

A major problem of the inverse design of radial impellers is the difficulty of defining an optimal velocity distribution that also satisfies the mechanical and geometrical constraints. Any change in the velocity distribution modifies the blade shape and the meridional contour because of the strong coupling between them. This is illustrated by redesigning an impeller where the original velocity distribution has been modified from fore to aft loading while keeping the average velocity unchanged (Figure 5.2):

$$(W_{SS} - W_{PS})_{imposed} = (W_{SS} - W_{PS})_{original} - 10. \, SF \sin\left(2\pi \, \frac{s}{s_{max}}\right) \tag{5.4}$$

As the velocity is specified along the absolute blade length, the energy input at the hub and shroud will be unchanged by imposing $SF = 1$ at the shroud and $SF = s_{max,S}/s_{max,H}$ at the hub. The impeller designed for this new velocity distribution is shown in Figure 5.3.

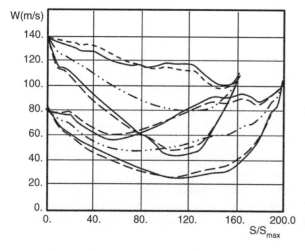

Figure 5.2 Velocity distribution with modified loading (——) versus original velocity distribution (- - - -).

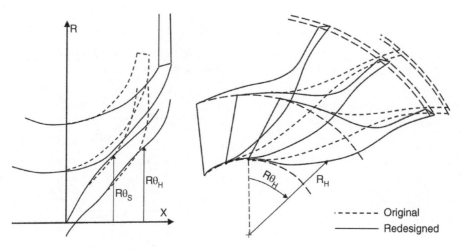

Figure 5.3 Comparison between the geometry for modified loading (——) and the original impeller (- - - -).

Although the hub and shroud average velocities are unchanged, the redesigned meridional contour is much less curved, resulting in a larger axial length and a smaller outlet radius. The smaller increase in the radius with an almost unchanged blade angle in the first part results from the smaller loading. A faster increase in the radius and more radial blades in the second half of the blade length are needed to achieve the increased blade loading and to compensate for the smaller radius.

The previous example demonstrates the difficulty of respecting the geometrical constraints when specifying the velocity distribution at the hub and shroud in a centrifugal impeller. It is difficult to control the loading and the average velocity at the same time without also changing the meridional contour. Many inverse design systems therefore keep the meridional contour fixed and specify only the blade loading. The streamwise variation of the average velocity at the hub and shroud is then an outcome of the design. This means that the diffusion is no longer controlled, which is known to have a major impact on the compressor performance (Lieblein 1960).

The most popular methods are those in which the blades are represented by sheets of vorticity whose strength is determined by a specific distribution of the circumferentially averaged swirl velocity $R\widetilde{V}_u$ along the prescribed hub and shroud meridional contours. The main advantages are that the energy input at the hub and shroud can be easily controlled and that by imposing

$$\partial(R\widetilde{V}_u)/\partial m = 0 \tag{5.5}$$

at the leading and trailing edge the blade will be at zero incidence and satisfy the Kutta condition at the trailing edge.

This method, initially developed for incompressible flow in axial turbomachines (Hawthorne et al. 1984), has been extended to compressible flow in centrifugal compressors by Borges (1990) and Zangeneh (1991), including splitter blades (Zangeneh 1998), and to transonic flows in centrifugal impellers (Dang 1992).

The quality of the results depends on the "optimality" of the swirl distribution at the hub and shroud and the stacking of the blade sections, for which not much guidance is available in the literature. Zangeneh et al. (1998) advocate swirl distributions and stacking line that minimize the hub to shroud pressure difference on the blade suction side in order to minimize the secondary flows and to achieve a more uniform spanwise distribution of the impeller outflow. This is in line with what was proposed by Baljé (1981) to obtain the optimal meridional contour as described in Section 3.2.1. This requires an aft loading at the hub and fore loading at the shroud. However, the latter may result in severe tip leakage flows and strong shocks when operating at high inlet Mach number.

Shibata et al. (2010) have studied the effect of the loading distribution in combination with a change in the meridional contour. They concluded that aft and front loading increased the surge and choking margin, respectively, but that this has little effect on the efficiency. A combination of aft loading with increased diffusion of the relative velocity and more backsweep showed an increase in the efficiency of 3.8% and an 11% increase in the surge margin.

Blade thickness and viscous effects have been accounted for by Zangeneh (1993) in an iterative procedure introducing a blockage distribution and extra vorticity terms related to the entropy gradients. Both are obtained from a viscous analysis of the flow in a previous approximate design of the geometry.

Viscous effects can also be accounted for in an iterative procedure whereby the new designs are analyzed by a Navier–Stokes solver. Comparing the calculated RV_u with the target whirl velocity $RV_u^\star$ allows the whirl distribution imposed in a new design iteration to be adapted. The procedure is repeated until the outcome of the Navier–Stokes solver is in agreement with the target (Tiow and Zangeneh 2000).

The imposed swirl distribution and stacking line may also be the outcome of an optimization procedure. Analyzing geometries corresponding to different parameterized load distributions by an accurate Navier–Stokes solver in combination with an optimization technique allows the parameter setting corresponding to minimum losses to be defined (Tiow et al. 2002; Yagi et al. 2008; Zangeneh et al. 2014). Compared to a classical optimization (Section 5.2), this approach facilitates guaranteeing the required energy input because it can be verified before starting the design by integrating the imposed load distribution. It is unclear what the advantage of such an indirect optimization is because there are many other parameters, such as meridional contour and splitter blade leading edge position, influencing performance. It is also more difficult to respect the mechanical constraints, which in classical optimization systems can already be verified in an early stage of the design before any analysis is started.

5.1.2 Inverse Design by CFD

The second type of inverse method combines a flow solver with a physical model to define the geometry modifications required for achieving the prescribed pressure or velocity distribution. This moving wall approach has shown to be very efficient and has been applied to a large variety of design problems (Léonard 1992; Léonard et al. 1992; Demeulenaere 1997; Demeulenaere et al. 1999).

The core of these methods is a modified time marching solution of the 3D Euler equations in a domain of which the walls are moving during the transient part of the calculation (Demeulenaere et al. 1998a). Two calculations are made at each time step (Figure 5.4).

In the first time step the flow, corresponding to the desired pressure distribution imposed on the walls of the present geometry, is calculated. The velocity will not be tangent to the walls except when the imposed pressure distribution equals the one resulting from a direct calculation around the same geometry. Permeable wall boundary conditions allow the velocity component normal to the blade walls to be calculated, which is then used to define a new blade geometry by means of the transpiration model.

The second time step makes an update of the flow field, taking into account the movement of the walls and mesh. These two time steps are separated by a geometry modification based on transpiration.

The transpiration model defines a new blade geometry as a function of the calculated normal and imposed tangential velocity along the initial blade suction and pressure side (Figure 5.5).

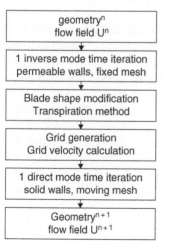

Figure 5.4 Flow chart of the moving wall inverse design method.

Figure 5.5 3D transpiration model (from Demeulenaere et al. 1998a).

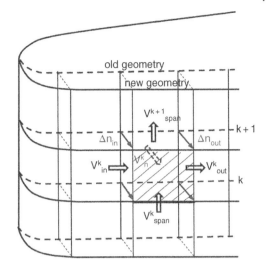

Starting at the leading edge stacking line the method progresses separately along the pressure and suction surface to the trailing edge. The stagnation line can be kept unchanged or modified to achieve a prescribed lean and sweep. All mesh points, including the ones on the blade wall, are displaced in the axisymmetric grid surfaces. The amplitude of the displacement Δn_{out} is obtained by applying continuity in the cells between the old and new blade walls. A grid-point velocity W_g is deduced from these displacements and is added as an extra term in the Euler solver with moving walls. After the blade shape modification, a new mesh is generated around the new profile.

Experience shows that after each time step the pressure distribution on the new geometry is closer to the target one, provided that the target corresponds to a feasible geometry. This procedure is repeated until the normal velocity and hence the transpiration flux is zero, i.e. when the pressure distribution obtained during the analysis step equals the target one.

The time marching procedure converges to the steady asymptotic solution simultaneously, with the geometry converging to the desired one. This progressive movement to the final geometry facilitates control of the blade geometry during the design process and, if needed, allows for an eventual adjustment of the target pressure distribution.

The existence of a converged solution with positive blade thickness depends on the design requirements and cannot be a priori guaranteed. The pressure to suction side average pressure can also be prescribed along the hub and shroud endwall, which gives rise to a distribution of velocities normal to those walls. The latter is input to the transpiration method applied to the shroud and hub contour.

The method is illustrated by the redesign of the centrifugal impeller shown in Figure 5.6. This impeller, with 20 twisted blades, rotating at 18 000 RPM, has been designed with a rapid 1D method. The grid lines on the blades define the 12 axisymmetric surfaces on which periodic H-grids have been generated to discretize the numerical domain and on which the displacement of the blades will be defined.

As already demonstrated in Section 5.1.1, it is particularly difficult to specify a completely new pressure distribution for a radial impeller that results in a realistic geometry. The design therefore starts from an existing geometry and target Mach number distributions at the hub and shroud that are smooth versions of the ones calculated in the starting geometry (Figure 5.7). The mean velocity of this target distribution is the outcome of an adjustment during the design process to assure a radial impeller outlet/diffuser inlet.

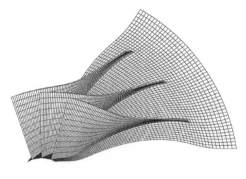

Figure 5.6 3D view of the initial centrifugal impeller geometry.

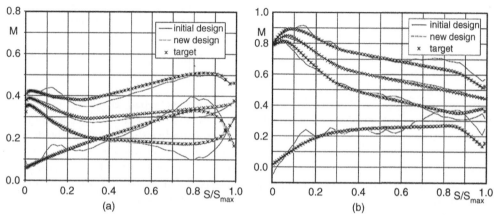

Figure 5.7 Initial and imposed isentropic Mach number distribution and loading at the (a) hub and (b) shroud (from Passrucker 2000).

Meridional contour and blade shape modifications are alternated in order to control the average velocity level and blade loading distributions in the blade channel, respectively. The 3D flow effects require the use of a 3D flow solver (Demeulenaere et al. 1998b) but normal velocities for the blade modifications may also be defined by the quasi 3D solver imposing a pressure distribution that accounts for the pressure difference between the target pressure distribution and the one obtained from the more accurate 3D solver (Passrucker et al. 2000). The thickness of radial impeller blades is usually fixed by mechanical considerations, and only the camberline of the blade is redesigned.

Figure 5.7 shows an almost perfect agreement between the target Mach number distribution and the one of the new design. The total modification of the hub endwall is shown in Figure 5.8 together with a comparison between the initial and modified blade shapes in four sections between the hub and the shroud. Analysis of the redesigned impeller with a Navier–Stokes solver confirms that smoothing the velocity distribution has a favorable effect on the efficiency (Demeulenaere et al. 1998b).

Design methods based on Euler solvers are unable to predict the secondary flows and flow separation usually occurring at the outlet of centrifugal impellers. It has, however, been shown (Ellis and Stanitz 1952) that secondary flows do not have much influence on the pressure distribution on the blade walls and therefore do not alter the validity of the method, at least not in the first half of the blade channel. The method is therefore particularly well suited to the design of transonic inducers, as illustrated by Demeulenaere et al. (1996, 1999) by the design of a shock-free transonic compressor (Figure 5.9).

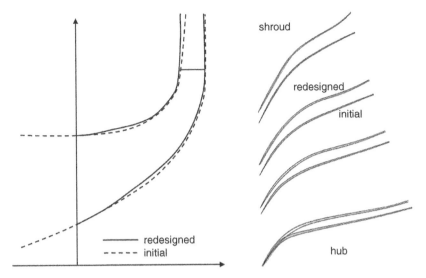

Figure 5.8 Initial and redesigned meridional contour and blade sections (conformal mapped) (from Passrucker 2000).

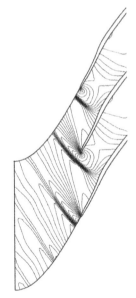

Figure 5.9 Shock-free transonic inducer (from Demeulenaere et al. 1999).

The extension of this inverse approach to a viscous solver with the zero slip condition on the walls is questionable, from both the physical and the mathematical points of view. Demeulenaere et al. (1997) and Léonard and Demeulenaere (1997) presented an extension whereby the normal velocity distribution W^n required by the transpiration model is no longer obtained from the compatibility equations but from the following simplified expression:

$$W_n^{n+1} = \frac{P - P^\star}{\rho a} \tag{5.6}$$

where $P^\star$ is the target pressure and P is the pressure calculated by a Navier–Stokes solver on the wall of the present geometry. An inverse time step of the solver is no longer required. The geometry modification can either be based on the transpiration principle, with the tangential velocity estimated at the edge of the boundary layer, or directly calculated from the normal velocity (Daneshkhah and Ghaly 2007):

$$dn = R_f W_n dt$$

where R_f is a relaxation factor. The blade modification procedure can be combined with commercial CFD programs as a user defined function. Arbabi and Ghaly (2013) observed a slower convergence than with the original Euler version for the redesign of subsonic blades.

Extensive testing for transonic and supersonic flows revealed serious convergence problems that can be explained by considering Equation (5.6) for the flow in a diverging Laval nozzle. If the required static pressure is higher than the computed one, it will indicate a negative W_n, i.e. fluid entering the numerical domain, and the nozzle will become narrower. This is correct only if the flow is supersonic because for the same pressure difference the nozzle should become wider at subsonic flows. The presence of a boundary layer does not invalidate this demonstration. Adapting the sign in Equation (5.6) to subsonic and supersonic flows could work for straight nozzles but it is unclear how the algorithm should be modified in more complex turbomachine applications where the boundary between subsonic and supersonic flow may move during the design process.

de Vito et al. (2003) avoided this ambiguity by combining a viscous solver with the inviscid inverse method. Two steps again alternate: an analysis step by means of a Navier–Stokes solver and an inverse step by means of an Euler solver where the imposed pressure distribution P^{imp} is defined by the difference between the target pressure and the one obtained from a Navier–Stokes analysis:

$$P^{imp} = P_{inviscid} + (P^* - P)_{viscous} \tag{5.7}$$

This second step defines the normal velocity used to redesign the blade shape by transpiration. The redesign of an axial turbine blade revealed the existence of more than one solution, i.e. more than one geometry satisfying the required Mach number distribution. This problem has not been observed in compressor design.

Blade designs in which the required pressure distribution is a modified version of that obtained from the analysis of an existing geometry do not show any particular problem. Under-relaxation may sometimes be needed if the required geometry is far from the initial one, but the CFD-based methods rapidly converge to the correct blade shape.

Starting from an arbitrary suction and pressure side pressure distribution may create convergence problems if the required pressure distribution is not compatible with the upstream and downstream boundary conditions or does not allow a realistic blade profile (closed with a positive thickness). Adjusting the pressure distribution to the free stream flow and satisfying mechanical constraints is facilitated by parameterizing the imposed pressure distribution. The parameters can be adjusted during the design process until the geometry satisfies the geometrical constraints. However, the "feasible" pressure distribution may be less optimal than the initially imposed one.

The most direct approach is by hybrid methods specifying the pressure only on one side of the blade together with the thickness distribution (Demeulenaere et al. 1998a).

The blade modification method based on CFD, presented previously, defines the blades by streamlines and should not suffer from the problem of contour intersection (negative thickness) except for numerical errors introduced when integrating the streamlines. The method normally converges to the closest physically possible geometry.

Euler solvers do not account for viscous effects, and the suction and pressure side boundary layer displacement thickness has to be subtracted from the designed blade to obtain the geometrical blade. This thickness can already be calculated before the blade geometry is known because the required Mach number distribution is known at the start. The sum of the geometrical thickness and boundary layer displacement at the trailing edge is imposed as a constraint on the designed geometry.

5.2 Optimization Systems

Optimization methods define the n_D design variables $\vec{X}(i = 1, n_D)$ of the geometry that minimizes an objective function $OF(U(\vec{X}), \vec{X})$, where $U(\vec{X})$ is the solution of the flow equations $R(U(\vec{X}), \vec{X}) = 0$ and subject to:

n_A performance constraints $A_j(U(\vec{X}), \vec{X}) \le 0$ $(j = 1, n_A)$

and

n_G geometrical constraints $G_k(\vec{X}) \le 0$ $(k = 1, n_G)$.

Figure 5.10 Optimization procedure.

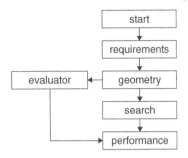

Numerical optimization procedures consist of the following basic components (Figure 5.10):

- A parameterized **definition of the geometry** by a choice of independent design parameters and a limitation of the addressable part of the design space according to the geometrical constraints.
- An **objective function** *OF* quantifying mathematically how far the design goals related to pressure ratio, mass flow, efficiency, noise, stress, manufacturing cost or any other design target are reached. When specifying the difference between a prescribed and achieved Mach number or pressure distribution as *OF*, the method becomes an inverse design method.
- A **performance evaluation** system, including an automatic grid generator and performance analysis, to provide the input for the *OF* of newly proposed geometries.
- A **search mechanism** to find the optimum combination of the design parameters, i.e. the one corresponding to the minimum *OF* while satisfying the geometrical, mechanical, or any other constraint.

5.2.1 Parameterized Definition of the Impeller Geometry

Defining a geometry by analytical functions allows a reduction in the number of unknowns. This is illustrated by a parametrized definition of an impeller, but similar systems can be used for diffusers, return channels or any other compressor component.

The hub and shroud meridional contours are easily defined by Bézier–Bernstein polynomials, between the leading and trailing edge, with extensions upstream and downstream (Figure 5.11):

$$x(u) = x_1(1 - u)^3 + 3x_2 u(1 - u)^2 + 3x_3 u^2(1 - u) + x_4 u^3 \tag{5.8}$$

$$R(u) = R_1(1 - u)^3 + 3R_2 u(1 - u)^2 + 3R_3 u^2(1 - u) + R_4 u^3 \tag{5.9}$$

Figure 5.11 Definition of the radial impeller meridional geometry by Bézier polynomials.

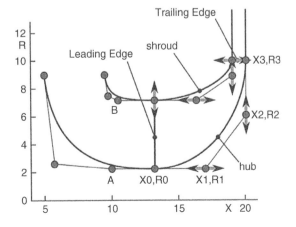

(x_1, R_1) to (x_4, R_4) are the design parameters. Higher order definitions are also possible but experience has shown that third-order equations can generate a wide range of hub and shroud contours. The points A and B are adjusted to obtain a smooth transition between the impeller and the environment defined by the axial or radial inlet section. The diffuser can be defined in a similar way.

The main advantages of this analytical definition are smooth contours with continuity of curvature and an easy calculation of all coordinates. The absolute value of the coordinate m along the meridional contour is defined by a stepwise integration of dx and dR along the contour from leading edge $u = 0$ to trailing edge $u = 1$.

The blade profiles are defined by the blade angle distribution $\beta(u)$ on the hub and shroud surface and eventually at intermediate sections. A typical distribution is shown in Figure 5.12a:

$$\beta_{bl}(u) = \beta_{1,bl}(1-u)^3 + 3\beta_{i1}u(1-u)^2 + 3\beta_{i2}u^2(1-u) + \beta_{2,bl}u^3 \tag{5.10}$$

$\beta_{1,bl}$ and $\beta_{2,bl}$ are the leading edge and trailing edge blade angles, respectively. β_{i1} and β_{i2} are control values at $u = 0.33$ and 0.66, respectively. More complex geometries are possible by increasing the order of Equations (5.8) to (5.10).

The blade angular coordinate θ is defined by integrating the β distribution along the hub and shroud contour (Figure 5.12b):

$$\theta(m) = \int_1^2 \frac{1}{R} \tan \beta_{bl}(m)\,dm \tag{5.11}$$

where m is the dimensional length of the meridional contour.

The wrap angle $\theta_2 - \theta_1$ can be influenced by changing the meridional contour or the β distribution at the hub and shroud. The circumferential position of the shroud versus hub section is an additional design variable commonly specified by a stacking line at the leading edge or by the trailing edge rake angle (Figure 5.13). Rake can be forward (+ in the direction of rotation) or backward and is normally limited to $\pm 45°$ because a too large rake will increase the whetted surface and hence the friction losses. The criteria are maximum efficiency, stress limitations or more uniform impeller outflow by influencing secondary flows (Section 3.2.4).

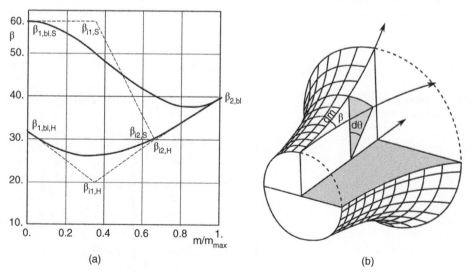

(a)   (b)

Figure 5.12 (a) Typical β distribution and (b) definition of a blade camberline as a function of the meridional contour.

Figure 5.13 Definition of impeller wrap angle $\Delta\theta$, lean, and rake angle.

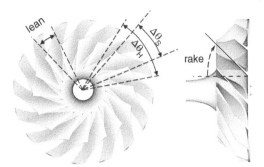

Figure 5.14 Thickness distribution along the camber line of the blade (not to scale).

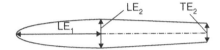

The blade thickness distributions at the hub and shroud of centrifugal impellers are mostly simple distributions with almost constant thickness and only a leading edge round-off, defined by two axes of an ellipse (LE_1 and LE_2) and straight trailing edge cut with thickness TE_2 (Figure 5.14). The values are different between the hub and the shroud according to vibration and stress considerations. Splitter blades can be defined in a similar way to the full blades and do not need to be short versions of the full blades.

Adding a blade thickness distribution to the camberline and connecting the corresponding points at the hub and shroud by straight lines defines blade shapes that can be manufactured by flank milling. The suction and pressure side generating lines being almost parallel to each other avoids the problem of undercutting. Non-linear stacking lines and additional β distributions in intermediate spanwise sections allow specifying blades that are curved from hub to shroud, including forward or backward sweep (Sugimura et al. 2012; Zheng and Lan 2014).

Some geometrical constraints can be imposed in the parameterization. However, it is important not to get stuck in a traditional suboptimal design but allow new and unconventional geometries that have not been considered in the past. Any limitation of the geometry may prevent the optimizer from finding the optimal geometry because the parameterization does not allow that geometry to be created.

5.2.2 Search Mechanisms

There are two main groups of search mechanisms:

- First-order methods calculate the required geometry changes in a deterministic way based on the output of the performance evaluations. A common one is the steepest descend or gradient method approaching the area of minimum *OF* by following the path with the largest negative gradient on the *OF* surface. This approach requires the calculation of the direction of the steepest gradient of the OF and of the step length.
- Zero-order or stochastic search mechanisms require only function evaluations. They vary from a random or systematic sweep of the design space to the use of evolutionary theories such as genetic algorithms (GA) or simulated annealing (SA) to find the optimum parameter combination. Zero-order methods may require more evaluations than gradient methods but can make use of well-proven existing flow solvers and have less risk of getting stuck in a local minimum.

5.2.2.1 Gradient Methods

Gradient methods are commonly used in many engineering applications to find the optimum combination of design parameters. The basic idea is illustrated in Figure 5.15a for a simplified problem with only two variables ($X(1)$ and $X(2)$).

The curved lines are lines of constant *OF*. The dashed lines are the constraints limiting the area of acceptable geometries. Starting from an initial geometry $(X(1)^0, X(2)^0)$ the method progressively approaches the optimum combination of $X(1), X(2)$, corresponding to the minimum value of the OF by marching in the direction of the steepest slope. The simplest way to find this direction is by evaluating the *OF* with a small perturbation of each variable. In addition to the function analysis in point $X(1)^0, X(2)^0$, additional analyses are needed in $X(1)^0 + \Delta X(1), X(2)^0$ and $X(1)^0, X(2)^0 + \Delta X(2)$ to define the gradient in the $X(1)$ and $X(2)$ plane. Each step requires $n_D + 2$ function evaluations, including an extra one in X^2 to estimate the optimal step length (Figure 5.15b). Fitting a parabolar tangent to the slope in the point X^0 and through the point X^2 provides an approximation of the *OF* along that direction and allows the approximate position of the minimum to be found. The same procedure is then repeated starting at the point of minimum *OF* defined previously, until the point with zero gradient is reached or when the line with steepest gradient is crossing a constraint.

Gradient methods perform poorly in the presence of noise in the function evaluation and may get stuck in a local minimum. The accuracy of the gradient calculation can be improved by means of the finite difference method or the complex variable method (Verstraete 2016a). Gradient methods cannot handle discontinuous variations of design parameters such as a change in the number of blades or vanes.

Assuming that N steps are needed, one optimization will require $N \times (n_D + 2)$ function evaluations. A real application on a geometry with 15 design parameters and 20 optimization steps thus requires 340 evaluations. Depending on the cost of one evaluation the required time and effort to find the optimum may be prohibitive.

Alternative methods to calculate the gradients are as follows:

- **Automatic differentiation** calculates the gradient of the *OF* based on the chain rule. This method acts at the programming level, where together with the value of a function the derivative of that function is also calculated based on the derivative of the input values of that function (Gauger 2008). Differentiation software is available for an automatic modification of the C++ (ADC; www.vivlabs.com/subpage ad v4.php) or FORTRAN code (ADF; www.vivlabs .com/subpage adf v4.php).

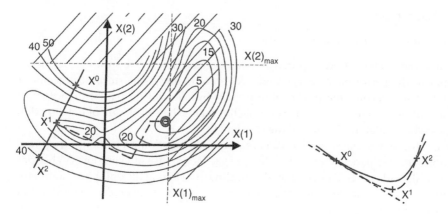

Figure 5.15 Steepest gradient method.

- **Adjoint methods** have their origin in control theory. The control variables are now the design parameters. This numerical technique provides the gradients of all variables by an effort that is comparable to that of an extra direct flow calculation (Giannakoglou et al. 2012). The technique has been successfully applied to the optimization of a radial turbine (Verstraete et al. 2017).

Both approaches to the gradient method require considerably less computer effort than the first one. However, they require a modification of the function evaluation code, which means that one can no longer profit from the accurate and readily available commercial flow and stress analysis codes.

5.2.2.2 Zero-order Search Mechanisms

Zero-order methods are an alternative to gradient methods. The simplest one is a systematic sweep of the design space by calculating the *OF* for a selected number of design variables between the maximum and minimum limits, as illustrated in Figure 5.16.

Only 3^{n_D} function evaluations are required to obtain a very good idea of the design parameters defining the optimal geometry. This approach is valid for problems with a small number of design variables but the number of required function evaluations increases to 14 348 907 for 15 design variables.

5.2.2.3 Evolutionary Methods

Different strategies have been developed to accelerate the procedure, i.e. by replacing the systematic sweep of the design space by a more intelligent selection of new geometries using in a stochastic way the information obtained during previous analyses. Such an approach is non-deterministic, i.e. it does not always converge to the same solution.

A single objective optimization is straightforward, as illustrated in Figure 5.17. Generations of geometries are replaced by new ones containing geometries that are hopefully performing better than the ones of the previous generation. The geometry with the lowest *OF* in the last generation is considered as the optimum.

The following are some techniques used to make the selection of new generations more efficient, i.e. converge faster to a minimum *OF*.

Genetic algorithms (GAs) is an evolutionary technique developed by Rechenberg (1973) that simulates Darwin's evolution theory stating that the fittest survives. According to this theory, individuals (geometries) with favorable genetic characteristics (low *OF*) will most likely produce better off-spring (with lower *OF*). Selecting them as parents increases the probability that individuals of the next generation will perform better than the previous ones.

Figure 5.16 Zero-order sweep of the 2D design space with three values of each design variable.

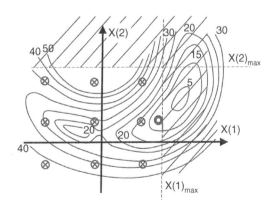

Figure 5.17 Single objective optimization.

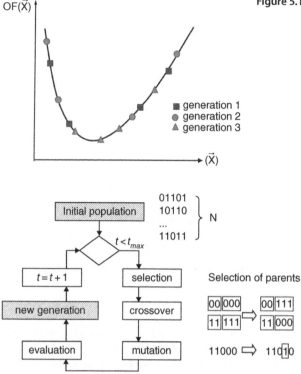

Figure 5.18 Schematic of a genetic algorithm.

In a standard GA, the n_D real values $X(i)$, defining a geometry, are represented by a binary string of length L_i. Low values of the substring length decrease the optimization effort by limiting the possible number of geometries, but the GA may not be able to accurately locate the minimum because of an insufficient resolution.

The operational principle of a standard GA is shown in Figure 5.18. Pairs of individuals (parents) are selected from an initial population of N geometries. Genetic material is subsequently exchanged between them (crossover) to create offspring, i.e. members of the next generation. Introducing new geometries into the population by switching a small number (typically 0.1%) of bits in the binary string of an offspring (mutation) avoids convergence to a sub-optimum geometry. After evaluation of the OF, the new geometry is added to the new generation. This process is repeated until the N new individuals are created, after which the procedure starts again from the new generation. Replacing a limited number of new geometries by geometries of the previous generation that have a lower OF is called elitism. It may accelerate the convergence, but increases the risk of converging to a suboptimal geometry.

An important issue of the GA is the selection scheme. A common one is the roulette (Figure 5.19a): a system in which the chance that an individual is selected increases proportional with $1/OF$. This scheme favors the best individuals as parents but avoids premature convergence to a local optimization by also allowing the selection of less optimal geometries.

In the tournament selection (Figure 5.19b), S individuals are chosen randomly from the population and the one with the lowest OF is selected as the parent. The same process is repeated for the second parent. The parameter S, called the tournament size, can take values between 1 and N. Larger values of S give more chances for the best samples to be selected and favor a

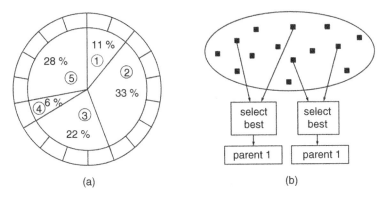

(a)   (b)

Figure 5.19 (a) Roulette and (b) tournament selection.

rapid, but maybe premature, convergence to a local optimum. Smaller values of S result in a more random selection of parents.

GAs are now widely used (Goldberg 1989; Bäck 1996). The quality of the GA optimizer is characterized by:

- GA efficiency: the required computational effort, i.e. the number of performance evaluations that are needed to find the optimum
- GA effectiveness: the value of the optimum OF. The tuning of the GA parameters (N, L_i) to accelerate the convergence is discussed by Harinck et al. (2005).

Simulated annealing (SA) is a technique derived from the annealing of solids where, at a given temperature, the state of the system varies randomly. This is simulated by perturbing an existing design with sufficient randomness to cover the design space (Van Laarhoven and Aarts 1987). A new geometry is accepted and replaces the previous one if the OF is lower. In order to avoid convergence to a local optimum a number of geometries with a higher OF are also accepted, but with a probability that decreases with the temperature such that the randomness slowly decreases and the search becomes more focused.

Particle swarm optimization (PSO) is a method based on the movement of individuals in a bird flock or fish school (Kennedy and Eberhart 1995). Each geometry of an initial database changes by moving in the design space with a velocity that depends on the properties of its own geometry as much as of the best geometry, by user defined constants and random numbers. The new geometry replaces the old one if it has a lower OF.

Differential evolution is a evolutionary method developed by Price and Storn (1997). For each geometry $\vec{X}$ a new one $\vec{Y}$ is defined by combining the design parameters of three other randomly selected geometries $\vec{A} \neq \vec{B} \neq \vec{C} \neq \vec{X}$. A new candidate geometry $\vec{Z}$ is obtained by following combination of $\vec{X}$ and $\vec{Y}$:

$$Z(i) = \begin{cases} Y(i) & \text{if} \quad r_i \leq C^{te} \\ X(i) & \text{if} \quad r_i \geq C^{te} \end{cases} \qquad i-1, n_D$$

where r_i is a uniformly distributed random number ($0 \leq r_i < 1$) and Cte is a user defined value between 0 and 1. The candidate geometry $\vec{Z}$ replaces $\vec{X}$ if its performance is superior (lower OF).

A rather large number of function evaluations are required before the optimal geometry is found and the method is useful only if the evaluation effort is low, i.e. in combination with analytical methods or fast approximated methods such as the ones described in Section 3.3 (Verstraete 2016b).

5.2.3 Metamodel Assisted Optimization

Zero-order search methods, even supported by evolutionary theory, still require a large number of performance evaluations. This becomes prohibitive in cases with expensive time-consuming performance evaluators. One way to reduce the computational effort is by working on different levels of sophistication and by making better use of the knowledge gained during previous designs. This is achieved by using fast but approximate prediction methods to find an optimum geometry, which is then verified by the more accurate, but also more expensive, function evaluator. Such a system is illustrated by the flow chart in Figure 5.20. The fast but less accurate optimization loop is to the right, the expensive but accurate one to the left.

Metamodels or surrogate models are interpolators using the information contained in the database to correlate the performance to the geometry similar to what is done by the function evaluator. They have the same input and output as the function evaluators they replace. In case of a Navier–Stokes analysis this will be the geometry and boundary conditions as input and the losses, Mach number distribution, flow directions, etc. as output. Once the metamodel has been trained on the data contained in the database, it is a very fast predictor of the *OF* of the many geometries generated by the GA, but with much less effort than the accurate function evaluators. As the metamodel is not always very accurate, the optimized geometry must be verified by means of a more accurate but time-consuming function evaluator. The new information resulting from this verification is added to the database and a new optimization cycle is started. It is expected that a new learning on the extended database will result in a more accurate metamodel and that the result of the next GA optimization will be closer to the real optimum. The optimization cycle is stopped once the metamodel results are confirmed by the accurate evaluator, indicating that the fast optimization has been made with an accurate performance predictor.

An important aspect of this iterative procedure is the fact that, once the system is converged, there will be no discrepancy between the results of the metamodel prediction and the one obtained by the accurate evaluator. Such an agreement is not obtained if a correlation or simplified solver (i.e. Euler or Navier–Stokes on a course grid) is used as the metamodel because the inaccuracy is not reduced during the design process.

The use of a metamodel constitutes an important gain in computational effort. In the case of a Navier–Stokes function evaluator the complete design iteration, including metafunction learning and GA optimization, requires typically 35% more time than that required for a Navier–Stokes analysis.

The main purpose of the **database** is to provide information to the metamodel about the relation between the geometry and performance. The more general and complete this information, the more accurate can be the metamodel and the closer the first optimum geometry, defined by the GA, will be to the real one.

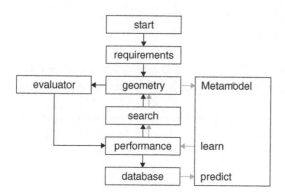

Figure 5.20 Flowchart of metamodel assisted optimization system.

Making a database is an expensive operation because it requires a large number of detailed function evaluations. Hence it is important to make the smallest possible database containing the maximum amount of information about the whole design domain, i.e. with a minimum of redundancy while covering the whole design space.

Design of experiment (DOE) refers to the process of planning for a reduced number of experiments such that the appropriate data, when analyzed by statistical methods, provide valid and objective conclusions. It is obvious that decreasing the number of samples in the database reduces the available information stored in it and hence decreases the accuracy of the artificial neural network (ANN) based on that information. Kostrewa et al. (2003) has demonstrated that, for an equal number of samples, the error of metamodel predictions trained on randomly selected databases is always larger than that obtained from a database defined by DOE.

The number of samples required in the initial database increases with the number of variables and at the same time the optimization effort. A careful selection of relevant design parameters is therefore recommended. Before starting the optimization, one can also verify the relevance of the design parameters used in the initial database by means of ANOVA (ANalysis Of VAriance), a statistical test procedure revealing the correlation between the input (design parameters) and outcome (*OF*). Disregarding the design parameters that have no relevant effect on the *OF* reduces the effort without compromising the outcome.

Any information missing in the database may lead to an erroneous metamodel that drives the GA to a non-optimal geometry. This is not a problem because the detailed performance analysis of that geometry will provide the missing information when it is added to the database. The critical case is when an incomplete database results in an erroneous extrapolation by the metamodel, predicting a poor performance (large *OF*) in that part of the design domain where in reality the *OF* is low. As a consequence, the corresponding geometry will not be selected by the GA and the real optimum may never be detected. It is therefore recommended that the quality of the database is verified during the design process by adding a number of detailed evaluations of geometries in regions where the uncertainty is largest. They can be selected on the basis of a merit function:

$$m(\vec{X}) = OF(\vec{X}) - w_m d_m(\vec{X})$$

where w_m is the weight factor that is given to points that are in unexplored areas of the design space, and $d_m(\vec{X})$ is the distance between a candidate geometry and the nearest known geometry.

Minimizing this function will favor the selection of geometries that are in less explored areas of the design space when extending the database. Figure 5.21 illustrates the evolution of the data point selection when a merit function is applied to a 1D design space.

The accuracy of the metamodel is a major factor in the convergence to the real optimum. The model needs to be fast, to shorten the time required to analyze a large number of samples, but at the same time accurate, to limit the number of iterations. Any approximating function can be used as a metamodel. Popular ones are response surface, ANN, radial basis functions (RBF), Kriging, etc. Only the most common ones are discussed here.

An ANN is composed of several elementary processing units called neurons or nodes. They are organized in layers and joined with connections (synapses) of different intensity, defined by the connection weight (*w*), to form a parallel architecture (Figure 5.22). Each node performs two operations:

- a weighted summation of all the incoming signals
- the transformation of the signal using a transfer function (*TF*) after a bias $b(j)$ has been added:

$$a(j) = TF\left(\sum_{i=1}^{n_D} w(i,j)X(i) + b(j)\right) \tag{5.12}$$

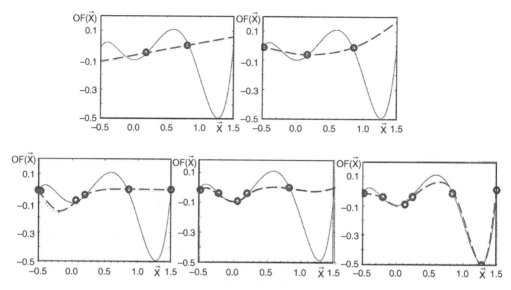

Figure 5.21 Selection of new database geometries based on a merit function.

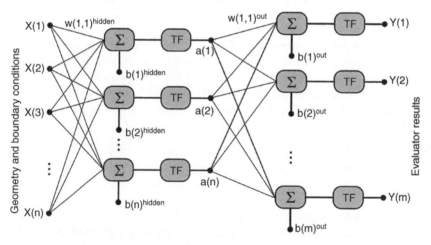

Figure 5.22 Artificial neural network architecture.

The same operations on the output of the hidden layer define the next hidden layer or the output layer.

Sigmoid functions are commonly used as the transfer function (Figure 5.23). They introduce power series (given implicitly in the form of an exponential term) and do not require any hypotheses concerning the type of relationship between the input and the output variables. In order to avoid saturation of the function, it is important to verify that the variation takes place in the central non-zero slope part of the curve.

The coefficients (weights and bias of the ANN) are defined by a "learning" procedure relating the input to the output layer for all the samples of the database. In case of a Navier–Stokes evaluator the ANN will be relating the η, β_2 and the Mach number distribution in N_M points on the blade surface to the boundary conditions and design parameters $\vec{X}$.

Figure 5.23 Sigmoid activation function.

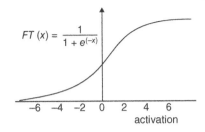

$$FT(x) = \frac{1}{1 + e^{(-x)}}$$

Figure 5.24 Early stopping method.

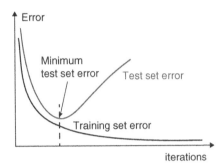

It is not the purpose of an ANN to reproduce the existing database with maximum accuracy, but to predict the performance of new geometries, i.e. to generalize. This is favored by dividing the available samples into "training", "test", and "validation" sets.

The training set contains the samples used for learning, i.e. to define the parameters (weights and bias) of the ANN. The test set contains the samples used only to assess the performance (generalization) of a fully specified ANN. The learning process starts with a rapid initial decrease of the training set error, which slows down as the network makes its way to a minimum (Figure 5.24).

Good generalization can be achieved by cross-validation of the ANN by a test set. In this procedure, the learning is periodically stopped (every so many training epochs) and the accuracy of the network is verified on the test set. The learning is stopped when the test set has reached its minimum error.

The validation set contains the samples that may be used to tune the architecture of the ANN (number of hidden nodes and layers).

A **radial basis function (RBF)** network is a three-layer network with a non-linear mapping from the input layer to the hidden layer and a linear mapping from the hidden layer to the output layer (Figure 5.25).

Each hidden neuron is associated with a so-called RBF center $C(j)$, a point in the n_D dimensional design space (Figure 5.26). The output of an RBF neuron is proportional to the distance between the input $X(i)$ and the RBF center. The proportionality is usually expressed by a Gaussian function with center $C(j)$ and amplitude $\sigma(j)$. The latter determines the activation range, i.e. the distance over which the neurons have a significant contribution to the output of the hidden layer. The output of the hidden neurons are input to a proportional sampling and transfer function defining the output layer, similar to the ANN. Training of the RBF network consists of finding the RBF centers $\vec{C}(j)$ and the amplitude σ_j for each neuron, such that the error on the predicted output is minimal for all the database samples. The advantage of

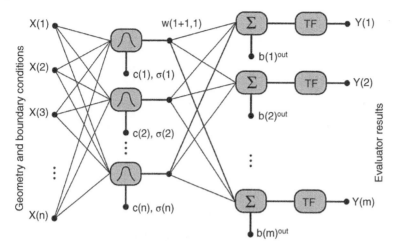

Figure 5.25 Radial basis function network architecture.

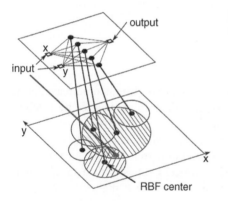

Figure 5.26 A 2D RBF interpolation network mapped on RBF space (from Verstraete 2008).

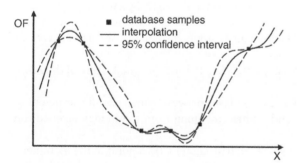

Figure 5.27 1D Kriging interpolation with exact prediction of data points and uncertainty between them.

the RBF is an increased accuracy due to the selectivity in the input parameters influencing the output.

Kriging is a model developed by geologists to estimate the concentration of minerals based on very scarce data that are available (Krige 1951). The main advantage of this technique is an estimation of the uncertainty of the prediction, as illustrated in Figure 5.27, which allows a better judgment of the geometry proposed by the optimizer.

Response surface models approximate the evaluation function by polynomials. The ζ values of the quadratic equation

$$OF_{meta}(\vec{X}) = \zeta_0 + \zeta_1 X(1) + \ldots + \zeta_{n_D} X(n_D)$$

$$+ \zeta_{1,1} X(1)^2 + \zeta_{1,2} X(1)X(2) + \zeta_{1,3} X(1)X(3) + \ldots + \zeta_{1,n_D} X(1)X(n_D)$$

$$+ \zeta_{2,2} X(2)^2 + \zeta_{2,3} X(2)X(3) + \ldots + \zeta_{2,n_D} X(2)X(n_D)$$

$$+ \ldots \zeta_{n_D,n_D} X(n_D)^2 \tag{5.13}$$

are defined by a least-squares regression to minimize $||\widetilde{OF}_{meta}(\vec{X}) - OF(\vec{X})||$. Hence the number of samples must be larger than the number of ζ coefficients to avoid that the system is under-determined. Different surfaces, each with their own coefficients, can be defined for the different components of the *OF*.

The main advantage of this type of metamodel is the analytical definition of the *OF* and the minimum can be found by a Newton method. However, using a quadratic or higher order function predefines the shape of the approximation function, which may be different from the real one. Another disadvantage is that the number of samples required increases rapidly with the number of design variables and order of the surface. At least $(n_D + 2)(n_D + 1)/2$ samples are needed for a quadratic equation, increasing to at least $(n_D + 3)(n_D + 2)(n_D + 1)/6$ samples for a cubic surface. For a ten-parameter design space this means more than 66 and 286 samples, respectively.

A metamodel optimization system can easily be extended to multidisciplinary by adding the corresponding function evaluators and metamodels to the system, as illustrated by the flow chart shown in Figure 5.28. The computational effort increases proportionally with the number of different performance evaluations that are needed.

The GA search for the multidisciplinary optimum geometry gets its input from the finite element stress analysis (FEA) as well as from the Navier–Stokes (NS) or heat transfer (HT) analysis or from any other discipline. The main advantages of such an approach are:

- The existence of only one "master" geometry, i.e. the one defined by the geometrical parameters selected by the GA optimizer. This eliminates possible approximations and errors when transmitting the geometry from one discipline to another.
- The possibility to do parallel calculations if each discipline is independent, i.e. if stress calculations do not need the pressure distribution on the blades or if the impact of the geometry deformations can be neglected on flow calculations.

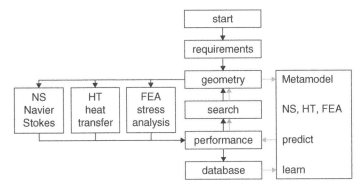

Figure 5.28 Multidisciplinary optimization flow chart.

- A concurrent convergence to a global optimum, considering simultaneously all aspects of a design, without time-consuming iterations between the aerodynamically optimal and the mechanically acceptable geometry. All aspects of the *OF* are considered when selecting new geometries leading to a optimization. The handling of possible conflicting objectives is explained in next section.

5.2.4 Multiobjective and Constraint Optimization

Multidisciplinary optimization gives rise to objectives and constraints that are often conflicting as there are minimum inertia of an impeller versus corrosion resistance, longer life and lower noise and stresses versus more compact high speed compressors, etc. Some of these constraints are readily satisfied by a simple limitation of the design space but most require a detailed analysis. Aerodynamic objectives may also be conflicting: maximum efficiency versus large operating range with a predefined choking mass flow, etc. The balance between the different objectives may not be clear, which complicates the selection of the "optimal" geometry. A ranking of the candidate geometries is needed.

5.2.4.1 Multiobjective Ranking

In case of two *OF*, the members of a population can be visualized in a 2D fitness space, allowing a trade off between them (Figure 5.29). Possible candidates for next generation are the non-dominated solutions, i.e. the collection of geometries for which there is no other geometry that has a lower value for both objectives. Geometry $\vec{A}$ dominates geometry $\vec{B}$ if

$$OF_1(\vec{A}) < OF_1(\vec{B}) \qquad \text{and} \qquad OF_2(\vec{A}) < OF_2(\vec{B}) \tag{5.14}$$

Geometry $\vec{B}$ is not dominated by $\vec{C}$ if at least one $OF(\vec{B}) < OF(\vec{C})$. Non-dominated geometries define the Pareto front and are candidate members of the next generation.

If the number of non-dominated geometries exceeds the population number *N* of the next generation, the individuals that are the closest to each other (most similar geometries) are removed to improve the diversity of the population. If the number of non-dominated solutions is smaller than required for the next population, the first series of non-dominated geometries is given rank 1 and removed from the population. The procedure starts again to select the members of rank 2, and so on until the complete population of the next generation is obtained.

Ranking geometries is more difficult when there are more than two objectives. Three *OF* can be visualized in a 3D graph and more objectives may be visualized in self-organizing maps

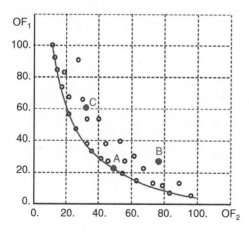

Figure 5.29 Pareto front ranking.

(SOMs) (Obayashi et al. 2005). Input for the SOMs are the geometries that have been analyzed and the value of the N_O objectives. The output is visualized in N_O 2D graphical maps. The conversion from the multidimensional space to the 2D space is based on the distance between the geometries. Similar geometries are located in neighboring positions in the 2D maps on which the value of each objective function, corresponding to these similar geometries, is visualized.

Figure 5.30 from Sugimura et al. (2008) visualizes the SOM technique applied to the optimization of a shrouded centrifugal fan. The four *OF* relate to the efficiency (OF_η), noise level (OF_{noise}), and the dependence of them on manufacturing uncertainty (OF_η^σ) and (OF_{noise}^σ). The color scales with the value of the respective *OF* and provides an intuitive understanding of how the specified *OF* vary with the type of geometry. Geometries (1) and (2) on the left side of each graph have longer blades with larger camber and the leading edge close to the shroud clearance gap. They have a lower efficiency and higher noise level but they are less affected by geometrical inaccuracies. Geometries (5) and (6) on the right side have shorter and less cambered blades, with the leading edge further away from the shroud clearance gap. They have the highest efficiency and lowest noise level but these characteristics are more sensitive to manufacturing inaccuracies.

Another way to handle multiple conflicting objectives is by defining a pseudo *OF*, i.e. a weighted sum of individual *OF*:

$$OF(\vec{X}) = w_1 OF_1(\vec{X}) + w_2 OF_2(\vec{X}) + \dots \dots \tag{5.15}$$

The balance between the different objectives is defined by the respective weight factors. The optimization thereby consists of finding the geometry that minimizes this pseudo *OF*.

The relation between the pseudo *OF* with two contributions and the Pareto front is illustrated in Figure 5.31. The optimization driven by a pseudo *OF* follows a path in the direction of the point where the line of constant pseudo *OF* is tangent to the Pareto front. The main advantage is a smaller number of geometries to be analyzed to find this optimum. The disadvantage is that the optimum depends on the relative weight given to the different penalties, as illustrated by

Figure 5.30 Self-organizing maps colored by the *OF* (from Sugimura et al. 2008).

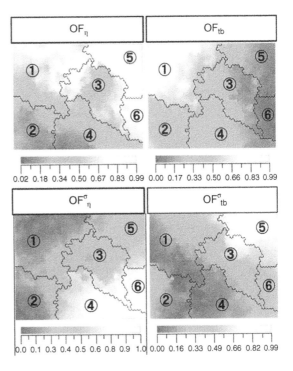

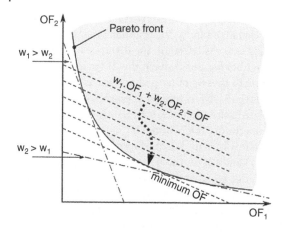

Figure 5.31 Pseudo objective function versus Pareto front.

the dash-dot line in Figure 5.31. However, once the pseudo *OF* has converged to a minimum it is still possible to modify the respective weights and to find other optima along the Pareto front.

5.2.4.2 Constraints

Constraints are expressed by inequalities that need to be respected. Geometrical constraints $G_k(\vec{X}), (k = 1, n_G)$ can be verified before any performance analysis is started and easily respected by limiting the design space. Most of the performance and mechanical constraints $A_j(U(\vec{X}), \vec{X}), j = 1, n_A)$ need input that is available only after a detailed function evaluation.

In a evolutionary optimization with multiobjective ranking, priority must be given to the constraints. Individuals of a new generation replace the ones of the previous generation if the constraints are better satisfied and if no new constraints are violated, even if the latter has an *OF* that is larger than that of previous generation.

Another way to account for constraints is by adding penalty terms to the pseudo *OF* that increase proportionally to the degree of violation:

$$OF(\vec{X}) = w_1 \cdot OF_1(\vec{X}) + w_2 \cdot OF_2(\vec{X}) + \ldots\ldots + \Sigma_{j=1}^{n_A} w_j \delta_j \cdot (A_j(U(\vec{X}), \vec{X}))^2 \qquad (5.16)$$

where δ_j is 0 inside the feasible domain and 1 in the domain where the constraint is violated (Figure 5.32).

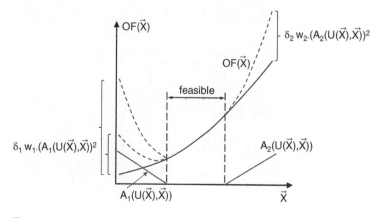

Figure 5.32 Pseudo objective function and constraints.

This will not guarantee that each individual constraint is satisfied but favors the convergence to the constrained optimum because the GA is discouraged to explore further outside the feasible part of the design space. Mechanical constraints that have a direct impact on the turbomachinery integrity, such as maximum stress and deformation, must be rigorously respected. Hence the corresponding weight factor should be sufficiently large to guarantee that the *OF* is always increasing outside the feasible domain, as illustrated for the constraint $A_1(U(\vec{X}), \vec{X})$ in Figure 5.32.

5.2.4.3 Multiobjective Design of Centrifugal Impellers

The use of pseudo *OF* functions is illustrated by the multidisciplinary and multiobjective optimization of a centrifugal compressor impeller of 20 mm diameter with splitter blades rotating at 500 000 RPM. The design mass flow is 20 g/s. The meridional geometry definition is the one shown in Figure 5.11. Only six coordinates of the Bézier control points are used to define the contours. The range in which they can vary is shown by arrows. The β distributions are of third order (Equation (5.10)) and have the same exit angle at the hub and shroud. The blade thickness is kept constant at the shroud (0.3 mm $< LE_2 = TE_2 <$ 0.6 mm) and LE_1 is fixed (Figure 5.14).

The splitter blades are a shorter version of the main blade and the only design parameter is the leading edge position between 20% and 35% of the non-dimensional length. The blade number is fixed at 7 + 7 for manufacturing reasons and tip clearance is 10% of the outlet width. The axial length and outlet diameter are fixed. The inlet geometry is automatically adapted to guarantee a smooth transition to the inlet and outlet sections. A fixed 0.25 mm fillet radius between the blade and hub is added for the stress calculations. This results in a total of 16 independent design parameters.

A Navier–Stokes solver is used to predict the aerodynamic performance and an FEA is used for the stress calculations. The optimization is driven by following pseudo *OF*:

$$OF(\vec{X}) = w_\eta P_\eta(\vec{X}) + w_{stress}P_{stress}(\vec{X}) + w_{massflow}P_{massflow}(\vec{X})$$
$$+ w_{Mach}P_{Mach}(\vec{X}) + w_{loading}P_{loading}(\vec{X}) + w_{geom}P_{geom}(\vec{X}) \tag{5.17}$$

The first term aims for maximum efficiency:

$$P_\eta(\vec{X}) = max(\eta_{req} - \eta, 0.0) \tag{5.18}$$

where η_{req} is set at a high unreachable value.

The constraint on maximum stress is replaced by following pseudo penalty:

$$P_{stress}(\vec{X}) = max\frac{\sigma_{max} - \sigma_{allowed}}{\sigma_{allowed}} \tag{5.19}$$

where σ_{max} is the maximum von Mises stress in the impeller. The penalty is zero when its value is below the limit value $\sigma_{allowed}$ and increases linearly when this limit is exceeded.

The mass flow penalty tolerates a 0.3% deviation from the imposed mass flow.

$$P_{massflow} = \left(max \left(\frac{|\dot{m}_{req} - \dot{m}|}{\dot{m}_{req}} - \frac{\dot{m}_{req}}{300}, 0.0 \right) \right) \tag{5.20}$$

The penalty on the Mach number favors Mach number distributions that are not only good at design point but are likely to be good also at off-design operation. It has two contributions. The first one penalizes negative loading and is proportional to the area between the suction and pressure side when the pressure side Mach number is higher than the suction side one (Figure 5.33):

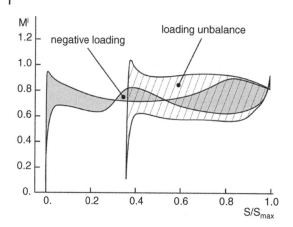

Figure 5.33 Negative loading and loading unbalance in a compressor with splitter vanes.

$$P_{Mach} = \int_0^1 max(M_{PS}(u) - M_{SS}(u), 0.0)du \tag{5.21}$$

The second part of the Mach penalty increases with the loading unbalance between the main and splitter blade. It penalizes the difference in area between the suction and pressure side Mach number distribution of the main A_{bl} and splitter A_{sp} blades, corrected for the difference in blade length:

$$P_{loading} = \left(\frac{A_{bl} - A_{sp}}{A_{bl} + A_{sp}} \right)^2 \tag{5.22}$$

Ibaraki et al. (2014) obtained a wide operating range by penalizing the Mach number peak near the leading edge and any reacceleration or non-smooth decelerations downstream of this peak.

The pseudo *OF* can further be complemented by $P_{geom}(\vec{X})$, a penalty for violating mechanical constraints such as maximum inertia and weight. The weight factors of the *OF* are determined based on the knowledge gained in previous optimizations and to the importance given to the different penalties. The values used in the present design are such that an efficiency drop of 1% is as penalizing as an excess in stress limit of 6.6 MPa.

The optimization starts from the outcome of a simple aerodynamic optimization, called the baseline impeller. Although this geometry has a high efficiency, it cannot be used because the stress analysis predicts von Mises stresses in excess of 750 MPa. The optimization starts with an initial database containing 53 geometries. Thirteen geometries out of the 2^6 initial ones defined by the DOE technique could not be analyzed because unfeasible geometry. Two additional geometries have been added: the baseline geometry and a geometry with all geometrical parameters at midrange of the design space.

Only ten iterations (metamodel optimizations followed by a detailed analysis) were needed to obtain a very good agreement between the metamodel and the Navier–Strokes analysis. It took another 15 iterations before the stress analysis confirmed that the proposed geometries satisfy the mechanical requirements.

The efficiency and stress level of the different database samples show a wide scatter and none of them satisfies the stress constraints (Figure 5.34). Most of the geometries created during the optimization process have a higher efficiency than the database samples and satisfy the stress constraint. The influence of the stress penalty on the optimization is clear by comparing the

Figure 5.34 Aero penalty versus stress penalty for baseline, database, and optimized geometries.

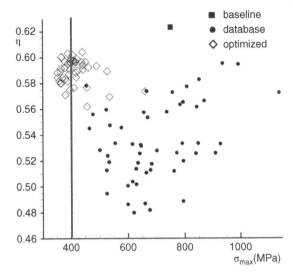

efficiency of the baseline impeller with the ones satisfying the stress constraint. The reduction of the maximum stress level with 370 MPa is at the cost of a 2.3% decrease in efficiency.

The drastic reduction in stress is the consequence of:

- the reduced blade height at the leading edge, resulting in lower centrifugal forces at the leading edge hub
- the increase in blade thickness at the hub
- the modified blade curvature resulting in lower bending stresses.

Results shown in Figure 5.43 are for geometries defined during that multipoint optimization and indicate a rather clear relation between blade lean, stress, and efficiency. Several geometries with good efficiency are found at $-12°$ lean angle. The drop in efficiency for lean angles above $-5°$ and below $-20°$ suggests that in that application, impellers perform better with approximately $-10°$ negative lean. This is a non-obvious outcome of the optimization.

5.2.5 Multipoint Optimization

Multipoint optimization aims to maximize the performance in more than one operating point. The simplest straightforward approach is to analyze every candidate geometry at the different operating conditions and compare a weighted value of the performance in the different operating points. Distinction should be made between designs with varying inlet conditions and those with varying outlet conditions. When optimizing a low solidity vaned diffuser, only the inlet conditions need to be specified because the target is to maximize the outlet static pressure. When optimizing compressors, the outlet pressure corresponding to a target mass flow is imposed and it cannot be guaranteed that this boundary condition allows convergence of the Navier–Stokes solver to a stable solution.

5.2.5.1 Design of a Low Solidity Diffuser

Widening the operating range with maximum pressure rise are the main objectives of low solidity diffusers (LSDs) (Section 4.2.1). Hence multipoint optimization with specified inlet conditions is the appropriate approach. The inlet flow angles listed in Table 5.1 are derived

Table 5.1 Diffuser inlet conditions at the three operating points.

	Low	Design	Choke
α_2	62.5°	52.8°	37.5°
$\dot{m}/\dot{m}_{ref}$	0.775	1.0	1.27

from the outlet conditions of a research compressor at the mass flow listed in the same table. The latter are only indicative and do not need to be achieved because it is believed that the diffuser pressure rise at subsonic flow depends mainly on the inlet flow angle and that small variations of the mass flow will have almost no impact on the non-dimensionalized values of CP_{2-4} and ω_{2-4}.

The diffuser blades are defined by a NACA thickness distribution ($th_{max}/c = 0.1$) superposed on a camber line defined by a four-parameter Bézier curve. The other design parameters are a scale factor for the thickness distribution (between 0.7 and 1.3) and the blade height. The latter is constant from leading to training edge but can be slightly different from the impeller outlet width.

The optimizer aims to maximize CP_{2-4} at the three operating points while minimizing the losses as expressed by following OF:

$$OF = -(w_{low}CP_{low} + w_{design}CP_{design} + w_{choke}CP_{choke})$$
$$+ w_{low}\omega_{low} + w_{design}\omega_{design} + w_{choke}\omega_{choke} \tag{5.23}$$

Weight factors are $w_{low} = w_{choke} = 0.25$ and $w_{design} = 0.5$.

The pressure rise and losses in channel diffusers depend mainly on the outlet to inlet area ratio and the way the corresponding diffusion is realized (opening angle and length) (Reneau et al. 1967). These geometrical characteristics do not exist in LSD because the vanes do not overlap. According to Equation (4.7) the divergence of a diffusing system can be characterized by the isentropic pressure rise coefficient $CP^i \approx CP + \omega$.

The variation of the components of the OF as a function of $CP + \omega$ is illustrated in Figure 5.35. The diffusers show minimum losses with an increase in CP for all geometries corresponding to $CP + \omega < 0.74$. All geometries below this limit are suboptimal in terms of pressure rise coefficient. Geometries with a higher divergence give rise to increasing losses and a decrease in

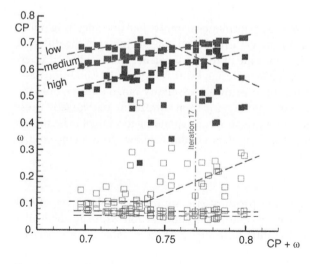

Figure 5.35 Performance criteria of LSD optimization (from Van den Braembussche 2010).

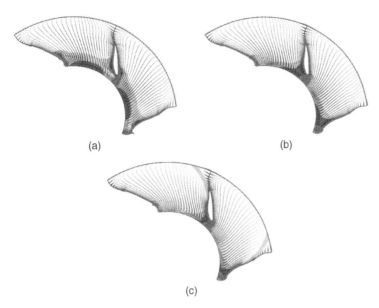

(a)

(b)

(c)

Figure 5.36 Velocity distribution in the optimized geometry at (a) maximum, (b) design, and (c) surge mass flow.

CP at the lowest mass flow. The scatter in *CP* at the same abscissa, hence similar divergence, is due to the variation in blade shape and setting angle. The geometry of iteration 17 is considered to be the optimal one because any further increase in the divergence results in a large increase in the losses at low mass flow, which is not compensated for by a corresponding increase in the pressure rise at the other operating points.

The Mach number and velocity vectors of the flow around these vanes are visualized in Figure 5.36. One can observe a small flow separation at minimum mass flow corresponding to the increase in the losses shown in Figure 5.35.

5.2.5.2 Multipoint Impeller Design

The main design objectives of a centrifugal compressor are to guarantee the imposed choking mass flow, stable flow up to a prescribed minimum mass flow, maximum efficiency near the design point or at least sufficiently far away from the surge point, and a pressure rise over mass flow curve with sufficiently negative slope to assure stability of the compression system (Figure 5.37).

Choking mass flow is a constraint and easily verified by a flow analysis at low back pressure. This is an objective of a first series of accurate analyses of the geometry optimized by the metamodel and GA. As metamodels are not always very accurate in this respect, not reaching the choking mass flow may be discovered only after the detailed Navier–Stokes analyses. One way to avoid loosing the large amount of optimization effort by the metamodel and GA is to scale the optimized geometry to make it satisfy the choking condition.

Adjusting the blade height from leading edge to trailing edge, also called trimming, is a common way of fine tuning the mass flow. The change in the impeller throat area dA_{th} as a function of dR is

$$dA_{th} = \frac{2\pi}{Z_r} k_{bl} \cos \beta_{1,bl} R dR \tag{5.24}$$

Assuming a linear variation of the leading edge blade angle from hub to shroud allows the new shroud leading edge radius to be defined as a function of the ratio between the calculated (before

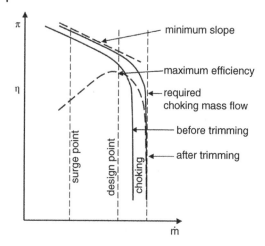

Figure 5.37 Requirements for a multipoint impeller optimization.

trimming line on Figure 5.37) and required choking mass flow. Any change in the inlet section requires an equivalent change in the impeller outlet width to conserve the same diffusion and work input, and to maintain the same velocity triangles at the diffuser inlet. It can be expected that this limited change in the geometry will not modify the optimality of the design.

The performance and flow stability have to be verified at different mass flows between choking and the target minimum mass flow. However, the mass flow in compressors is an outcome of a compressible flow calculation with an estimation of the corresponding pressure ratio as outlet boundary condition. It is not a priori known at what pressure ratio or mass flow a compressor will surge. It cannot be guaranteed that the NS solver will be able to provide a converged solution at each of the required operating points. A simple shift of the original performance curve to larger mass flows in accordance with the shroud trimming provides a good approximation of the new performance curve and allows feasible pressure ratios to be defined with more confidence for the final detailed analysis of the scaled geometry.

This procedure has been used for the multipoint optimization of a turbocharger compressor where the splitter blade geometry is independent of the full blades (Van den Braembussche et al. 2012). The trailing edge position of the splitter at the impeller exit is fixed halfway between the full blades to minimize the circumferential distortion of the diffuser inflow. The meridional contour is defined by fourth-order Bézier curves with 12 independent control points. The β distribution at the hub and shroud of the full and splitter blades is defined by third-order polynomials. The same $\beta_{2,bl}$ is imposed at the hub and shroud of all blades. The blade number and thickness distribution are fixed at their initial value. Together with the variable length of the splitter blade this results in 31 design parameters.

The *OF* is the same as Equation (5.17) with the following extra penalty terms. The diffuser inflow uniformity is further enhanced by penalizing the difference in mass flow on both sides of the splitters. As confirmed by Ibaraki et al. (2014), this helps to equalize the blade loading between the splitter and the full blade:

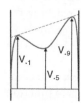

$$P_{\Delta massflow}(\vec{X}) = w_{\Delta massflow}\left(\frac{\dot{m}_{left} - \dot{m}_{right}}{\dot{m}_{left} + \dot{m}_{right}}\right)^2 \qquad (5.25)$$

The following additional term penalizes the distortion and skewness of the impeller outflow. It aims for enhanced diffuser stability and increased pressure rise (Senoo 1984):

Figure 5.38 Distortion and skewness of impeller outflow.

$$P_{dist-skew}(\vec{X}) = w_{dist}|dist| + w_{skew}|skew| \qquad (5.26)$$

where the distortion and skewness are defined by:

$$dist = \frac{2 . V_{0.5}}{V_{0.1} + V_{0.9}} - 1 \qquad skew = \frac{2(V_{0.1} - V_{0.9})}{V_{0.1} + V_{0.9}} \qquad (5.27)$$

The subscripts 0.1, 0.5, and 0.9 indicate the non-dimensional spanwise position where the velocities are taken (Figure 5.38).

In this multipoint optimization the mass flow penalty is replaced by one that increases when the pressure ratio at design mass flow is below the required value:

$$P_\pi(\vec{X}) = max(0, \pi_{req} - \pi) \qquad (5.28)$$

The pressure slope penalty

$$P_{slope}(\vec{X}) = max(0, slope + tol) \qquad (5.29)$$

starts increasing when the negative slope of the pressure rise curve

$$slope = \frac{P_{2,surge} - P_{2,design}}{P_1^o} \frac{\dot{m}_{design}}{\dot{m}_{2,design} - \dot{m}_{2,surge}} \qquad (5.30)$$

is less than a prescribed value *tol*.

Bertini et al. (2014) provide a clear analysis of the vibrations in radial impellers with vaned diffusers and demonstrate that the vibrational energy increases when both

$$nZ_D\Omega = \omega_m \qquad (5.31)$$

and

$$\frac{nZ_D \pm d_m}{Z_r} = integer \qquad (5.32)$$

The first condition verifies if the nth harmonic of the excitation frequency coincides with one of the modal frequencies of the impeller (ω_m), which is normally visualized on a Campbell diagram. The second one verifies that the excitation force has the same shape as the associated mode shape of the vibration, which is a necessary condition for adding vibrational energy to the system. The latter provides the critical excitation frequencies for each number of modal diameters and can be visualized in a Singh's advanced frequency evaluation (SAFE) diagram.

Perrone et al. (2016) add the following penalty to the *OF* to avoid designs that have an eigenfrequency within the 5% range around the excitation frequencies:

$$P_{vibr}(\vec{X}) = w_{freq} \left(0.05 - \left| \frac{\omega_m}{\omega_{exc}} - 1 \right| \right) \qquad (5.33)$$

if

$$\left| \frac{\omega_m}{\omega_{exc}} - 1 \right| < 0.05$$

The leading edge lean is the outcome of maximizing the efficiency while respecting maximum stress, but a limitation to a maximum rake of ±45° is appropriate to avoid too large friction losses near the exit. The results of this optimization procedure, shown in Figure 5.39, required the analysis of 150 geometries: 50 samples of the initial database and two sets of 50 optimized geometries, before and after the scaling. The complete optimization required 600 Navier–Stokes analyses but they were performed completely automatically without the manual intervention of the designer.

The good agreement between the scaled and verified choking mass flow illustrates the accuracy of the scaling procedure. A better alignment of the full- and splitter blade leading edge to the local flow direction, together with a more uniform loading distribution, is at the origin of the considerable gain in efficiency and the larger operating range.

The full and splitter blade β distribution of four high performance geometries are compared to the one of the starting geometry in Figure 5.40. The results in Figure 5.39 are for the β distribution of IT 49. They all show an initial increase of β near the shroud leading edge of the full blades followed by an almost linear decrease towards the trailing edge. The splitter blades are shorter than in the baseline impeller and their β angles have largely increased near the shroud and decreased near the hub, to better capture the flow swirling from suction to pressure side on the shroud and in the opposite direction at the hub, as illustrated in Figure 3.14.

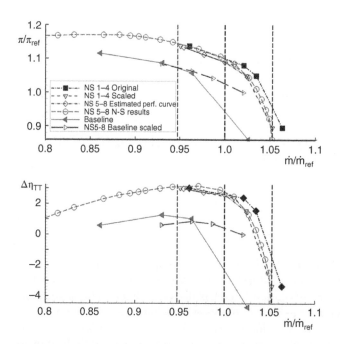

Figure 5.39 Results of the multipoint optimization scheme (from Van den Braembussche et al. 2012).

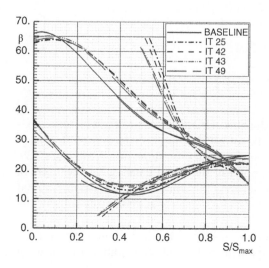

Figure 5.40 Optimized β distributions at the hub and shroud of the full and splitter blades (from Van den Braembussche et al. 2012).

5.2.6 Robust Optimization

The robustness of a design characterizes the insensitivity of its performance to small changes in the design parameters, operating conditions, manufacturing inaccuracies, geometrical variations during operation or inaccuracies of the analysis programs. Those variations may trigger large changes in the performance of turbomachinery components designed at the verge of stall (limited diffusion factor) or operating at near sonic inlet flow. Lack of robustness limits the practical use of those geometries unless very accurate and expensive analysis and manufacturing techniques are used.

Optimizers, aiming only to minimize the *OF*, should arrive at the point X_D of the example shown in Figure 5.41. A small variation in the design parameter X of this geometry results in a large deterioration in the performance. An optimization focusing only on minimizing the *OF* is commonly called "deterministic" even if it is obtained by a non-deterministic technique such as evolutionary theory.

Robust optimizers seek for a minimum that is less sensitive to geometrical and operational variations. They may arrive at geometry X_R where the *OF* is larger than the deterministic one but where a variation of the design parameters has a much smaller impact on *OF*.

A first way to enhance robustness is by adding extra terms to the pseudo *OF*, which increase for Mach number or pressure distributions that are known to be at the limit of stall, or by penalizing any other performance limiting phenomenon. The extra terms anticipate the effect of possible performance deteriorations resulting from small changes in the design parameters (Pierret 1999; Pierret et al. 1999). Such a modified *OF*, shown by the dash-dot line in Figure 5.41, will drive the optimizer to a robust design.

A rigorous approach to robust design is to evaluate the impact of the uncertainty of the design parameters on the *OF*. The possible variation of a design parameter can be defined by a rectangular (Figure 5.41), normal or any other probability density function (PDF). The impact on the *OF* is statistically evaluated by calculating the *OF* for a representative number of geometries and operational conditions in the range of uncertainty. Once all simulations have been performed the mean value $\widetilde{OF}$ and standard deviation σ_{OF} can be defined by a Monte Carlo technique, taking into account the respective weights of the geometrical and operational uncertainties. The PDF of the *OF* may be very skewed and more moments of the output may be needed to correctly reconstruct it (Wunsch et al. 2015).

A linear sampling of the uncertainty domain is representative only in the case of a uniform distribution of uncertainty (Figure 5.42a). In the case of an arbitrary probability density distribution of uncertainty, the representative geometries and operational conditions should be equally distributed on the cumulative density function, $CDF = \int PDF dx$ (Figure 5.42b).

Both $\widetilde{OF}$ and σ_{OF} must be minimized and a single *OF* optimization becomes a two-objective one. Visualizing the two objectives in a Pareto plot allows a trade-off between performance and robustness. Multiobjective optimizations require either to specify weight factors in the

Figure 5.41 Deterministic versus robust design.

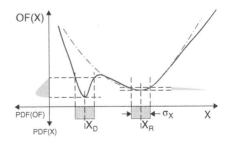

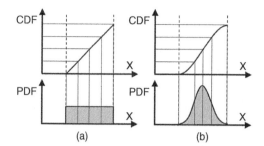

Figure 5.42 Sampling of the uncertain variable as a function of the probability density distribution.

pseudo $OF = w_1\widetilde{OF} + w_2\sigma_{OF}$ or to use special sorting techniques, such as the one proposed by Sugimura et al. (2008) and illustrated in Figure 5.30.

Evaluating the impact of uncertainties requires a large number of additional function evaluations. Operational uncertainties demand running the function evaluator on the same geometry, but at different boundary conditions. Geometrical uncertainties also require a modification of the geometry and mesh. Doing this with detailed CFD and FEA may be prohibitively expensive, even after reducing the number of required function evaluations by means of a DOE or other statistical techniques. An ANOVA sensitivity analysis of the database allows uncertainties that have little or no influence on the solution to be identified and a reduction in the number of design dimensions.

The use of a metamodel is again an efficient way to reduce the required computational effort if it is sufficiently accurate. Defining such a metafunction may be more cumbersome than for a deterministic optimization because extra samples in the database may be required to account for the uncertainty of the geometrical and operational parameters. The advantage of using a metafunction for a robust optimization of a centrifugal impeller is demonstrated by Sugimura et al. (2008). In addition to minimizing the $\widetilde{OF}$ for noise and efficiency, the two σ_{OF} are also

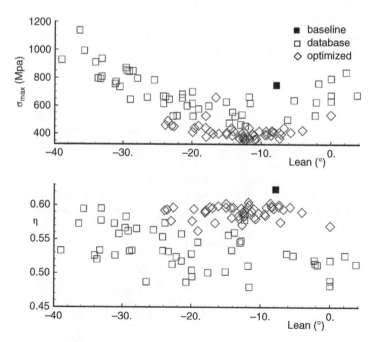

Figure 5.43 Stress and efficiency versus blade lean for the database and optimized geometries (from Verstraete et al. 2008).

minimized. Each of the four *OF* shown in Figure 5.30 have been calculated by a separate Kriging model, trained on an extended database obtained by a linear sampling over the uncertainty domain of the two design variables, i.e. blade length and camber.

The availability of a trained metamodel at the end of a deterministic optimization allows an a posteriori estimation of the robustness of that design. The metamodel has repeatedly been improved during the design process by training on the results of accurate analyses of geometries that are close to the optimum. One can expect that it will be sufficiently accurate to estimate the variation of *OF* for small changes of the design variables around the optimum and allows an evaluation of σ_{OF}. Such an analysis requires only a minimal effort and in the case where it indicates insufficient robustness the optimization can be continued either with modified weight factors or after a redefinition of the design space. This approach does also not account for a variation in the boundary conditions or uncertainties resulting from the inaccuracies of the solvers used to build the database.

The data shown on Figure 5.43 are the results from the geometries available in the database at the end of a radial impeller optimization (Verstraete et al. 2008). They allow the conclusion that a variation of the leading edge lean angle by ±5° around the optimum value has almost no impact on the stress or efficiency. Similar graphs can be produced for other design parameters and contributions to the *OF*.

6

Volutes

This chapter provides a short overview of the different non-axisymmetric inlet and outlet geometries for radial impellers. It describes the advantages and disadvantages of the different types, and explains the relation between the geometry and the flow in terms of performance and circumferential pressure distortion. Inlet and outlet volutes are discussed separately because of the fundamental differences between the flows.

The roll-up of the flow in outlet volutes, in combination with the internal shear, results in a forced vortex type flow structure (Figure 6.1a). As explained in Section 6.2 the flow structure and losses change with impeller operating point. Off-design operation is an important issue.

Depending on the upstream geometry, the flow in a suction pipe of a compressor may have parallel streamlines with a nearly constant total pressure. Irrotational flows entering an inlet volute remain irrotational up to the impeller inlet, except for the vorticity created in the boundary layers and in areas of separated flow (Figure 6.1b). The velocities change proportional to the mass flow with only small variations due to compressibility. Losses scale proportional to the inlet dynamic pressure with a dependence on the Reynolds and Mach numbers (Koch et al. 1995; Flathers et al. 1996). Inlet volutes that perform well at design point are likely to perform equally well at other mass flows and impeller rotations per minute (RPM). Off-design operation is not a big issue.

In this chapter reference is made also to intensive research made on pump volutes because of the similarity of the geometry and flow structure. This link is further justified by the low Mach number in compressor volutes. Hence most of the knowledge obtained for pump volutes is equally valid for compressors.

6.1   Inlet Volutes

The main consequence of a non-axisymmetric inlet duct is a spanwise and circumferential distortion of the pressure, velocity, and flow angle at the inlet of the impeller. The spanwise distortion of the inlet flow causes a variation in incidence over the blade height. This may change the performance, but the impeller flow remains steady because the inflow conditions do not vary during a rotation. A non-axisymmetric inlet flow causes circumferentially varying incidence and flow patterns in the impeller channels. In addition to increased losses, this also results in unsteady blade loading, hence blade vibrations and noise (Reichl et al. 2009). It also changes the radial forces on the shaft and may reduce the operating range because of premature stall or a modified choking limit.

An accurate prediction of the impact of an inlet volute on the impeller performance requires the simultaneous prediction of the flow in the volute and impeller by means of a 3D unsteady Navier–Stokes solver. This is computationally expensive and time-consuming. Numerical flow

Design and Analysis of Centrifugal Compressors, First Edition. René Van den Braembussche.
© 2019, The American Society of Mechanical Engineers (ASME), 2 Park Avenue, New York, NY, 10016, USA (www.asme.org).
Published 2019 by John Wiley & Sons Ltd.

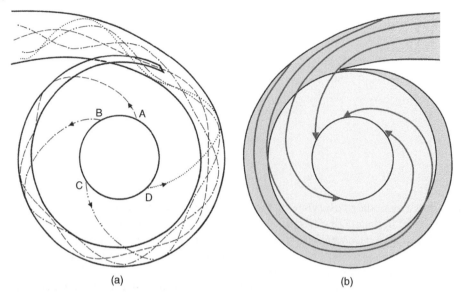

Figure 6.1 Streamlines inside (a) an outlet and (b) an inlet volute.

predictions are therefore often limited to steady flow in the non-rotating parts. However, the presence of an impeller can significantly modify the flow in the nozzle when return flow is occurring (Tomica et al. 1973). The same warning applies to experimental studies of only the inlet volute.

Inlet volute calculations are complicated by the complexity of the geometry, requiring a flexible grid generator. Symmetric geometries allow the numerical domain to be limited to half the inlet. An efficient exploitation of the numerical results requires good insight into the flow and loss mechanisms and their dependence on the geometry. The following sections intend to contribute to that.

6.1.1 Inlet Bends

The simplest non-axial inlet geometry is a curved channel of circular cross section, connecting the inlet pipe to the impeller inlet section (Figure 6.2). The outlet flow is characterized by a variation of the throughflow velocity between the inner and outer wall of the bend and two streamwise counter-rotating vortices. Depending on the ratio of the bend radius over the pipe diameter this may result in flow separation, i.e. a zone of low or negative throughflow velocity with low total pressure in position 5. Because of the blockage by this recirculation zone, the flow accelerates on the other side (positions 1 to 3). The consequence is a circumferential variation of the meridional velocity and hence of the impeller relative inlet flow angle and incidence.

The second consequence of the inlet bend are two counter-rotating vortices in positions 4 and 6. One of them rotates in the same direction as the impeller. Yagi et al. (2010) explain the impact of this asymmetry of the impeller inlet flow on the performance (Figure 6.3). Preswirl in the same direction as the impeller rotation ($\alpha_1 > 0$) has only a small impact on the impeller inlet flow conditions. The inlet relative velocity is smaller and the change in relative flow angle results in a decrease or only a small increase in the incidence. This constitutes an unloading of the impeller and unchanged or even lower losses can be expected. The flow is counter swirling ($\alpha_1 < 0$) on the opposite side (position 4). The incidence and inlet relative velocity increase above the average value. Both have a negative impact on the impeller performance, particularly

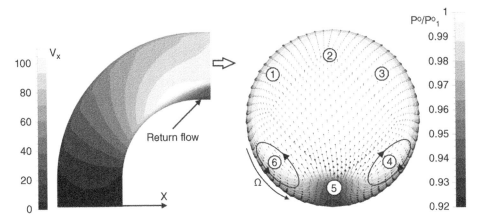

Figure 6.2 Axial velocity distribution in an inlet bend and flow structure in the outlet section.

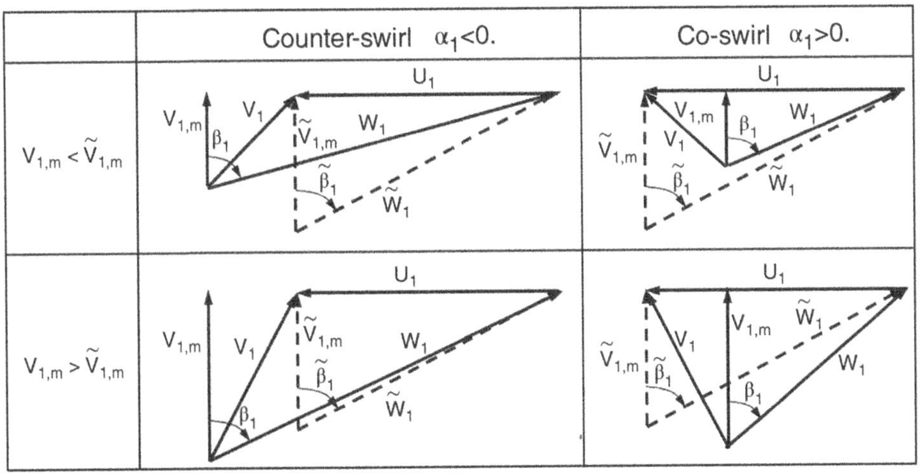

	Counter-swirl $\alpha_1 < 0.$	Co-swirl $\alpha_1 > 0.$
$V_{1,m} < \tilde{V}_{1,m}$		
$V_{1,m} > \tilde{V}_{1,m}$		

Figure 6.3 Impact of meridional velocity and prerotation on the impeller inlet conditions.

when the increase in the inlet Mach number gives rise to higher shock losses and an eventual reduction in the choking mass flow, as confirmed by the numerical results of Ueda et al. (2015).

Inlet bends differ by the cross section area and curvature radius of the central line. The favorable impact of a streamwise decrease in the cross section area on the impeller inlet velocity distortion is quite substantial, as shown in Figure 6.4 (Matthias 1966). The constant cross section inlet shows a velocity variation between 0.6 and 1.15 of the average value, with a 15° variation in the flow angle. The acceleration of the flow on the convex side, followed by a deceleration toward the impeller inlet, is at the origin of the flow separation and large velocity deficit on the inner wall. Increasing the convergence rate drastically improves the uniformity of the impeller inlet velocity (Figure 6.4b,c). Results shown here are for thin inlet boundary layers. Thicker boundary layers would enhance the two counter-rotating vortices shown in Figure 6.2.

The design approach for inlet bends is to minimize the boundary layer growth and hence the secondary flow effects by a continuous acceleration of the fluid along the inner and outer walls. Inlet bends of constant area show an acceleration of the flow on the convex side followed by a deceleration towards the impeller inlet (Figure 6.5). The latter is the main cause of flow separation upstream of the impeller leading edge. The concave side shows first a velocity deceleration

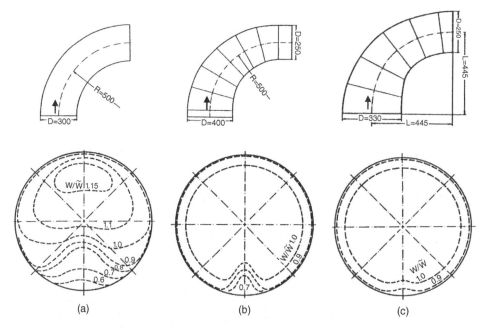

Figure 6.4 Different types of curved inlet ducts and impact of convergence on inlet flow distortion at the impeller inlet (from Matthias 1966).

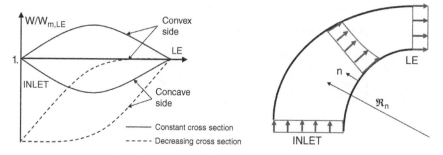

Figure 6.5 Influence of the convergence of a curved channel on the inner and outer wall velocity distribution.

followed by a reacceleration towards the average velocity at the exit. Flow separation is less likely to occur because of the increased turbulence on this concave surface. A streamwise increase in the average throughflow velocity by a gradual reduction of the cross section area compensates for these local decelerations and may allow a continuous velocity increase on both sides up to the impeller inlet, as shown by the dashed lines.

Another possibility to decrease the flow distortion and losses in an inlet bend is to increase the curvature radius of the bend or add a straight axial section in front of the impeller (Kim et al. 2001). However, both require a larger axial length (Figure 6.6). The impeller inlet flow uniformity can also be improved by installing deflection vanes in the inlet bend. The corresponding friction losses are very small in comparison with the gain in impeller performance because of a more uniform inlet flow. However, vanes should extend sufficiently far upstream of the convex side of the bend to avoid flow separation (Pinckney 1965a,b).

The impact of an upstream extension of the hub on losses and flow distortion has been studied by Wilbur (1957). An optimized non-symmetric nose, as shown in Figure 6.7a, cannot rotate

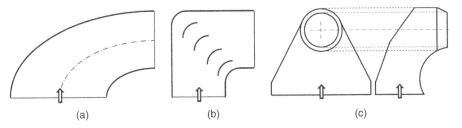

Figure 6.6 Inlet bend with (a) increased curvature radius, (b) cascade, and (c) large inlet cavity.

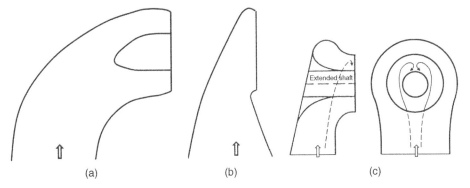

Figure 6.7 Inlet bend with (a) optimized impeller nose (from Pinckney 1965b) and inlet volute (b) without and (c) with shaft extension.

Table 6.1 Impact of inlet volute geometry on losses and distortion.

	$\Delta H/(\rho V_0^2/2)$	$\Delta NPSHR$ (%)
Constant section bend (Figure 6.4a)	0.040	30.0
Convergent bend (Figure 6.4c)	0.006	0.0
Inlet cascade (Figure 6.6b)	0.008	0.0
Inlet volute with weak inlet convergence (Figure 6.11)	0.03	0.12
Inlet volute with strong inlet convergence (Figure 6.11)	0.02	2.0
Large inlet cavity (Figure 6.6c)	0.004	0

with the impeller. Extra struts (and losses) are required unless the shaped nose can be supported by the inlet guide vanes (IGVs).

Kovats (1979) has estimated the impact of different inlet geometries on losses and net positive suction head required (NPSHR; a pump characteristic sensitive to inlet flow distortion). The results are summarized in Table 6.1. Some geometries (Figures 6.4c and 6.6c) have very low losses and almost no flow distortion, but losses may increase by an order of magnitude and NPSHR by 30% in constant cross section inlet bends

Kim et al. (2001) experimentally evaluated the impact of the inlet distortion on the compressor performance and observed a 10% efficiency drop at minimum mass flow. The energy input is almost unchanged because of the symmetry of the prerotation but the extra impeller losses impact on the total pressure ratio and efficiency. The static pressure ratio is almost identical for a straight inlet and curved bend. Kim et al. conclude from this that the inlet distortion affects only the impeller and not the diffuser because the inlet conditions of the latter are almost unchanged.

Yang et al. (2013) investigated the clocking effect between the inlet bend and outlet volute and observed a non-negligible variation in the pressure ratio at high mass flow when changing the azimuthal position of the inlet bend relative to the volute.

6.1.2 Inlet Volutes

A long axial length is the main disadvantage of curved inlet pipes. Inlet geometries frequently need to be designed within such space limitations that the optimum proportions cannot be respected. Both the convex and concave wall curvature radii may become very small (Figure 6.7b). The flow may further be distorted by an eventual upstream extension of the hub contour hosting the shaft (Figure 6.7c), resulting in an annular inlet section distributing the flow circumferentially before turning it axially towards the impeller inlet and an eventual flow separation downstream of the shaft protection.

Most inlet volutes are symmetric, with an intake rib separating the flow going to each side of the extended shaft (Figure 6.8). Opposite to the inlet ($\theta = 180°$) is a flow splitter, sometimes called "suction baffle". It may be a flat plate, preventing collision of the flow coming from both sides in order to decrease the tangential flow oscillations in the plenum, or may be shaped to create a more gradual decrease of the volute cross section and avoid excessive flow separation downstream of the extended hub.

Experimental and numerical studies by Lüdtke (1985) and Yagi et al. (2010) provide the following picture of the flow. When approaching the shaft at $\theta = 0°$ (full lines on Figure 6.8), the flow sharply turns from radial to axial. Position A is therefore the first location with a risk of flow separation for reasons already explained for inlet bends. This risk is enhanced by the wake of the intake rib. A second location where separation may occur is near the inlet upstream of the extended hub (position B), where the inlet duct diverges to minimize the velocity in the circumferential duct and hence the preswirl at the impeller inlet. The corresponding deceleration is partly compensated for by the local acceleration around the convex shroud and by blockage of the extended hub. A third zone where separated flow is likely to occur is the wake behind the extended hub at position C, as visualized by the dashed lines in Figure 6.7c and confirmed by the computational results of Han et al. (2012).

The outcome is a non-uniform velocity at the impeller inlet (Figure 6.9). Zero swirl flow is observed at $\theta = 0°$ where most of the flow enters the impeller near the hub because of flow

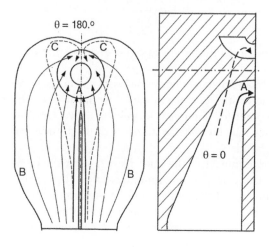

Figure 6.8 Flow structure in a radial inlet volute (from Lüdtke 1985).

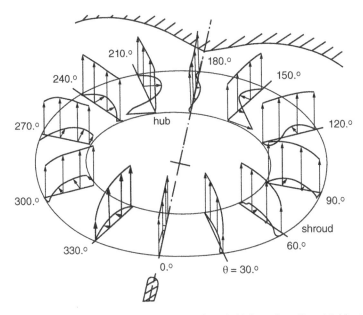

Figure 6.9 Flow distribution at the exit of a radial inlet volute (from Lüdtke 1985).

separation on the shroud side (position A on Figure 6.8). The bulk of the flow enters the impeller with a tangential velocity component on each side of the shaft. This is the consequence of the transport of the fluid around the volute circumference. Separation on the shroud may give rise to small opposite swirl at $\theta = 60$–$90°$ and 270–$300°$.

The rest of the flow turns around the shaft and fills the bottom part between $\theta = 150°$ and $210°$. Flow separation downstream of the extended hub creates a dead-water zone near the hub which drives the incoming streamlines towards the shroud (Figure 6.11).

After impinging on the suction baffle, the flow turns backward and fills the dead-water zone near the hub. The tangential velocity component is opposite, between the hub and the shroud. This gives rise to two counter-rotating vortices near the baffle. The impeller inlet swirl distribution at $\theta = 150°$ and $210°$ is opposite to the one at $\theta = 90°$ and $270°$. Numerical calculations by Yagi et al. (2010) indicate that the orientation and intensity of the vorticity near $\theta = 150°$ and $210°$ depend on the casing geometry.

Figure 6.10a shows a calculated variation of the absolute inlet flow angle at the impeller eye which is in line with the flow model shown in Figure 6.9. Prerotation angles decrease from $\pm 25°$ at hub (25% span) to less than $\pm 10°$ near the shroud (99% span) for $60° < \theta < 120°$ and $270° < \theta < 330°$. The larger swirl angle at the hub is the consequence of a lower meridional velocity than at the shroud, because of meridional curvature, in combination with a larger tangential velocity because of conservation of momentum ($RV_u = C^{te}$) towards the smaller hub radius. The opposite sign of the absolute flow angle at the hub and shroud for $\theta = 170°$ and $190°$ corresponds to the streamwise counter-rotating vortices in the bottom dead centers.

The static pressure (Figure 6.10b) increases from shroud to hub, and from the inlet, where the flow accelerates around the shroud before entering the impeller, to the bottom dead center. The highest static pressure is near $\theta = 180°$, where the circumferential transport velocity is stopped by the bottom rib.

If the axial length allows, the uniformity of the flow can be improved by adding a bellmouth to the shroud contour (Figure 6.11). The concept of a bellmouth goes back to Stepanoff (1957) and has a double purpose. The first one is to avoid flow separation at $\theta = 0°$ because of a too small

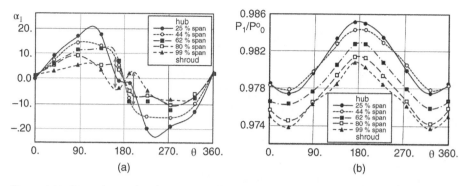

Figure 6.10 Circumferential and spanwise variation of (a) the absolute flow angle and (b) static pressure at the impeller eye (0% span at hub) (from Flathers et al. 1996).

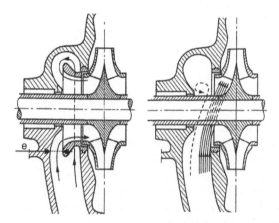

Figure 6.11 Impact of the bellmouth on the impeller inlet flow (from Neumann 1991).

curvature radius near the shroud. A second one is to prevent the flow from entering directly into the impeller with a large tangential velocity at $\theta = 90°$ and $270°$, as shown by the full lines on Figure 6.11.

The inlet volute impacts on the compressor performance by the friction and separation losses in the volute and by the increased impeller losses resulting from the distorted inlet flow. Based on experimental data and numerical calculations, Flathers et al. (1996) estimated the total pressure losses in the inlet volute at about 45% of the volute inlet dynamic pressure with a concentration in the zones with risk of separation.

Yagi et al. (2010) show experimental evidence that the pressure losses in the inlet nozzle are rather small in comparison with the extra impeller losses resulting from the inlet distortion and that a more uniform flow at the eye of the impeller results in an increase in efficiency that largely compensates for the pressure losses in the inlet volute. Optimizing the shape of the radial inlet volute, they achieved a 3.8% increase in efficiency back to a level comparable to the one for an axial inlet. The improvements also resulted from a reduction in the volute losses as well as from an improved flow uniformity by:

- avoiding flow separation in position B on Figure 6.8 by a smoother transition from the inlet pipe to the volute
- decreasing the fluid velocity in the section around the bellmouth by decelerating the flow at the outlet of the inlet nozzle and keeping it low by increasing the cross section of the annular duct outside the bellmouth. A smaller circumferential transport velocity contributes to

a smaller preswirl velocity at the impeller inlet because the two are related by the conservation of the tangential momentum. A smaller transport velocity further allows a proper acceleration of the flow towards the impeller eye to reduce or even avoid separation losses.

- a contracting section just upstream of the impeller eye. Accelerating the flow towards the impeller eye not only helps to decrease the losses, but also reduces the flow angle distortions because it increases the meridional velocity component without influencing the swirl velocity component.

Pazzi and Michelassi (2000) have numerically evaluated the impact of the axial distance e between the lip of the bellmouth and the hub wall (Figure 6.11). Decreasing this gap creates an obstacle for the flow at $\theta = 0°$ and contributes to a circumferentially more uniform distribution of the mass flow and impeller inlet Mach number. However, the total pressure losses may increase because of the increased flow deceleration and subsequent flow separation when turning the flow from radial to axial along the shroud. A larger gap e results in lower separation losses because of a stronger acceleration of the flow towards the impeller eye. However, increasing the gap shortens the axial length of the shroud and reduces the curvature radius, which may provoke separation. Furthermore, the circumferential non-uniformity of the meridional velocity increases because more flow is entering the impeller at $\theta = 0°$. A circumferential increase in e from $\theta = 0°$ to $\theta = 180°$ contributes to a more uniform inlet flow and lowers the losses, but at the cost of a more complex geometry.

Lüdtke (1985) and Koch et al. (1995) emphasized the possible impact of the flow distortion resulting from the geometry upstream of the inlet volute. An inlet bend in a plane perpendicular to the shaft may guide the flow towards one or the other side of the volute. The higher velocity on the convex side of the inlet bend (Figure 6.12) directs the flow towards the upper part of the inlet volute, resulting in an increase in the prerotation at the impeller eye and hence extra impeller losses. The larger flow does not reach the baffle at $\theta = 180°$ but ends in a strong vortex. An inlet bend in the plane of the shaft has a much smaller impact. The flow is directed towards the hub or shroud side of the volute, which does not alter the symmetry, and the results are comparable to those for a straight radial inlet pipe.

6.1.3 Vaned Inlet Volutes

A more complex and expensive way to uniformize the impeller inlet flow is by installing guide vanes. Vanes can be limited to the radial part (Tan et al. 2010) or extend up to the impeller eye (Koch et al. 1995). Placing straight vanes in the bellmouth limits the circumferential extend and amplitude of the swirl. However, the high incidence and subsequent separation of the flow on the vane suction side results in large vortices near the shroud (Figure 6.13).

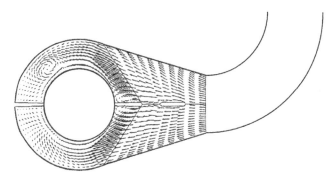

Figure 6.12 Impact of upstream bend on the flow in an inlet volute (from Koch et al. 1995).

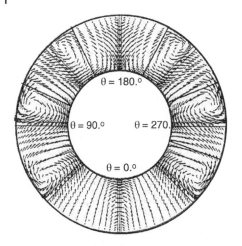

Figure 6.13 Calculated velocity vectors at the exit of an inlet section with straight vanes (from Koch et al. 1995).

Flathers et al. (1996) and Xin et al. (2016) designed cambered guide vanes with the leading edge adapted to the local spanwise averaged inlet flow angle (Figure 6.10a). The rounded leading edge allows a variation of the incidence from hub to shroud (Figure 6.14). The camber of each vane aims for zero outlet swirl. The flow downstream of the vanes is very much equalized, with only local pressure distortions in the vicinity of the vanes and swirl angle variations below $\pm 2°$. The overall total pressure losses are only 11% higher than for the vaneless inlet and are more than compensated for by a much larger improvement in the impeller performance because of more uniform inlet flow.

6.1.4 Tangential Inlet Volute

Tangential volutes have a centerline that does not cross the impeller axis but follows a spiral, tangent to the shroud of the inlet annulus (Figure 6.15). Non-symmetric or tangential volutes may be a valid alternative for radial ones when there are space restrictions or to facilitate mounting and dismounting of horizontally split compressors. Another argument to replace a radial volute by a tangential one is to avoid the circumferential asymmetry of the flow by replacing the opposite swirl velocities on both sides of the inlet by a more uniform one in only one direction. This, however, is likely to be at the cost of an increased prerotation and a change in pressure ratio (Neumann 1991). When required, such a change in energy input could be compensated for by a modification of the impeller. Frozen rotor type calculations by Michelassi et al. (2001)

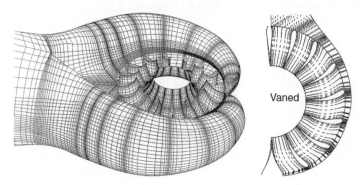

Vaned

Figure 6.14 Accelerating volute with cambered vanes and velocity vectors calculated in the vaned inlet section (from Flathers et al. 1996).

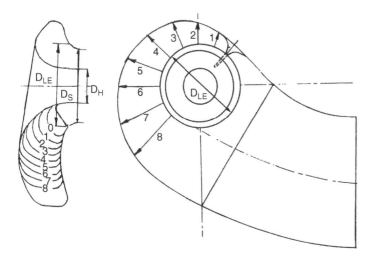

Figure 6.15 Typical tangential inlet volute geometry (from Neumann 1991).

indicated up to a 10% decrease in the impeller enthalpy rise depending on the amount of swirl generated by the volute, but with no effect on the stall limit. Their calculations also indicated that the amount of swirl can be drastically reduced by changing the position of the suction baffle, but at the expense of a 50% increase in total pressure loss.

Tangential inlet nozzles differ by the orientation of the inlet pipe, the shape of the curved intake rib separating the two sides, and by the shape and circumferential position of the suction baffle (Figure 6.16). The latter has a strong impact on the tangential flow component. There are two distinct flow zones. The first one is located near the convex side of the turn where the fluid approaches the annulus almost radially. The second one has a quasi spiral form and is located at the outside of the turn, feeding the top of the annulus. The cross section of this part decreases gradually to compensate for the decrease in mass flow around the circumference. The purpose is to keep the stagnation pressure and tangential velocity constant around the impeller eye.

An optimization by Lüdtke (1985) to minimize the overall swirl and losses resulted in a rather unconventional geometry (Figure 6.16) providing a circumferentially more uniform tangential and meridional flow and lower losses than in the original radial inlet.

The impact of a bend upstream of a tangential volute is similar to that of symmetric volutes, i.e. flow may shift to one or the other side depending on the orientation of the bend.

Figure 6.16 Tangential inlet volute optimized for minimum swirl (from Lüdtke 1985).

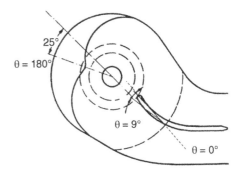

6.2 Outlet Volutes

An outlet volute can be considered as a stator with only one vane and the complete circumference as pitch (two vanes at 180° pitch in case of double volutes). The layout depends on the leading edge (tongue) suction side angle and hence is well adapted for only one mass flow per speed line. At off-design operation, outlet volutes generate a circumferential variation of the static pressure. The main consequence is an unsteady impeller flow with a circumferential variation of mass flow and blade loading.

The main purpose of the present section is to provide insight into the flow structure and to present geometrical modifications that can improve the performance or remediate problems. A description of the flow in different types of outlet volutes is followed by a critical reflection on performance prediction models and design procedures. The flow models presented here may help to get a better understanding of the CFD output.

Outlet volutes influence the compressor and pump performance mainly in two ways:

- A direct reduction of efficiency because of friction and non-isentropic deceleration of the fluid. The corresponding losses are proportional to the kinetic energy available at the volute inlet. The origin of these losses and prediction models are the main issue in the present chapter.
- Indirectly by a reduction of the impeller efficiency and operating range due to the circumferential pressure distortion resulting from the off-design operation of the volute. The origin of this pressure distortion is presented but its influence on the impeller flow is the main topic of Chapter 7.

6.2.1 Volute Flow Model

There is only one operating point per constant speed line for which a well-designed volute imposes a circumferentially uniform pressure at the diffuser outlet (Figure 6.17b). At smaller mass flows, the volute is too large and acts like a diffuser, resulting in a static pressure rise between the volute inlet (tongue) and outlet (Figure 6.17c). At larger than design mass flows, the volute is too small and the flow accelerates towards the exit, resulting in a decreasing pressure from the volute inlet to the outlet (Figure 6.17a). The main consequences are extra losses and a circumferential pressure distortion at the diffuser and impeller exit.

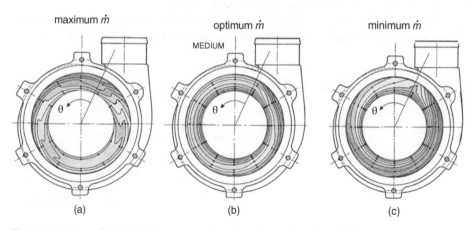

Figure 6.17 Circumferential static pressure distortion due to the volute off-design operation.

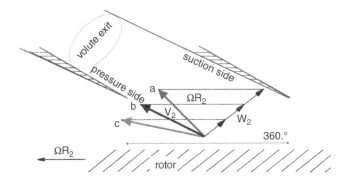

Figure 6.18 Transformation of an impeller–volute combination into a rectilinear cascade.

The real cause of the circumferential pressure distortion is best understood by transforming the impeller (11 blades) and volute (1 blade) conformally into an equivalent cascade (Figure 6.18). At a given peripheral speed, ΩR_2, the flow enters the stator tangent to the vane (zero incidence) for a mass flow corresponding to the velocity shown by arrow b on Figure 6.18. The flow is not deviated by the volute wall and the pressure remains constant from the tongue to the exit. This operating point is called volute design mass flow. At higher mass flow (arrow a on Figure 6.18) the flow enters the stator with a negative incidence. The flow decelerates on the leading edge suction side followed by a reacceleration further downstream on that side. The negative incidence creates an expansion around the leading edge pressure side, resulting in a low pressure on that side. The combination of the two is a pressure decrease from suction side to pressure side, i.e. along the volute. At a lower mass flow (arrow c on Figure 6.18) the flow approaches the tongue with a positive incidence. The flow accelerates around the leading edge suction side and decelerates on the pressure side of the tongue, which explains the static pressure rise from the volute tongue to the volute exit.

The volute static pressure distortion applies to the diffuser exit. In the case of a vaneless diffuser, this pressure variation propagates upstream and results in periodic outlet conditions for the rotating impeller. In the absence of a diffuser, the distortion applies directly to the impeller exit. Vaned diffusers act like filters and attenuate or even eliminate the distortion.

The main consequences of the impeller outlet distortion are as follows:

- **Radial forces:** The circumferential variation of the impeller outlet pressure results in a circumferential change in the velocity and blade loading, and hence in a net radial force acting on the impeller shaft. A second contribution to this radial force is the non-axisymmetric static pressure acting on the impeller outlet section and on the shroud and hub outer walls.
- **Losses:** The cyclic variation of the flow in the impeller channels at each rotation results in an additional energy dissipation. The circumferential non-uniformity of the flow at the diffuser inlet imposes an unequal momentum around the periphery that produces mixing losses and extra shear in the diffuser and volute.
- **Noise and vibration:** The pressure fluctuations caused by the rotating impeller are at the origin of noise and vibrations. The amplitude is largest near the volute tongue, indicating that the volute tongue–impeller interaction is the main source of the noise and vibrations.

6.2.2 Main Geometrical Parameters

Most volutes are complex tri-dimensional geometries that are often difficult and expensive to manufacture. The following geometrical parameters have a major influence on the flow:

- **Volute size:** Smaller volutes give rise to higher pressure ratios and steeper performance curves than standard ones, while larger volutes result in lower pressure ratios and flatter performances curves (Whitfield and Robert 1983; Mishina and Gyobu 1978).

  Figure 6.19 shows a comparison of the performance obtained with the same impeller and vaneless diffuser, but two different volutes. The volute of Figure 6.19a was optimized for a pressure ratio of 3.8, whereas the volute of Figure 6.19b is optimized for a pressure ratio of 6 and 30% smaller. The largest volute resulted in a larger operating range at low RPM but the flow becomes unsteady at higher pressure ratios, which is reflected by the wavy shape of the pressure rise curves. The smaller volute had a stable flow at all pressure ratios but a smaller range because of choking at the lower pressure ratios.

- **Cross section shape and location:** Cross sections can be symmetric or asymmetric, circular, elliptic or rectangular with constant, increasing or decreasing central radius (Figure 6.20). It has been shown by Mishina and Gyobu (1978) that this has an important influence on the volute losses (Figure 6.21). This will be discussed in more detail in the next sections.

- **Circumferential variation of the volute cross section:** Even in a 2D volute with circumferentially and axially uniform inlet conditions, the flow will not be uniform over a cross section (Figure 6.22a). Continuity defines the change in the radial velocity component along a streamline:

$$\rho b R V_R = C^{te} \tag{6.1}$$

The decrease in the tangential velocity is defined by the conservation of momentum

$$R V_T = C^{te} \tag{6.2}$$

Both change with radius. The local velocity at every point depends on the tangential velocity at the starting point of the streamline (the impeller outlet) and the radius change along the

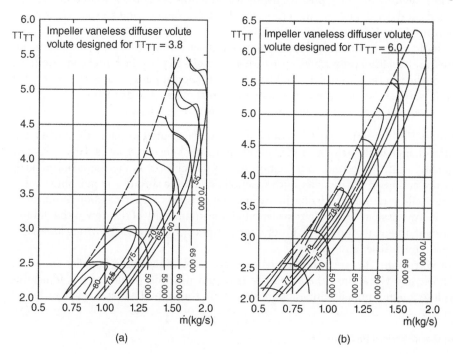

(a)   (b)

Figure 6.19 Influence of the volute size on the radial compressor performance map: (a) larger volute designed for a pressure ratio of 3.8 and (b) smaller volute designed for a pressure ratio of 6.0 (from Stiefel 1972).

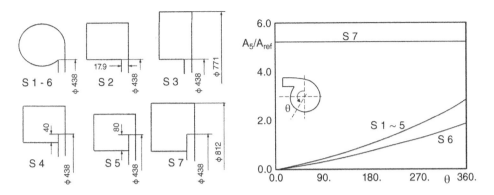

Figure 6.20 Definition of different volute cross sectional shapes (from Mishina and Gyobu 1978).

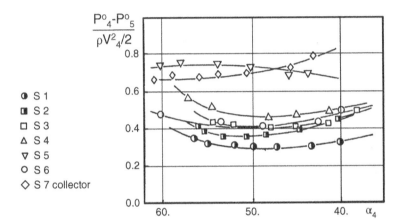

Figure 6.21 Influence of the volute cross sectional shape and radial location on the volute loss coefficient (from Mishina and Gyobu 1978).

streamline. A uniform V_{u2} distribution at the impeller trailing edge will therefore result in a free vortex through flow velocity(V_T) over a 2D cross section A, as shown in Figure 6.22b. Circumferential velocity distributions at the impeller outlet, corresponding to lower or higher than optimum mass flow, result in the velocity distributions shown in Figure 6.22a and 6.22c, respectively.

At high mass flow, the larger tangential velocity near the volute inlet results in a large throughflow velocity near the outer wall in section A (Figure 6.22c) in spite of the large increase in radius. One should account for this variation of the velocity with radius when defining the volute cross sections.

- **Volute tongue:** A detailed examination of the influence of the volute tongue on performance is given by Lipski (1979). Flow visualizations (Brownell and Flack 1984; Elholm et al. 1992) have shown that the stagnation point on the volute tongue moves from the discharge side to the impeller side when the mass flow increases. The distance between the tongue and the impeller exit has an important influence on the circumferential pressure distortion. Large gaps between them provide more space for the attenuation of an eventual flow and pressure distortion. However, the amplitude of the pressure distortion may increase from the vaneless diffuser exit to the impeller exit, as shown in Figure 7.6 (Sideris et al. 1987a,b; Yang et al. 2010).

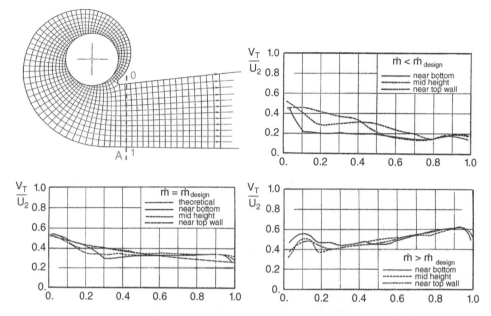

Figure 6.22 Crosswise velocity distribution in cross section A of a 2D volute (from Bowerman and Acosta 1957).

It seems that the perturbed velocity also respects the laws of mass and momentum conservation (Equations (6.1) and (6.2)) and that the amplitude of the pressure perturbations changes inversely proportional to the square of the velocity perturbation.

6.2.3 Detailed 3D Flow Structure in Volutes

The following is a detailed description of a volute flow model and its validation with flow measurements in simplified and real volutes of different shape. The main purpose is to clarify the relation between the volute flow, the volute geometry, and the incoming flow conditions. This will be the basis for the volute prediction models.

The flow in volutes is very different from the flow in a classical vortex tube, where the migration of the fluid from the outer radius to the center results in an increase in the swirl velocity V_S. Vortex tubes, therefore, have a free vortex circulation near the walls $rV_S = C^{te}$ and a forced vortex circulation $V_S = rC^{te}$ in the center. A free vortex swirl occurs also in conical diffusers (Senoo et al. 1978), when the swirling fluid migrates from a small radius at the inlet to a large radius at the outlet.

The structure of the vortices in a volute, however, depends on the circumferential variation of the inlet radial velocity and can be more easily explained by a linear model of a volute (Figure 6.23). The fluid entering the volute at a small radius close to the tongue fills the center of the volute. After one rotation it is absorbed by new fluid, entering the volute further downstream at a larger radius, which starts swirling around the upstream fluid. Vortex tubes of increasing radius wrap around each other and each part of the fluid remains at almost the same radius it has entered the volute. Hence the swirl velocity V_S at a given crosswise position depends on the radial velocity V_R of the incoming fluid. The other properties of the flow inside the volute, such as T^o and P^o, depend on the incoming flow and on the changes in the vortex structure by shear forces inside the volute.

The model is derived from the experimental investigation of the 3D flow in a simplified large-scale model of a typical overhang volute by Van den Braembussche and Hände (1990). The

Figure 6.23 Flow structure in an centrifugal compressor exit volute.

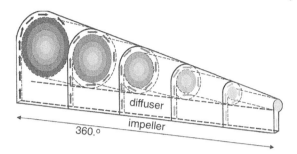

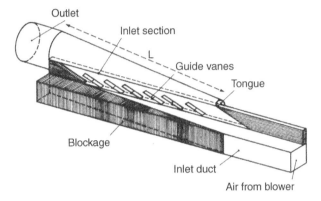

Figure 6.24 3D view and dimensions of a straight diffuser and volute.

parallel walled inlet and conical channel, connected to one side of the inlet duct (Figure 6.24), represent a radial diffuser followed by an asymmetric volute. The main differences between the model and a real volute are a constant inlet total pressure and the absence of circumferential curvature between the volute tongue and outlet. However, this curvature radius (defined by the volute center line radius) is normally much larger than the volute cross section radius, and its influence on the flow will be discussed later.

A blower provides air to a settling chamber upstream of the rectangular inlet duct. The variable blockage in the inlet duct enables adjustment of the inlet pressure distribution at the volute inlet. An empirical procedure is used to adjust the vane setting angle and blockage until the required volute inlet flow conditions, corresponding to maximum, optimum or minimum mass flow, are obtained. The static pressure in four crosswise and 11 longitudinal positions on the volute wall is shown in Figure 6.25a–c for three operating conditions.

Figures 6.28, 6.31, and 6.34 show the crosswise variation of the swirl velocity V_S and through-flow velocity V_T together with the static and total pressure distribution at five longitudinal positions for three operating points. Only one traverse is made at each cross section because one can assume that the flow is axisymmetric in a conical channel.

6.2.3.1 Design Mass Flow Operation

A circumferentially constant impeller exit radial velocity and total pressure along the volute inlet (Figure 6.26) theoretically results in a vortex with constant swirl velocity V_S and constant total pressure, as shown in Figure 6.27a. However, such a flow cannot exist because of the large shear forces it generates in the center. Kinetic energy is dissipated until a forced vortex structure, corresponding to a friction free solid body rotation, appears in the center (Figure 6.27b). The shear forces generate losses and are responsible for a decrease in the total pressure in the center.

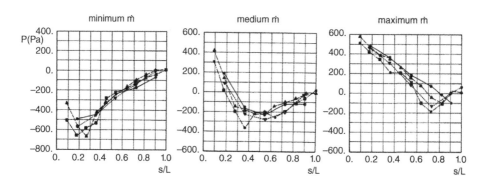

Figure 6.25 Longitudinal and circumferential static pressure variation corresponding to (a) minimum, (b) optimum, and (c) maximum mass flow (from Van den Braembussche and Hände 1990).

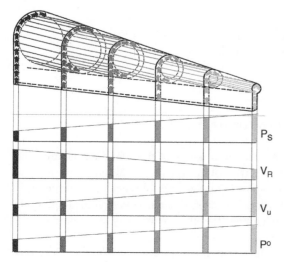

Figure 6.26 Volute inlet conditions at design mass flow.

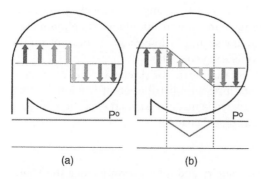

Figure 6.27 Flow structure in a volute at design mass flow: (a) isentropic and (b) real.

This flow structure is confirmed by the experimental results shown in Figure 6.28. The swirl velocity is nearly constant close to the walls and has a forced vortex type structure in the center. The total pressure is almost constant except for the small zone in the center corresponding to the local energy dissipation resulting from the creation of the forced vortex.

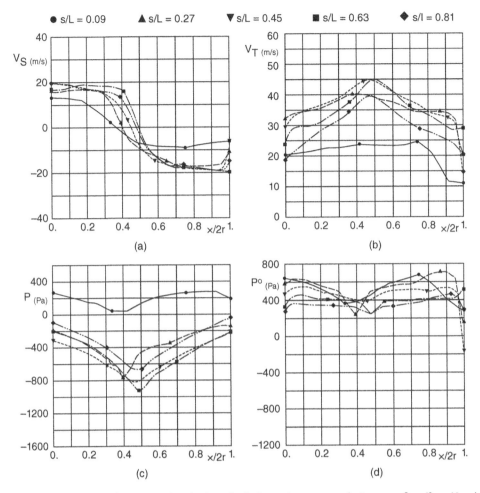

Figure 6.28 Measured crosswise distribution of velocity and pressure at design mass flow (from Van den Braembussche and Hände 1990).

The radial equilibrium between the swirl velocity and the static pressure creates a zone of low static pressure in the center defined by

$$\frac{dP}{dr} = \rho \frac{V_S^2}{r} \tag{6.3}$$

The throughflow velocity is defined from what is left after the swirl energy is subtracted from the local kinetic energy:

$$\rho \frac{V_T^2}{2} = P^o - P_S - \rho \frac{V_S^2}{2} \tag{6.4}$$

Except for a distortion near the inlet, the spanwise averaged value of V_T is almost constant along the volute inlet, which is expected at design mass flow. Any deceleration or acceleration of the throughflow velocity would result in an increase or decrease in the static pressure and be in conflict with the definition of design point operation. The nearly constant total pressure in combination with a decreasing static pressure and low swirl in the center results in a substantial increase in V_T towards the center of the volute.

6.2.3.2 Lower than Design Mass Flow

At less than optimal mass flow, the circumferentially decreasing radial and increasing tangential velocity and total pressure (Figure 6.29) along the volute inlet result in a nearly free vortex flow with maximum swirl velocity in the volute center and maximum total pressure near the outer wall (Figure 6.30a). Extra kinetic energy is dissipated in the center of the volute resulting from the transformation of the free vortex with the maximum velocity in the center into a forced vortex with zero velocity in the center. The corresponding shear losses are responsible for a further decrease in total pressure in the center. The small swirl velocity is responsible for the more uniform static pressure distribution over the cross section (Equation (6.3)). In combination with the low P^o in the center this results in a decrease in V_T towards the center of the volute (Equation (6.4)). The experimental results at less than design mass flow (Figure 6.31) confirm this. The swirl velocity first increases from the walls toward the center after which it decreases again by the shear forces creating the forced vortex in the center.

At low mass flow, the fluid enters the volute with an increasing tangential velocity component V_u. Continuity, however, requires a small V_T inside the volute. A deceleration of the tangential velocity takes place and the volute acts like a diffuser. As observed in Figure 6.31, the average throughflow velocity V_T decreases and the average static pressure increases from volute tongue to the outlet. The swirl component keeps the flow close to the walls and the zone of low throughflow velocity and low P^o remains in the center. This reduces the effective cross section, decreases the static pressure recovery, and avoids flow separation at the walls. The low static pressure gradient, because of the small swirl velocity, in combination with the low total pressure in the center, may even result in negative values of V_T in the center of the outlet diffuser.

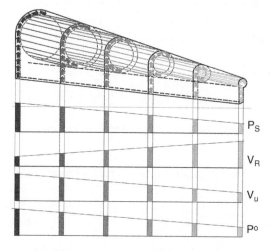

Figure 6.29 Volute inlet conditions at less than design mass flow.

P_S

V_R

V_u

P^o

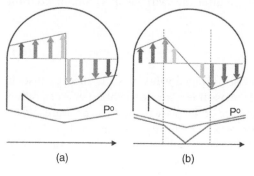

Figure 6.30 Flow structure in a volute at less than design mass flow: (a) isentropic and (b) real.

P^o

P^o

(a)   (b)

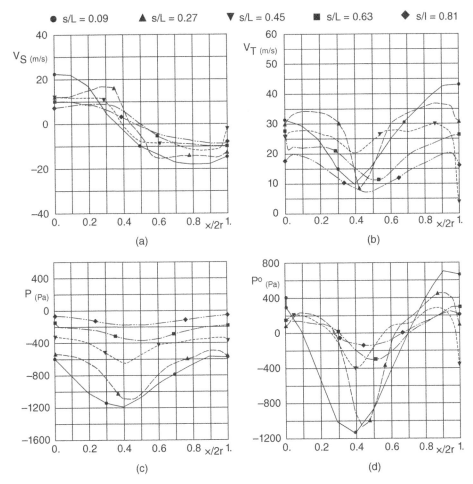

Figure 6.31 Measured crosswise distribution of velocity and pressure at less than design mass flow (from Van den Braembussche and Hände 1990).

The strong asymmetry observed at minimum mass flow at the first two positions is an incidence effect on the tongue because there is no tongue leakage flow, as in real volutes, and it takes some distance to fill the volute with fluid when the incoming flow is very tangential. The corresponding high losses in the center gradually disappear by mixing with new incoming fluid.

6.2.3.3 Higher than Design Mass Flow

The circumferentially increasing radial and decreasing tangential velocity and total pressure along the volute inlet (Figure 6.32), when operating at higher than optimal mass flow, theoretically results in a nearly forced vortex with maximum swirl velocity near the outer wall and maximum total pressure in the center (Figure 6.33a). Only a small amount of kinetic energy must be dissipated in the center of the volute to create a forced vortex. Hence the losses are very small and the total pressure remains high.

This theoretical flow structure is again confirmed by the experimental results shown in Figure 6.34. The average static pressure is decreasing from the tongue to the outlet. The increase in the swirl velocity from $s/L = 0.09$ to $s/L = 0.81$ results from an increase in the radial velocity along the diffuser exit/volute inlet. This is a consequence of the decreasing static

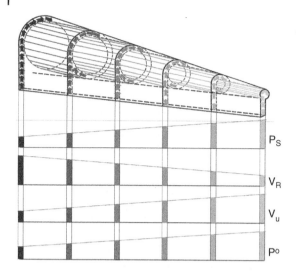

Figure 6.32 Volute inlet conditions at larger than design mass flow.

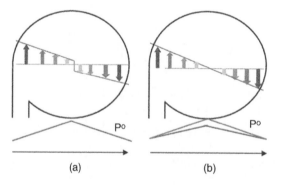

Figure 6.33 Flow structure in a volute at larger than design mass flow: (a) isentropic and (b) real.

pressure along the circumference. At all locations the swirl velocity decreases towards the center because of the specific inlet conditions.

The total pressure increase expected in the center and indicated on Figure 6.33 is not observed because the linear test facility behaves differently from a real volute/impeller combination. The decrease in the static pressure along the volute inlet does not give rise to a decrease in total pressure as would be the case with an impeller. Hence the inviscid total pressure is constant along the inlet and over the whole section.

The radial distribution of the throughflow velocity V_T increases from a small value at the walls towards a much larger one at the volute center (Figure 6.34). This is the consequence of the nearly constant P^o and the large decrease in P towards the center because of the large swirl. The high throughflow velocity in the volute center results in a mass flow which is larger than the one predicted when using the pressure measurements at the volute walls. This is similar to a negative blockage in the volute numerically observed by Hübl (1975) and experimentally by Stiefel (1972) who stated that "the optimum scroll operation is achieved for a scroll area 10–15% smaller than the frictionless computed one".

Experimental results show very thin boundary layers near the walls. This is the consequence of the continuous absorption of the boundary layers after each rotation in the cross section by new fluid coming out of the diffuser (Figure 6.35) and the enhanced mixing of the boundary layer flow on concave surfaces as described in Section 3.2.3. This absorption of the boundary layers lowers the total pressure in the center of the volute.

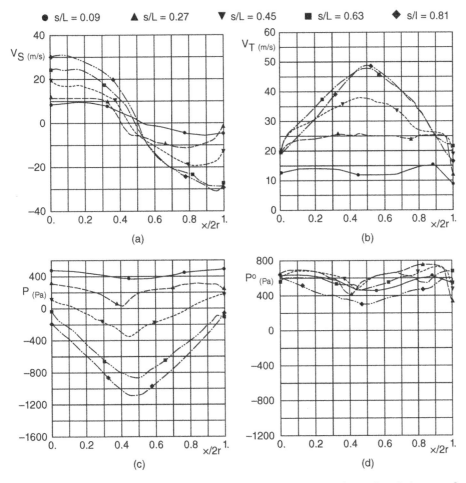

Figure 6.34 Measured crosswise distribution of velocity and pressure at larger than design mass flow (from Van den Braembussche and Hände 1990).

Figure 6.35 Absorption of volute boundary layers by new incoming fluid.

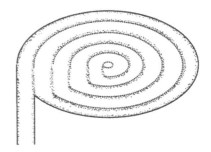

Previous considerations lead to the following volute flow model:

- The forced vortex distribution of the swirl velocity results from internal friction and/or from the radial velocity distribution at the volute inlet. The first mechanism results in high losses and low total pressure in the center of the cross sections. The second one shows much lower losses and the total pressure is a function of the total pressure distribution at the volute inlet.

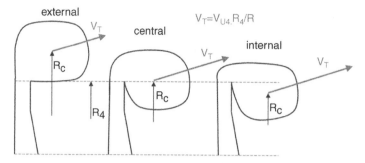

Figure 6.36 External, central, and internal volute geometry.

However, the complete swirl velocity will ultimately dissipate by friction on the walls, which explains the higher overall total pressure losses at maximum mass flow.

- The static pressure distribution depends on the centrifugal forces resulting from the swirl velocity in the cross section. In a real volute, curved around the diffuser exit, there will also be a radial pressure gradient due to the centrifugal forces resulting from the throughflow velocity along the volute periphery.
- The crosswise throughflow velocity distribution results from the total and static pressure variation over a cross section and can be very different from the average one.

In what follows the model will be validated on different types of real volutes. They differ by the cross section shape and by the radial location of the center of the cross section (Figure 6.36). External volutes have the inner wall at constant (diffuser exit) radius. They show a higher pressure ratio and efficiency because of the extra diffusion that takes place between the diffuser exit and the volute centerline. However, they are much larger because the outer radius increases circumferentially with the increasing volute cross section.

The maximum volume of centrifugal compressors is often limited by the available space and by the material and manufacturing costs, proportional to size. This favors the use of internal volutes, i.e. volutes with a constant outer radius and the center of the cross sections at a smaller radius than the diffuser outlet. The main disadvantages of internal volutes are a lower static pressure rise and larger total pressure losses. The pressure recovery in the diffuser due to velocity deceleration with increasing radius is annihilated by the subsequent reacceleration of the flow in the volute due to the decreasing volute centerline radius.

6.2.4 Central Elliptic Volutes

The previous picture of the 3D flow in volutes has been confirmed by a series of detailed measurements in a central volute with elliptic cross section, designed by the method of Brown and Bradshaw (1947) (Figure 6.38). Measurements are described in detail in Ayder et al. (1991, 1993, 1994). The flow is produced by a low specific speed impeller with an outer diameter of 256 mm and 19 radial ending blades in combination with a narrow vaneless diffuser of radius ratio 2.24. The performance of the compressor at standard inlet conditions is shown in Figure 6.37. Measurements have been made at the three operating points indicated by high, medium, and low mass flow.

The spanwise distribution of the radial and tangential velocity and the total temperature and pressure at the diffuser exit is measured at eight circumferential positions (Figure 6.38). A five-hole pressure probe is used to measure the 3D flow field by three to five radial traverses and one axial traverse in the seven cross sections. The total and static pressures are measured

Figure 6.37 Characteristic of compressor with an elliptic central volute and a rectangular internal volute.

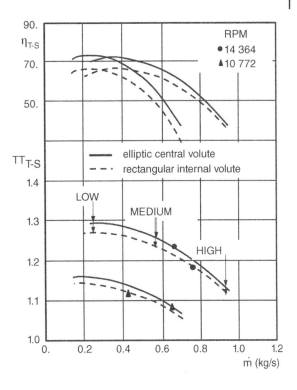

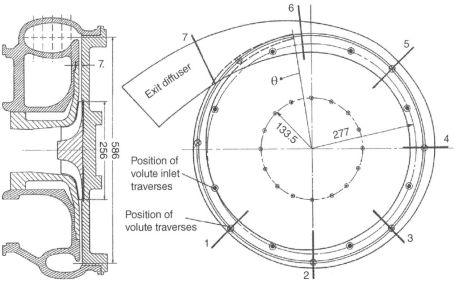

Figure 6.38 Geometry of the elliptic volute and the location of the measurements (from Ayder et al. 1993).

relative to the volute exit static pressure. Positive values indicate pressures that are higher than the atmospheric outlet pressure.

The circumferential variation of the mass averaged flow characteristics at the three operating points are shown in Figure 6.39. The decreasing static pressure (Figure 6.39b) corresponds to higher than optimum mass flow. The constant static pressure at low mass flow indicates that

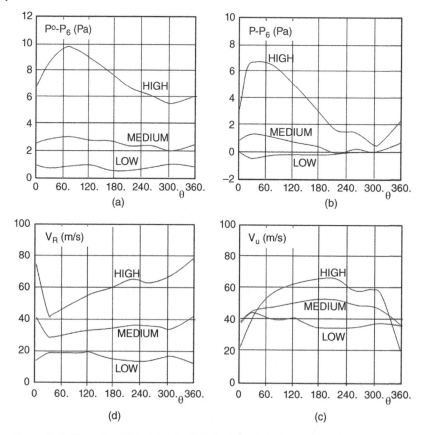

Figure 6.39 Circumferential variation of (a) the spanwise averaged total pressure, (b) the static pressure, (c) the tangential velocity, and (d) the radial velocity at the volute inlet (from Ayder et al. 1993).

the volute is well adapted to the impeller at this mass flow. Measurements could not be made at lower mass flows because of impeller instability.

The results of the detailed volute flow measurements over the cross sections are shown in Figures 6.40 to 6.47.

6.2.4.1 High Mass Flow Measurements

In accordance with the flow model described in the previous section, the fluid entering the volute close to the tongue starts swirling around the center of the cross section and gives rise to large shear stresses at the center. A well-defined spot of high total pressure losses, resulting from this dissipation, is observed in the center of the first cross section (Figure 6.40).

The circumferential increase in the radial velocity at the volute inlet from the tongue to the volute outlet (Figure 6.39) results in an increase in the swirl velocity with cross section radius. Hence the solid body rotation in the downstream sections of the volute (Figure 6.41) is the consequence of the particular volute inlet flow velocity distribution and is only slightly influenced by the shear stresses in the center. As a consequence, the total pressure distribution will depend mostly on the incoming flow. The decreasing total pressure along the volute inlet, from the tongue to the volute outlet, results in a decrease in total pressure from the volute center to the outer radius at the downstream sections of the volute (Figure 6.40). Hence the total pressure over the volute cross section is higher than the one locally entering the section.

Figure 6.40 Total pressure variation over the cross sections at high mass flow (Pa) (from Ayder et al. 1993).

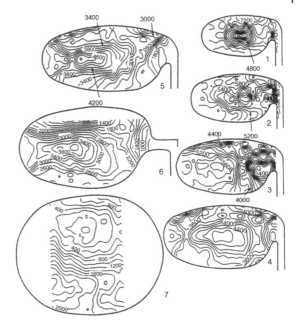

Figure 6.41 Swirl velocity distribution over the volute cross section at high mass flow (from Ayder et al. 1993).

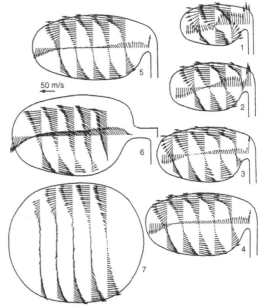

The internal shear stress is zero in a solid body (forced vortex) swirl. Any further decrease in the swirl velocity is by the friction on the volute walls or by an increase in radius in the exit diffuser. Figure 6.41 shows lower swirl velocities perpendicular to the longer horizontal axis and larger ones perpendicular to the short vertical axis of the elliptic cross sections. This results from the conservation of angular momentum of the flow swirling in an elliptic section.

The change in location of the vortex center between section 5 and section 6 is due to an axial shift of the volute centerline and an increase in the centerline radius. Because the measurements are made in planes parallel to the x axis, the probe measures an additional axial and radial velocity component and an apparent shift of the vortex center towards the lower left corner.

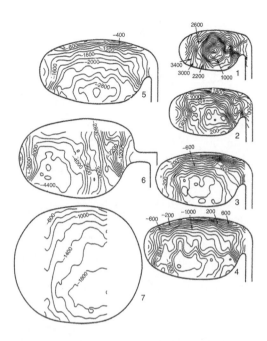

Figure 6.42 Static pressure variation over the volute cross section at high mass flow (from Ayder et al. 1993).

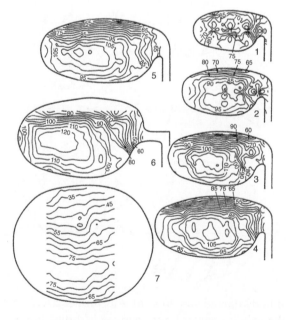

Figure 6.43 Throughflow velocity variation over the volute cross section at high mass flow (m/s) (from Ayder et al. 1993).

The measured static pressure variation over the cross section results from the swirl and the circumferential curvature of the volute channel. The largest static pressure gradient is due to the swirl velocity (Equation (6.3)). The pressure gradient between the inner and outer walls is due to the throughflow velocity and volute central radius R_C:

$$\frac{dP}{dR} = \rho \frac{V_T^2}{R_C}$$

(6.5)

Figure 6.44 Total pressure variation over the cross sections at low mass flow (Pa) (from Ayder et al. 1993).

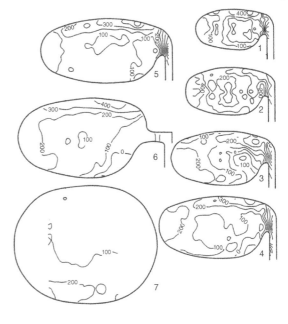

Figure 6.45 Swirl velocity distribution over the cross sections at low mass flow (from Ayder et al. 1993).

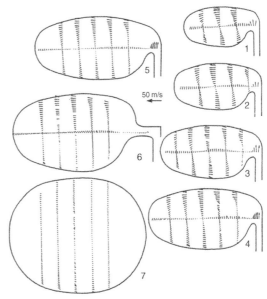

The measurements (Figure 6.42) indicate that the static pressure variation over the small volute cross sections near the tongue is mainly determined by the large swirl velocity and small radius r (Equation (6.3)) in combination with a low throughflow velocity. This gradually changes into an increased pressure variation between the inner and outer walls at the downstream cross sections as a consequence of the increasing throughflow velocity (Equation (6.5)) and decreasing swirl velocity in combination with a larger distance between the inner and outer walls of the volute.

The variation of throughflow velocity over the cross section (Figure 6.43) is function of the static and total pressure distribution (Equation (6.4)). In the sections near the inlet of the volute

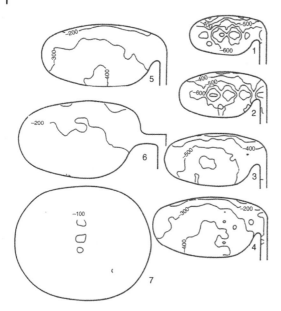

Figure 6.46 Static pressure variation over the cross sections at low mass flow (Pa) (from Ayder et al. 1993).

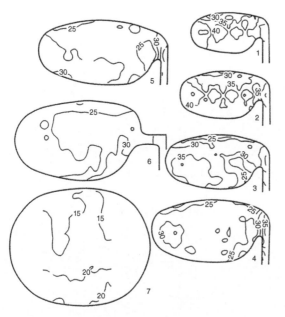

Figure 6.47 Throughflow velocities over the volute cross section at low mass flow (m/s) (from Ayder et al. 1993).

both P^0 and P decrease from the volute wall towards the center, resulting in a nearly uniform throughflow velocity distribution. At the downstream volute sections, the total pressure increases towards the center whereas the static pressure is decreasing. This results in a throughflow velocity at the center which is twice that at the volute inlet. The corresponding kinetic energy, which is four times larger than that at the inlet, is partially dissipated between sections 6 and 7 and is responsible for the high exit cone losses. As will be explained in Section 6.2.7, 1D volute models that assume a uniform flow over the cross sections require unrealistic loss coefficients to account for this.

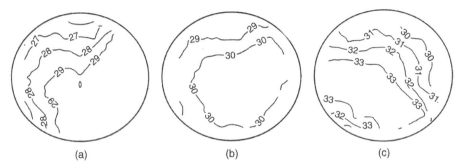

Figure 6.48 Total temperature ($T^o - T_1^o$) variations over the exit section of the compressor at (a) high, (b) medium, and (c) low mass flow (from Ayder 1993).

6.2.4.2 Medium and Low Mass Flow Measurements

The low circumferential pressure distortion at medium and low mass flow corresponds to almost design operation of the volute. The lower radial velocity at the volute inlet for the latter leads to smaller swirl velocities (Figure 6.45) and smaller crosswise static pressure gradients. The nearly constant swirl velocity near the walls results from the uniform radial velocity at the volute inlet. The forced vortex distribution at the center is due to internal shear. The corresponding energy dissipation results in only a small local decrease in total pressure at the center (Figure 6.44). Downstream of section 3, the size of this forced vortex core gradually increases to cover the whole cross section at the downstream measurement planes.

The small swirl velocity has only a small impact on the static pressure and throughflow velocity. The latter is almost exclusively determined by the circumferential curvature of the volute channel (Figures 6.46 and 6.47) and is zero in the straight last section.

6.2.4.3 Volute Outlet Measurements

The impeller exit circumferential pressure distortion at the volute off-design operation results in a circumferential variation of the energy input and hence of the total temperature. Streamlines do not mix very much inside a volute, as illustrated in Figure 6.1, and the variation of the inlet temperature is conserved up to the volute exit (Figure 6.48a,b). This non-uniformity of the flow at the volute exit is important in terms of performance measurement because the latter will change with the location where outlet pressure and temperature are measured. The largest temperature rise occurs at low mass flow due to a higher energy transfer from the impeller to the fluid. The uniform static pressure and temperature at the impeller exit excludes the possibility that the temperature variation measured at the low mass flow (Figure 6.48c) could be the consequence of a variation in work input. It can only be due to the heat loss in areas with very low velocity or separated flow, concentrated near the walls. This unphysical variation emphasizes the need for a thermal insulation of the model during tests.

6.2.5 Internal Rectangular Volutes

The same impeller has been combined with an internal volute of rectangular cross section downstream of a shorter vaneless diffuser (Figure 6.49). Internal volutes are used to reduce the dimensions and weight of the compressor. Rectangular volutes are easier to manufacture (eventually by welding). The following explains the influence of the cross sectional shape and reduction in the radius between the diffuser exit (R_4) and the volute center line (R_C) on the volute losses shown on Figures 6.20 and 6.21. As the diffuser is much shorter ($R_4/R_2 = 1.75$) than for the elliptic volute ($R_4/R_2 = 2.24$), the volute inlet velocities will be larger.

Figure 6.37 illustrates the impact of these geometry changes on the performance. The lower efficiency is due to the lower static pressure rise in the diffuser and the higher losses in the volute. The larger radial velocity at the diffuser exit is at the origin of larger swirl losses. The reduction of the cross section centerline radius in an internal volute creates an increase in the fluid throughflow velocity that partially annihilates the static pressure rise achieved in the vaneless diffuser. The larger throughflow and swirl velocity in the rectangular cross sections, with a smaller hydraulic diameter, further result in larger friction losses.

The larger cross section over diffuser exit radius ratio and the circumferentially decreasing central radius result in a design- and smaller than design mass flow operation. This is illustrated by the circumferential variation of the static pressure at the inlet and outlet of the diffuser (Figure 6.50). Except for a perturbation near the exit pipe ($\theta = 300–360°$), a nearly constant pressure distribution is observed at high mass flow operation (Figure 6.50a) indicating design point operation of the rectangular volute. The continuous static pressure rise along the volute circumference, at medium mass flow (Figure 6.50b), results from the flow deceleration taking place between the volute tongue and $\theta = 300°$. At low mass flow (Figure 6.50c) the static pressure increases only over the first third of the volute length. The nearly constant static pressure further downstream in the volute is the consequence of a flow separation, as will become evident from the detailed measurements.

The pressure perturbation observed at all three mass flows, for θ values between 300° and 360°, is because that part of the diffuser outlet is directly connected to the exit pipe (Figure 6.49). The local diffuser outlet pressure is fixed by the pressure variation between the outer and inner walls of the volute exit pipe. The higher pressure at $\theta = 270°$ is due to the centrifugal forces in a curved duct that have disappeared at $\theta = 360°$ where the static pressure is uniform over the cross section. The static pressure decrease is further enhanced by the additional fluid that enters the channel of constant cross section between the two sections.

The 3D flow field in the volute section is measured at the six circumferential positions shown in Figure 6.49. Radial traverses at five axial positions cover nearly the whole flow area. The detailed results for high to low mass flow are shown in Figures 6.52 to 6.56, respectively.

6.2.5.1 High Mass Flow Measurements

The velocity and pressure distributions, measured in cross sections 1 to 5, have a flow structure that is very similar to the one observed in the volute with elliptic cross section, i.e. the flow swirls around the center with a nearly forced vortex type of velocity distribution (Figure 6.52) and the centrifugal forces corresponding to the swirl are in equilibrium with a static pressure increase from the center towards the walls.

The following observations deserve special attention:

- The flow approaches the corners with a throughflow component V_T (perpendicular to the cross section) which is often much larger than the swirl velocity component V_S. Hence the flow experiences turning over an angle that is much larger than 90° (Figure 6.51) and the corner vortices are very weak.

- There is an apparent contradiction between the circumferentially uniform static pressure at the diffuser exit (Figure 6.50a) and a static pressure decrease in the cross sections from the volute inlet to outlet (Figure 6.53a). The change in minimum pressure from −1200 Pa in section 1 to −3000 Pa in section 5 corresponds to an increase of throughflow velocity from 60 m/s in section 1 to 95 m/s in section 5. The velocity increase from the outer to the inner wall of the volute is defined by the conservation of the angular momentum (Equation (6.2)). The corresponding decrease in the static pressure is a first cause of the lower efficiency of internal volutes.

- There is an accumulation of low energy fluid and a clockwise vortex exists in the hub–inner wall corner. This is a consequence of the secondary flow resulting from the circumferential

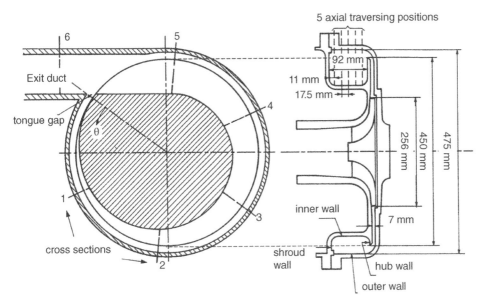

Figure 6.49 Geometry of the compressor with internal volute of rectangular cross section (from Ayder et al. 1991).

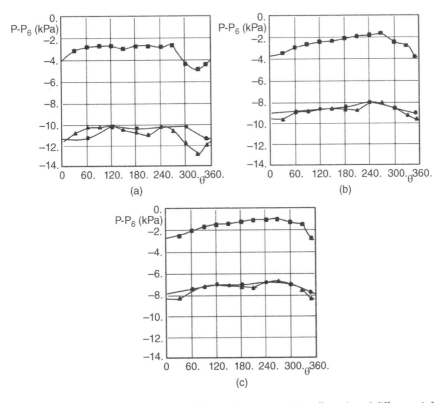

Figure 6.50 Circumferential variation of the static pressure at impeller exit and diffuser exit for (a) high mass flow, (b) medium mass flow, and (c) low mass flow (from Ayder et al. 1991).

curvature of the volute. The radial pressure gradient in the volute is a function of the inviscid throughflow velocity V_T and the circumferential curvature of the volute R_C. The fluid in the boundary layers has a smaller V_T but is subjected to the radial pressure gradient imposed by the larger free stream V_T. Hence it will move to the inner wall along streamlines with a smaller curvature radius. The low velocity fluid, on the hub wall, migrates backwards to a smaller radius and creates a vortex in the hub–inner wall corner (sections 2 to 5 on Figure 6.52a) that is opposite to the main swirl. This is at the origin of an accumulation of low total pressure fluid in that corner. On the shroud wall this motion is in the same direction as the main swirl velocity, which facilitates the transport of this low velocity fluid along that wall and further towards the hub–inner wall corner (sections 4 to 5 on Figure 6.53b).

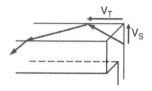

Figure 6.51 Flow in the corners of rectangular volutes.

The direction of the leakage flow under the tongue depends on the pressure difference between cross sections 5 and 1 (Figure 6.49). As the static pressure is much lower on the inner wall between sections 5 and 6 than in cross section 1 (Figure 6.53b), fluid will escape from the volute inlet towards the exit duct. Hence nothing prevents the diffuser outflow from filling the empty inlet section of the volute with high energy fluid from the diffuser. The result is an asymmetric vortex and a high total pressure in section 1.

The fluid leaving the diffuser at $315° < \theta < 360°$ directly enters the volute on the hub side between sections 5 and 6 (Figure 6.49) and blows the zone of high total pressure loss and low throughflow velocity towards the shroud–inner wall corner of section 6 (Figures 6.53b and 6.52b).

6.2.5.2 Medium Mass Flow Measurements

The medium mass flow is less than optimal for the volute. The circumferential increase of the inlet static pressure and lower radial component of the inlet velocity have a substantial impact on the flow pattern. The static pressure gradient is almost exclusively from the inner to the outer wall of a cross section because much less influenced by the smaller swirl velocity than by the trough flow velocity and circumferential curvature (Figure 6.55a).

A large zone of low total pressure fluid (Figure 6.55b) with large throughflow velocity but almost no swirl (Figure 6.54a,b) is already present near the shroud side of section 1. This structure originates from the leakage flow entering the volute through the tongue gap because the static pressure on the inner wall of section 5 is larger than in section 1. This low energy fluid migrates to the hub–inner wall corner (sections 2 to 5) because the inner to outer wall static pressure gradient prevents it from moving further around and, as already explained, is at the origin of the clockwise vortex in the hub–inner wall corner.

This low energy fluid has completely disappeared in section 6 (Figure 6.55a). It is aspirated through the tongue gap into the volute inlet because the average static pressure is lower in section 1 than on the inner wall of section 5. This is the origin of the very low P^o and absence of swirl at the shroud side of section 1 (Figure 6.54a).

The expected static pressure decrease, because of the increasing through flow velocity at decreasing radius (Equation (6.2)), is annihilated by a diffusion of the flow along the volute. The outcome is a circumferential increase of the static pressure along the volute outer wall (hence also at the diffuser outlet) and a small increase along the volute inner wall. The concave curvature prevents separation of the fluid on the outer wall. The small acceleration of the flow along the inner wall prevents separation of the fluid in spite of the convexity of that wall.

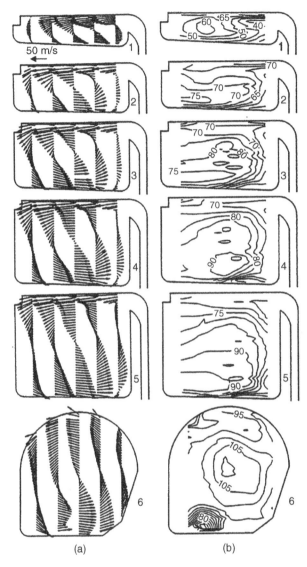

Figure 6.52 (a) Swirl velocity and (b) throughflow velocity variation over the volute cross section at high mass flow (from Ayder et al. 1991).

6.2.5.3 Low Mass Flow Measurements

The flow at low mass flow is characterized by a very low swirl velocity and large leakage flow at the tongue in combination with a large incoming tangential velocity. The latter decreases along the outer wall and is the origin of the static pressure increase up to section 3, from whereon the pressure remains nearly constant because of flow separation on the convex inner wall. Centrifugal forces push the high velocity fluid towards the outer wall, resulting in a growing zone of low energy fluid near the inner wall, filled by leakage flow at the tongue (Figure 6.56). A large zone of separated flow, with no structure and negative total pressure (because it is opposite to the probe direction), is clearly visible in section 6. It results from the extra diffusion between sections 5 and 6 because of the large flow leakage through the tongue gap.

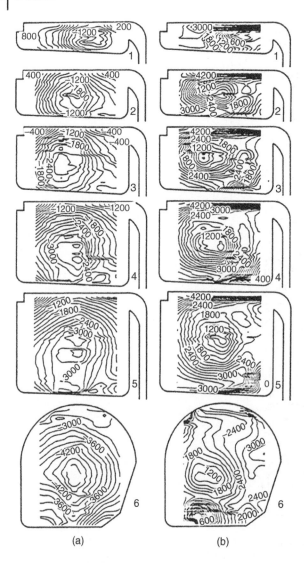

Figure 6.53 (a) Static pressure and (b) total pressure distribution over the volute cross section at high mass flow (from Ayder et al. 1991).

(a) (b)

In conclusion the main differences between the rectangular internal and central elliptic volute are as follows:

- A counter swirling vortex, generated in the hub/inner wall corner by the radial pressure gradient, not present in the central and external volutes, favoring the accumulation of the low energy fluid on the inner wall. Together with the decreasing curvature radius of that wall it is at the origin of flow separation at lower mass flows.
- Friction on the wall separating the diffuser from the internal volute creates extra losses in internal volutes. As will be discussed in Section 6.3.1, these losses can be reduced by shortening the vaneless diffuser upstream of the volute.
- The special way the outlet duct of an internal volute is connected to the diffuser exit results in a local perturbation of the circumferential pressure distribution for $300° < \theta < 360°$, even at design mass flow.
- The cross section area should be adapted to the change in velocity when changing the volute centerline radius (Mojaddam et al. 2012).

Figure 6.54 (a) Swirl velocity and (b) throughflow velocity variation over the volute cross section at medium mass flow (from Ayder et al. 1991).

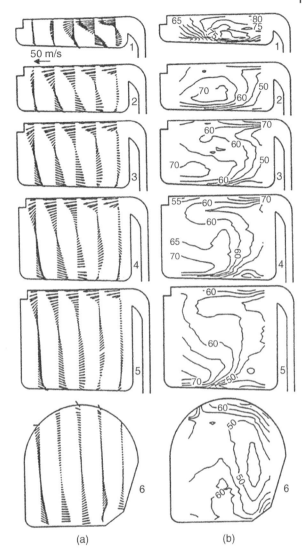

(a)　　　　　(b)

6.2.6　Volute Cross Sectional Shape

The 3D flow in symmetric volutes depends on the volute geometry as well as on the spanwise variation of the impeller outlet flow (Peck 1951; Hübl 1975). Figure 6.57 shows the vortex structures for different operating points. The weak asymmetric vortex observed at shut-off mass flow (a) is likely to change into two symmetric vortices at design mass flow (b) and back to a strong asymmetric vortex (c) at large mass flow. The direction of rotation of the asymmetric vortex depends on the spanwise variation of the inlet conditions. The one on (c) is the consequence of the larger radial velocity at the hub side of the impeller exit and remains attached to that wall by the Coanda effect. The different flow patterns have different losses and the switch from one type to another may lead to discontinuities in the performance curve.

A symmetric volute can be approximated by two asymmetric ones (Figure 6.58) with the same structure on each side of the volute. The impact on energy input is very small, but the efficiency is 2.5% lower than for eccentric volute (Hübl 1975).

Figure 6.55 (a) Static pressure and (b) total pressure distribution over the volute cross section at medium mass flow (from Ayder et al. 1991).

Previous observations are confirmed by Heinrich and Schwarze (2017), who also made a numerical optimization of the volute cross sectional shape. The best performance was obtained with the slightly eccentric volute shown on Figure 6.59. The flow leaving the diffuser is not squeezed together by the primary vortex and the lower radial velocity results in lower radial velocity dump losses.

6.2.7 Volute Performance

The overall performance of a volute is characterized by the static pressure recovery coefficient

$$CP = \frac{\widetilde{P}_6 - \widetilde{P}_4}{\widetilde{P}_4^o - \widetilde{P}_4}$$

(6.6)

and total pressure loss coefficient

$$\omega = \frac{\widetilde{P}_4^o - \widetilde{P}_6^o}{\widetilde{P}_4^o - \widetilde{P}_4}$$

(6.7)

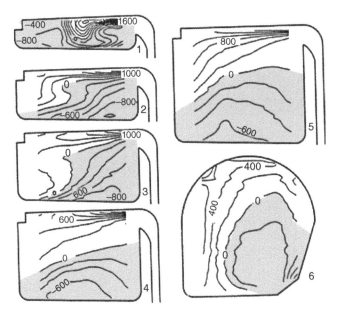

Figure 6.56 Total pressure distribution over the volute cross section at low mass flow (from Ayder et al. 1991).

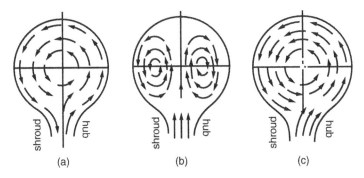

Figure 6.57 Vortex structures in a symmetric volute at different operating points (from Peck 1951).

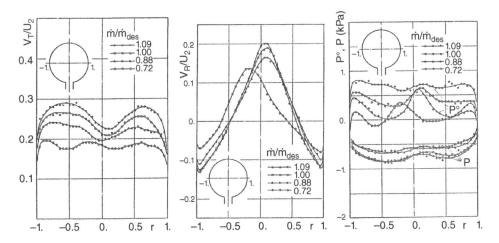

Figure 6.58 Flow and pressure distribution in a symmetric volute at different operating points (from Hübl 1975).

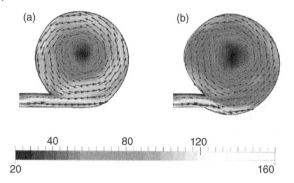

(a)   (b)

Figure 6.59 Velocity distribution in (a) a conventional and (b) an optimized eccentric volute cross section shape (from Heinrich and Schwarze 2017).

calculated as a function of the averaged flow quantities at the volute inlet and outlet.

Equation (4.8) shows that for a given inlet section and flow conditions and for a fixed outlet section the sum of $CP + \omega$ is constant and independent of the flow structure between the inlet and outlet. It provides the following approximation of the volute outlet over inlet velocity ratio:

$$1 - (\omega + CP) = \frac{V_6^2}{V_4^2} \tag{6.8}$$

6.2.7.1 Experimental Results

The overall performance of the elliptic volute, based on the mass averaged flow quantities at sections 4 and 6, is shown in Figure 6.60a. One observes a decrease in the static pressure ($CP < 0$) at all mass flows. This is due to the high friction losses and the absence of diffusion in a volute that is too small at all operating points. At high mass flow, the losses in the elliptic volute are even larger than the kinetic energy available at the inlet ($\omega > 1$). Together with the increase in the kinetic energy ($\omega + CP < 0$) in Equation (6.8), this results in an important decrease in static pressure ($CP = -2$).

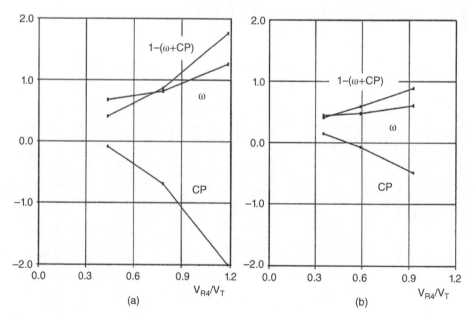

Figure 6.60 Overall performance of (a) elliptic and (b) rectangular volutes.

The overall performance of the rectangular internal volute is shown in Figure 6.60b. The static pressure rise is higher than in the elliptic volute, resulting in a positive *CP* at minimum mass flow. The total pressure losses are lower and the flow decelerates at all operating points ($\omega + CP > 0$). One should not conclude from the higher pressure rise coefficient that internal rectangular volutes are superior to central elliptical ones. The present rectangular volute is only better adapted to the impeller and vaneless diffuser. The elliptic volute is too small for the test impeller due to the large diffusion in the (too) long vaneless diffuser.

6.2.7.2 Performance Predictions

Another way of looking at the volute impeller interaction that helps us to understand the losses is illustrated in Figure 6.61. The flow leaving the impeller or vaneless diffuser at lower than optimal mass flow has a small radial and a large tangential velocity component. Transporting that small amount of fluid in the volute requires only a small throughflow velocity V_T. The deceleration from V_{u4} at the diffuser exit to a smaller V_T at the center of the volute cross section results in a pressure rise and diffusion losses. The small radial velocity gives rise to a small swirling motion in the volute. The latter is dissipated by internal shear and wall friction, and a second contribution to the volute losses.

The flow leaving the impeller or vaneless diffuser at higher than optimum mass flow has a large radial and smaller tangential velocity. Transporting that large amount of fluid in the volute requires a large throughflow velocity. The corresponding acceleration from V_{u4} at the diffuser exit to V_T at the volute cross section center results in a pressure decrease. This partial destruction of the diffusion that took place in the diffuser results in extra friction losses. The dissipation of the swirl energy, resulting from the large radial velocity component, creates losses that are much larger than at low mass flow.

The experimental results, discussed in previous sections, make clear that there is not much interest in verifying loss prediction models that assume uniform flow without swirl over the volute cross section. One of the models that accounts for the main features of the flow in volutes is that of Japikse (1982), extended by Weber and Koronowski (1986). It accounts for following contributions to the losses:

- meridional velocity dump losses ($\Delta P^o_{MV\,DL}$)
- friction losses (ΔP^o_f)
- tangential velocity dump losses ($\Delta P^o_{TV\,DL}$).

The meridional velocity dump losses are based on the assumption that the kinetic energy associated with the swirl resulting from the meridional velocity at the volute inlet is dissipated.

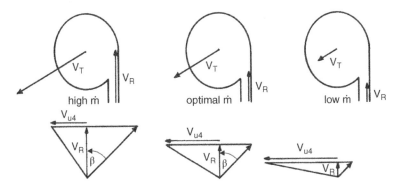

Figure 6.61 Acceleration and diffusion of tangential and radial velocity between impeller exit and volute.

This amount of losses is expressed by

$$\Delta P^o_{MV\,DL} = \frac{\rho\,V^2_{R4}}{2} \tag{6.9}$$

Senoo and Kawaguchi (1983) mention the possibility of reducing these swirling losses in a collector by adding circumferential fences, turning the flow in the circumferential direction.

The friction losses are a function of the wall roughness, the hydraulic diameter of the volute channel (DH), the path length of the fluid (LH) within the volute (which is assumed to be equal to the length of the volute channel), and the throughflow velocity (V_T). Only the latter is considered for the friction losses since the dissipation of the swirl velocity is already accounted for by the meridional velocity dump losses:

$$\Delta P^o_f = C_f \frac{LH}{DH} \frac{\rho V^2_T}{2} \tag{6.10}$$

The friction coefficient (C_f) depends on the Reynolds number and relative surface roughness. It can be obtained from the standard friction charts for pipes (Equation (1.16)). Javed and Kamphues (2014) studied the impact of volute surface roughness on turbocharger compressor performance. Although the upstream components (impeller and diffuser) turn out to be the main contributors to entropy production by friction, they experimentally verified that an increase in the volute roughness from 45 μm to 100 μm resulted in an efficiency drop of 1%, a value that is twice that predicted by CFD.

The tangential velocity dump losses are assumed to be zero at high mass flow, when the tangential velocity accelerates from the diffuser outlet to the volute outlet ($V_{u4} < V_{T5}$). At low mass flow ($V_{u4} > V_{T5}$) the total pressure losses are assumed to be equivalent to those of a sudden expansion mixing process:

$$\Delta P^o_{TV\,DL} = \omega_T \frac{\rho(V_{u4} - V_{T5})^2}{2} \tag{6.11}$$

where $\omega_T = 1$ for decelerating flows and $\omega_T = 0$ for accelerating flows.

The exit cone losses, between the volute and compressor outlet, are those for a conical diffuser:

$$\Delta P^o_{EC} = \omega_{EC} \frac{\rho}{2} \frac{(V_{T5} - V_{T6})^2}{2} \tag{6.12}$$

where ω_{EC} varies from 0.15 for a gradual expansion at an opening angle of 10° to a value of the order 1.1 for an opening angle of 60° when the flow is fully separated. Since the opening angle of a well designed volute exit cone should not exceed 10°, a constant value of 0.15 is proposed by Weber and Koronowski (1986). Larger opening angles may be possible when the remaining swirl keeps the flow against the exit cone wall. However, too large a swirl may result in a very low static pressure and even return flow in the center.

Variation of *CP* and ω as a function of the inlet swirl parameter V_{u4}/V_{R4}, calculated and measured in turbocharger volutes with different outlet over inlet area ratios (*AR*), is shown in Figure 6.62. The parameter *AR* is defined by

$$AR = r^2_5/2R_4 b_4 \tag{6.13}$$

It appears from Figure 6.62 that the previous model provides a useful basis for the prediction of the volute static pressure rise but is less accurate in terms of losses. The overestimation of loss coefficient (ω) for high mass flows (small values of V_{u4}/V_{R4}) is likely to be due to the

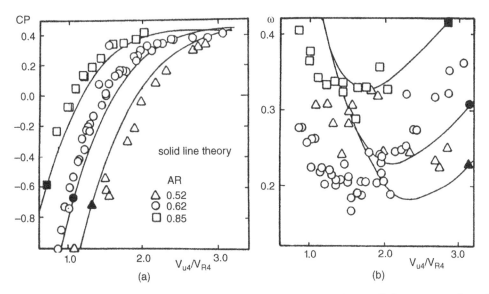

Figure 6.62 (a) Volute pressure recovery coefficient *CP* and (b) pressure loss coefficient ω, comparison of theory and experimental data (from Japikse 1982).

remaining swirl at the exit of the volute not included in the experimental losses and assumed fully dissipated in the model. The swirl losses are less important at lower mass flows, which explains the higher accuracy of the predictions for large values of V_{u4}/V_{R4}.

The experimental and theoretical investigations of Weber and Koronowski (1986) reveal that the modeling of the tangential velocity dump losses without taking into account the variation of the central radius of the volute channel causes an incorrect prediction of the losses, especially for the internal type of volutes ($R_C < R_4$). They therefore propose to introduce an intermediate station in the volute channel at the 50% collection point (point where 50% of the mass flow has entered the volute). The throughflow velocity at this intermediate station (section 4.5) and at the volute exit (section 5) are calculated from continuity, assuming a uniform velocity over the cross section:

$$V_{T4.5} = \dot{m}/(2\rho A_{4.5}) \qquad\qquad V_{T5} = \dot{m}/(\rho A_5) \qquad\qquad (6.14)$$

The fictitious tangential velocities of the flow entering both halves of the volute ($V_{u(4-4.5)}$ and $V_{u(4-5)}$) are calculated from the conservation of angular momentum between the stations (4.5 and 5) and the volute inlet (4):

$$V_{u(4-4.5)} = V_{T4.5}R_{4.5}/R_4 \qquad\qquad V_{u(4-5)} = V_{T5}R_5/R_4 \qquad\qquad (6.15)$$

In this extended model, the tangential velocity dump losses for the first half of the mass flow are defined as follows:

- If the tangential velocity at the volute inlet (V_{u4}) is larger than the fictitious one ($V_{u(4-4.5)}$), the flow is decelerating in the first part of the volute with following contribution to the losses:

$$\Delta P^o_{TV\,DL4.5} = \omega_T \frac{\rho(V^2_{u4} - V^2_{u(4-4.5)})}{4} \qquad\qquad (6.16)$$

where $\omega_T = 0.5$ for volutes and 1.0 for a plenum.

- If V_{u4} is smaller than $V_{u(4-4.5)}$, i.e. when the flow is accelerating, the following losses are introduced:

$$\Delta P^o_{TV\,DL4.5} = \frac{\rho(V_{u4} - V_{u(4-4.5)})^2}{4} \qquad (6.17)$$

The same expressions are used for the second half of the mass flow entering the volute further downstream where the pressure losses $\Delta P^o_{TV\,DL5}$ are a function of V_{u4} and $V_{u(4-5)}$. The total tangential velocity dump losses are given by

$$\Delta P^o_{TV\,DL} = \Delta P^o_{TV\,DL4.5} + \Delta P^o_{TV\,DL5} \qquad (6.18)$$

The exit cone losses are as specified by Equation (6.12). Weber and Koronowski (1986) compared this model with several experimental data available for a turbocharger and an industrial compressor. The prediction of the tangential velocity dump losses showed a better agreement than the simple model of Japikse (1982), but they concluded that the method is subject to further improvements. Eynon and Whitfield (2000) proposed a correction for the extra diffusion between the diffuser exit and volute tongue and for non-zero incidence on the volute tongue.

6.2.7.3 Detailed Evaluation of Volute Loss Model

The main discrepancies between predictions and experiments, however, result from the assumptions of uniform flow conditions at the volute inlet and outlet, conditions that are not satisfied at off-design operation. The detailed experimental data of the elliptic volute presented in Section 6.2.4 have been used to evaluate the previous model and to find the possible sources of discrepancies between experiments and predictions (Ayder et al. 1993). The results are summarized in Table 6.2.

The losses shown in the first column at each mass flow are the ones calculated according to Equations (6.9) to (6.12) assuming circumferentially uniform inlet flow predicted by a 1D analysis model. The second column differs from the first one because the meridional velocity dump losses are obtained by circumferentially integrating the losses defined by Equation (6.9) using the spanwise mass averaged velocities measured at the volute inlet. The losses are considerably higher than the ones obtained with circumferentially uniform velocity (column 1). Both

Table 6.2 Summary of measured and calculated volute and exit cone total pressure losses (Pa).

	High mass flow			Medium mass flow			Low mass flow		
	Calculated		Measured	Calculated		Measured	Calculated		Measured
	1D	Data	Measured	1D	Data	Measured	1D	Data	Measured
ΔP^o_f	2483	2483	2483	982	982	982	281	281	281
ΔP^o_{MVDL}	1453	2486	1939	492	791	674	94	157.5	126
ΔP^o_{TVDL}	98	98	98	9	9	9	215	215	215
ω_T	0.2			0.2			0.5		
ΔP^o_{4-5}	4034	5067	4520	1478	1777	1660	590	653	622
ΔP^o_{MC}			547			117			31
ΔP^o_{EC}	2414	1381	1381	696	397	397	89	26	26
$(V_5 - V_6)^2$	2230	2230	5654	754	754	1517	145	145	329
ω_{EC}	1.1	0.62	0.24	0.92	0.53	0.26	0.61	0.18	0.08
ΔP^o_{TO}	6448	6448	6448	2174	2174	2174	679	679	679
CP_{EC}			0.578			0.483			0.55

Figure 6.63 Impact of the circumferential distortion on the meridional velocity dump losses.

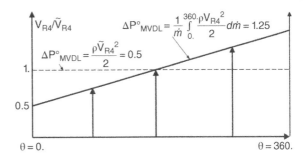

radial velocity distributions correspond to the same mass flow but the non-uniformity of the flow results in higher swirl losses, as illustrated in Figure 6.63. Regions of larger than average V_R have not only higher losses but also higher mass flow. This is not compensated for by the lower swirl losses in regions with low V_R because less mass flow is involved. The difference between the two approaches can be very large.

The third column contains the measured volute swirl losses. They are smaller than the ones in the second column by the amount ΔP^o_{MC} corresponding to the residual swirl remaining at the volute exit. The swirl is not yet fully dissipated at the volute exit and dissipation continues further downstream in the exit cone and piping.

The friction and tangential velocity dump losses have been quantified by subtracting the measured (column 3) ΔP^o_{MVDL} from the measured total volute losses ΔP^o_{4-5}. The friction losses are defined by Equation (6.10) with $C_f = 0.019$ based on the average Reynolds number ($Re = 3.10^5$) and relative surface roughness ($k_s/D_H = 0.001$). The same value of C_f is used in all three calculations. The remaining losses are assumed to be the tangential velocity dump losses ($\Delta P^o_{TV\,DL}$). According to the theory the latter should be zero at high and medium mass flow because the velocity is increasing from the volute inlet to outlet. However, as explained in Section (6.2.3), the strong swirling flow at high mass flow creates a large distortion of the throughflow velocity and the average value of V_T may not provide correct loss evaluations. Substituting the average V_T in Equation (6.11) provides a value of $\omega_T = 0.2$, which is reasonable for the accelerating fluid at high and medium mass flow. These losses are anyway less than 1.5% of the total volute losses and could well be the consequence of an inaccurate estimation of the other losses. In the case of decelerating flows (low mass flow) the resulting tangential flow losses are well predicted by a coefficient $\omega_T = 0.5$, which is identical to the one proposed by Weber and Koronowski (1986) for diffusing flows.

The exit cone losses concern only the diffusion losses, exclusive of the remaining swirl losses. They are obtained by subtracting the volute losses from the measured total losses and the accuracy depends on the predicted value of the other volute losses. Equation (6.12) does not account for the non-uniformity of the flow. Losses have therefore been related to the change in the mass averaged kinetic energy between inlet and outlet. Very high loss coefficients are needed to predict the exit cone losses by the 1D model (column 1) because of an underestimation of the upstream losses and an underestimation of the kinetic energy at the inlet of the exit cone.

Calculating the experimental exit cone losses as a function of the inlet dynamic pressure, predicted by the improved volute loss prediction (column 2), shows more reasonable values ($\omega_{EC} = 0.62$ and 0.53 at maximum and medium mass flow, respectively). They are higher than in column 3 because they are based on a lower dynamic pressure by neglecting the non-uniformity of the volute exit, i.e. cone inlet, flow.

The exit cone losses are very small at low mass flow. The values of ω_{EC} based on the measured data disagree with the value at the other operating points. This is likely to be due to perturbed inlet flow and/or a reduced measurement accuracy at very low values of V_T.

Consistently large values of static pressure rise coefficient CP_{EC} are observed at all operating points. This may be attributed to the stabilizing effect of the swirling flow in conical diffusers. Swirl creates a velocity component on the concave wall with radius r and, as explained in Section (3.2.2), this increases the turbulence in the boundary layer and reduces the risk of flow separation.

As a conclusion one can state that the main shortcomings of the 1D volute prediction models result from the lack of knowledge about the real velocity distribution at the different cross sections. Depending on the operating point, both the swirl and throughflow velocity may not be uniform and the kinetic energy available at each cross section is larger than the one evaluated with a uniform flow. Considering these distortions in the loss modeling results in a considerable improvement of the predictions.

The main difficulty is an accurate prediction of the circumferential pressure and flow distortion at off-design operation of the volute. This requires an impeller response model. A more complete model, taking into account the circumferential pressure distortion and crosswise non-uniformity of the flow, is discussed in Section 7.4.

6.2.8 3D analysis of Volute Flow

The large number of parameters influencing the performance of a centrifugal compressor volute and the large manufacturing cost of the complex 3D geometries prohibits a systematic experimental investigation of the flow. Reliable analysis methods are therefore of great help in determining the influence of the different design parameters on the volute flow and losses.

Modern CFD solvers exhibit no problems in the calculation of the rotational flows described in previous sections. They all show solutions, with the swirling velocity going to zero at the center of the vortices, and finite values for the pressure, density, and axial velocity (Reunanen 2001; Steglich et al. 2008; Javed and Kamphues 2014; Giachi et al. 2014). The main problems are the definition of the volute geometry and the numerical grid.

Experimental results have shown that the flow in volutes is affected more by the losses in the core than by the wall boundary layers. It is therefore important to have a sufficiently fine grid also in the center of the cross section to capture correctly the local shear forces. Boundary layers are very thin and their influence may be limited to the shear forces on the walls.

The following results have been obtained by a simplified flow solver in which viscous effects are simulated by second-order dissipation terms and wall shear forces (Denton 1986). A detailed description of the method, boundary conditions and artificial dissipation is given by Ayder et al. (1994). As will be demonstrated, the method provides a correct image of the flow and emphasis will be on specific problems related to volutes.

The numerical calculations have been validated by calculating the 3D swirling flow inside the elliptic volute, for which the experimental results are described in Section 6.2.4. The flow field is extended upstream to the impeller outlet to minimize the influence of the inlet conditions on the tongue and volute flow (Figure 6.64). The circumferential variation of the spanwise uniform inlet boundary conditions (P_2^o, T_2^o, and V_{u2}/V_{R2}) are derived from the measured values at the volute inlet, corrected for the change in radius between diffuser inlet and outlet. The circumferential shift of total pressure and temperature, due to the tangential velocity in the diffuser, is accounted for by the diffuser model described in Section 7.4. The non-uniform static pressure at the outlet of the exit cone is defined by the radial equilibrium between the pressure and centrifugal forces.

The circumferential variation of the radial velocity (V_R) and static pressure (P) at the diffuser and volute inlet are a result of the calculation. Spanwise averaged results at the volute inlet are compared with experimental ones in Figure 6.65 and show a very good agreement for all operating conditions.

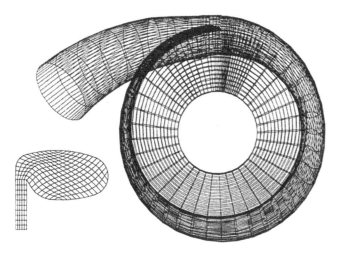

Figure 6.64 3D view of the discretized elliptic volute and diffuser.

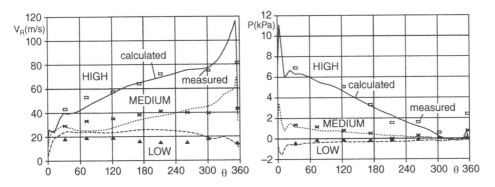

Figure 6.65 Comparison between calculated and measured radial velocity and static pressure at the volute inlet for high, medium, and low mass flow (from Ayder et al. 1994).

The calculated swirl velocity (V_S), throughflow velocity (V_T), total pressure (P^o), and static pressure (P) over the sections 1, 5, and 7 (Figures 6.66 and 6.67) show a good agreement with the experimental results in Figures 6.40 to 6.43. The total and static pressures are slightly overpredicted in cross section 5 (Figure 6.67). Measurements at section 7 do not cover the whole cross section and it is therefore not possible to find out what is the origin of the higher experimental losses. They may be due to separation at the tongue.

Calculated results have been used to visualize streamlines at medium mass flow inside the diffuser and volute, starting in four circumferential positions at the diffuser inlet (Figure 6.1a). One observes that the fluid entering the volute at a position close to the tongue (A) remains in the center of the volute until the compressor exit. Fluid entering the volute further downstream (B, C, and D) swirls at a larger radius, which confirms the volute flow model described in Section 6.2.3.

6.3 Volute-diffuser Optimization

The strong interaction between the diffuser and volute in terms of circumferential pressure distortion and losses suggests that both should be designed/optimized simultaneously.

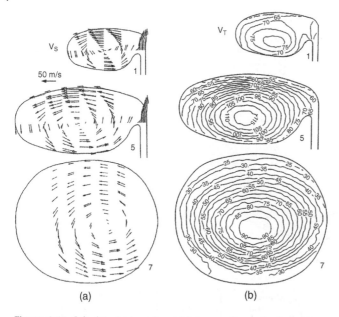

Figure 6.66 Calculated (a) swirl and (b) throughflow velocity at high mass flow (from Ayder et al. 1994).

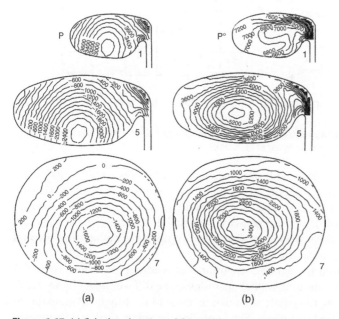

Figure 6.67 (a) Calculated static and (b) total pressure at high mass flow (from Ayder et al. 1994).

Furthermore, the boundary conditions are far from obvious when the volute is operating at off-design with large circumferential distortions. Two geometry modifications for performance improvement are discussed in the context of a more integrated approach to compressor design (Steglich et al. 2008). Both have been applied to the compressor with the internal volute described by Hagelstein et al. (2000).

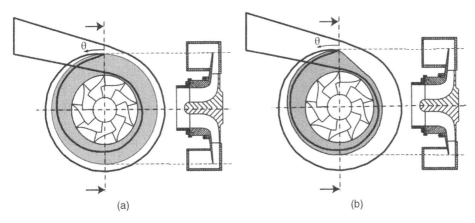

Figure 6.68 Geometry of a compressor with (a) axisymmetric and (b) non-axisymmetric diffuser (from Steglich et al. 2008).

6.3.1 Non-axisymmetric Diffuser

A major source of losses in a diffuser/internal volute combination is the reacceleration of the fluid in the volute after it has been decelerated in the diffuser. The wall between a vaneless diffuser and an internal volute does not contribute to a net increase in the pressure but the friction on both sides results in extra losses. They can be eliminated by removing that wall where geometrically possible. The result is an asymmetric vaneless diffuser as shown in Figure 6.68b, where the vaneless diffuser radius ratio R_4/R_2 decreases from 1.5 for small values of θ to 1.07 at $\theta = 300°$. In addition to a reduction in the friction losses, this also results in a small increase in the volute cross section area without a change in the overall dimensions. In absence of a wall between the two, the radial pressure gradient in the diffuser and volute will coincide.

According to Equation (6.8) $(CP + \omega)_{2-6}$ depends only on the diffuser inlet and volute outlet velocity and is independent of the geometry between them. The area of the compressor exit flange, where the performance is measured, is identical for all geometries, and any decrease of the diffuser or volute loss coefficient will give rise to an equal increase in the static pressure coefficient.

Shortening the vaneless diffuser transforms the internal volute into a partially external one (Figure 6.68b) in which the additional pressure rise in the volute compensates for the lower pressure rise in the diffuser. A combination of both effects results in performance improvements at high and medium mass flow, as illustrated in Figure 6.69. The efficiency and pressure ratio at the design point are considerably higher than those measured with a concentric diffuser. The higher volute inlet tangential velocity at minimum mass flow and the increase in the volute cross section area lead to a flow separation downstream of the shorter diffuser. The lower pressure recovery in the diffuser is no longer compensated for by a larger pressure rise in the volute, and at the origin of the positive slope of the pressure rise/mass flow curve at low mass flow. Limiting the useful operating range to the mass flows between maximum pressure rise and choking, it turns out that the higher efficiency and pressure ratio is at the expense of a smaller operating range.

Another consequence of the non-axisymmetry of the diffuser is a more pronounced circumferential static pressure distortion when operating away from the nominal operating point, causing slightly higher dynamic loads on the blades and bearings.

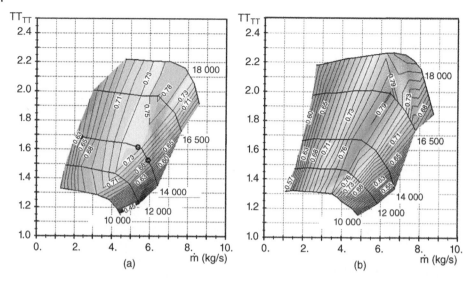

Figure 6.69 Experimental performance map of a compressor with (a) an axisymmetric diffuser and (b) a non-axisymmetric diffuser (from Steglich et al. 2008).

6.3.2 Increased Diffuser Exit Width

The 1D loss model explained in Section 6.2.7 indicates that the friction losses, tangential velocity dump losses, and radial velocity dump losses are the main contributions to the volute losses. The first two are proportional to the square of the diffuser exit tangential velocity and influenced by the circumferential volute cross section area. The latter is proportional to the square of the radial velocity at the diffuser outlet. Its contribution to the volute losses directly depends on the diffuser exit width and is particularly important at higher mass flows. However, an increase in the diffuser exit width to minimize the radial velocity component has a negative effect on vaneless diffuser stability (Section 8.2).

Vaned diffusers have enhanced stability because the pressure gradient is better aligned with the velocity. They achieve a larger pressure rise by decreasing the tangential velocity component, i.e. by turning the flow more radially. At unchanged diffuser width, the radial exit velocity (V_{R4}) and hence the radial velocity dump losses will be unchanged, except for a small impact of the increased density.

A smaller reduction of the tangential velocity component results in a smaller pressure rise in the diffuser but may allow a larger decrease in the radial velocity component and lower radial velocity dump losses by increasing the diffuser width. A larger volute inlet tangential velocity allows a smaller volute cross section, but with increased friction losses.

An increase in the diffuser exit width without reducing the operating range is possible by installing low solidity vanes, as shown in Figure 6.70. The convergent vaneless diffuser with exit radius ratio $R_4/R_2 = 1.5$ but only 13% area increase is replaced by a parallel one with 50% area increase. These modifications result in a decrease in both the tangential and radial velocity components at the diffuser exit without compromising on diffuser stability. There is room for an optimization between a decrease of V_T or V_R

The experimental results (Figure 6.71) show an increase in the static pressure rise between the rotor and diffuser exit CP_{2-4}. However, this increase in pressure rise is associated with a reduction in the outlet kinetic energy and an increase in the total pressure loss coefficient ω_{2-4} at low $\dot{m}$ due to the incidence losses on the low solidity diffuser (LSD) vanes.

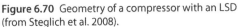

Figure 6.70 Geometry of a compressor with an LSD (from Steglich et al. 2008).

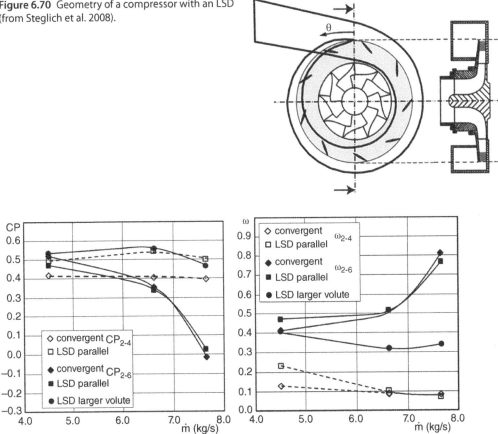

Figure 6.71 Predicted static pressure and total pressure loss coefficient for different geometries and mass flows (from Steglich et al. 2008).

The lower tangential velocity V_{T4} at the diffuser exit is insufficiently compensated for by the higher fluid density. Not adapting the volute cross section makes the fluid reaccelerate, resulting in a circumferential decrease in the static pressure and an increase in the friction losses in the volute. The extra static pressure rise achieved in the LSD is then annihilated in the volute. The consequence is shown in Figure 6.71 where, in spite of the larger diffuser pressure rise CP_{2-4}, the overall static pressure rise CP_{2-6} is almost equal to that of the convergent vaneless diffuser. Adapting the volute size to the LSD exit flow conditions requires in the present application a 34% increase in the cross section. Comparing the experimental performance in Figure 6.72a with the ones of both vaneless diffusers (Figure 6.69) one observes a considerable improvement when using an LSD in combination with a larger volute. The pressure ratio and operating range approach those obtained for a concentric diffuser with a larger external volute (Figure 6.72b).

The higher performance of the external volute (Figure 6.72b) is due to the extra deceleration, resulting from the increase in radius between the diffuser exit and the center of the volute cross section. A similar deceleration has already taken place upstream of the volute when the flow is deflected by the LSD vanes. This explains how a pressure ratio, comparable to the one obtained with an external volute, could be obtained with a smaller LSD–internal volute combination. The weaker secondary vortex in the volute cross section also results in a lower blockage and losses.

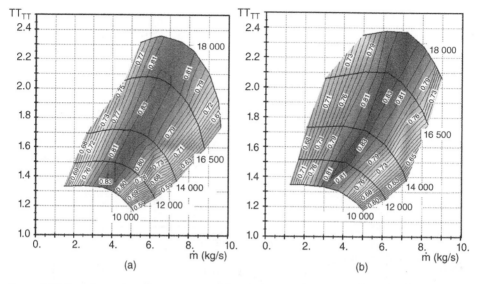

Figure 6.72 Experimental performance map of the compressor with (a) an LSD and larger internal volute and (b) a baseline diffuser and external volute (from Steglich et al. 2008).

The shift in surge limit to higher mass flows, when increasing from 14 000 RPM to 16 500 RPM, is due to a narrowing of the LSD incidence range with increasing inflow Mach number. The LSD and internal volute have a higher efficiency at low RPM and a lower efficiency at high RPM for the same reasons. The apparent increase in choking mass flow is the consequence of a modification of the test facility allowing measurements at lower outlet pressure.

The performance can be further improved by optimizing the circumferential position of the LSD vanes relative to the volute tongue (Iancu et al. 2008).

Increasing the vaned diffuser solidity allows a further decrease in the velocity components at the inlet of the volute, but at the cost of increased diffuser friction losses and a reduced operating range. Maximizing the diffuser and volute performance separately leads to conflicting design changes and a simultaneous optimization of both is recommended.

7

Impeller Response to Outlet Distortion

The flow perturbation by the diffuser vanes, a volute operating at off-design, and diffuser rotating stall are the main sources of the circumferential pressure distortion downstream of a centrifugal impeller. The direct consequences are the time varying flow and forces in the rotating blade passages (relative frame of reference) and a circumferential variation of the velocity at the diffuser inlet (absolute frame of reference). This chapter concentrates on the unsteady impeller response to the downstream pressure distortion, imposed by volutes operating at off-design. After a short overview of some experimental data, the following topics will be discussed:

- the flow mechanisms governing unsteady incompressible and compressible flow
- 1D and full 3D unsteady flow calculation models, including the frozen rotor model
- the forces acting on the impeller shaft and blades, and the influence of the impeller geometry.

The forces generated by a flow distortion are proportional to the fluid density and hence are much larger in pumps than in compressors. This explains the intensive research that has been done on pumps in contrast to more sporadic work on compressors. However, the larger forces in compressors operating at high inlet pressure and the use of magnetic bearings and squeezed film damper bearings with limited radial and axial trust have been an incentive to also study the forces in compressors.

Although unsteady phenomena are different in compressible and incompressible fluids, some experimental data obtained in pump volutes are also presented here. Geometrical similarities help to clarify some aspects and compensate for the lack of information about these unsteady phenomena in compressors. Except if specified otherwise, all data and discussions are for compressible fluids.

Unsteady stage interaction is important because it gives rise to:

- extra losses due to the unsteady rotor flow and additional vorticity in the diffuser because of non-uniform inflow
- a reduced operating range due to a local increase in blade loading and destabilization of the diffuser flow by the extra vorticity
- radial forces on the shaft because the circumferentially varying blade forces do not compensate each other
- noise and vibrations due to time varying flow and blade forces.

The appropriate method to study these phenomena depends on the degree of unsteadiness. For incompressible fluids, unsteadiness is characterized by the reduced frequency (Figure 7.1):

$$\Omega_R = \frac{fL}{V} \tag{7.1}$$

Design and Analysis of Centrifugal Compressors, First Edition. René Van den Braembussche.

Figure 7.1 Schematic representation of a blade to blade channel.

where L is the length of the flow channel under consideration, V is the velocity by which a perturbation is transported through the channel, and f is the frequency of the perturbation.

The acoustic Strouhal number is more appropriate to characterize unsteady effects in compressible fluids, where the pressure perturbations propagate at the speed of sound. It is defined by the product of the reduced frequency and Mach number:

$$S_r = \Omega_R M = \frac{fL}{a} \tag{7.2}$$

It relates the time needed for a pressure wave to travel at the speed of sound a over a distance L, to the period $1/f$ of the pressure perturbation.

In an impeller, L is defined as the length of a blade passage where f is defined by

$$f = (\Omega - \omega_\sigma)N_s \tag{7.3}$$

where Ω is the rotational speed of the impeller and ω_σ is the rotational speed of the perturbation (rotating stall cells) and is zero in case of a vaned diffuser or volute pressure distortion. N_s is the number of stall cells or number of stator vanes ($N_s = 1$ for a volute).

Unsteady effects are small for $S_r < 0.1$ and the flow can be approximated by a steady calculation. For $S_r > 0.1$ accurate results can be obtained only by means of unsteady calculations. The amplitude of the flow perturbations is likely to grow up to $S_r = 0.5$ and decrease again for larger values. Vaned diffuser impeller interactions give rise to high frequency pressure oscillations at Strouhal numbers that are much larger than 1. The time such an outlet perturbation acts on an impeller flow channel is much shorter and the impact is much smaller and limited to the area near the interface (Baghdadi 1977; Dawes 1995).

7.1 Experimental Observations

Many papers present the circumferential pressure distortion, measured at the impeller exit. Only a few also present the circumferential variation of velocity, flow angle, etc. (Binder and Knapp 1936; Pfau, 1967; Thomas and Salas 1986; Sideris 1988).

Comparing the static and total pressure with the total temperature variations in different circumferential locations at the exit of a compressor impeller (Figure 7.2), Pfau (1967) claimed that the out of phase changes of P^o and T^o near the tongue are a consequence of two counter-rotating vortices at the impeller inlet (counter-rotating because of zero total vorticity upstream of the impeller). At lower than optimum mass flow, one of the vortices creates a negative swirl resulting in an increased incidence, hence a higher temperature rise but lower pressure rise because of increased incidence losses. The other vortex results in a decrease in incidence (lower temperature rise but increased efficiency and higher pressure rise). Hence the change in total pressure rise does not follow the same trend as the change in temperature rise (Figure 7.2).

The upstream propagation of the distortion is confirmed by Uchida et al. (1987) (Figure 7.3) who show at the diffuser outlet:

- a typical, almost linear, static pressure rise from the volute tongue to the volute outlet at minimum mass flow
- a nearly constant pressure at medium mass flow, corresponding to the volute design point

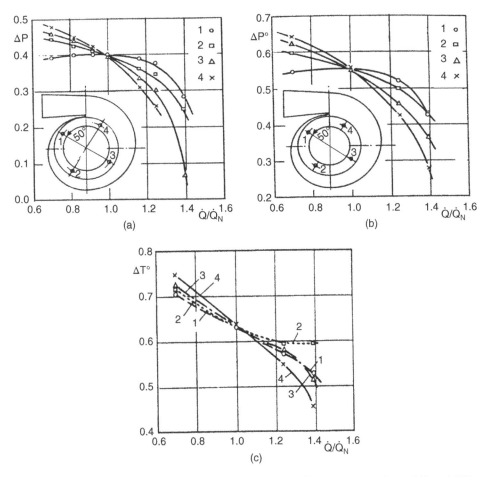

Figure 7.2 Variation of (a) static pressure, (b) total pressure, and (c) temperature rise at $R/R_2 = 1.15$ in a centrifugal impeller volute combination (from Pfau 1967).

Figure 7.3 Circumferential static pressure distribution at the impeller inlet, outlet, and vaneless diffuser outlet (from Uchida et al. 1987).

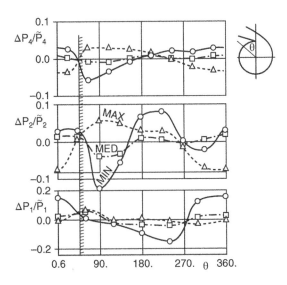

- a nearly linear pressure decrease at maximum mass flow, followed by an abrupt pressure rise in the tongue region.

The wavy variation of the static pressure at the rotor exit has a larger amplitude than at the diffuser exit and is in line with the strong vortices described by Pfau (1967). The inlet distortion is observed only at lower than optimum mass flows. It is likely that inducer choking at higher mass flows acts as a filter.

Worster (1963) concludes from theoretical considerations that the volute flow distortion results in local return flow in the impeller. The main consequences are an increase in the losses and less guidance of the flow because the trailing edge becomes a leading edge. The incidence angle of the return flow is a function of the absolute flow angle imposed by the volute shape or tongue and is no longer defined by the blade trailing edge shape ($\beta_{2,bl}$) and the slip factor.

Based on theoretical considerations about the pump impeller response to a circumferential pressure distortion, Bowerman and Acosta (1957) concluded that a perfect guidance of the flow by the blades ($\beta_{2,fl} = C^{te}$) is not possible at all circumferential positions for backward leaning blades. Assuming a constant relative flow angle ($\beta_{2,fl}$) a local increase in the radial velocity would result in an equivalent increase in the mass flow but with reduced work input ($U_2 V_{u2} < U_2 \tilde{V}_{u2}$ on Figure 7.4). A smaller part of the mass exits the impeller at the area with lower radial velocity, where at constant $\beta_{2,fl}$ there would be a local increase in work input ($U_2 V_{u2} > U_2 \tilde{V}_{u2}$). The product of the mass flow and enthalpy rise is the energy input. The sum of the contribution in the areas with lower and higher mass flow is smaller than the one corresponding to the (uniform) average velocities $\widetilde{V}_{u2}$ and $\widetilde{V}_{R2}$. Hence a reduction in the torque should occur at distorted outflow. This is in contradiction with the measurements of Bowerman and Acosta (1957), who observed an increase in torque at circumferentially distorted outflow and concluded that $\beta_{2,fl}$ cannot remain uniform along the circumference. As the total energy input increases, the measured decrease in the manometric head outside the volute design point operation must be due to increased losses.

Sideris et al. (1987a) made crossed hot wire measurements at the exit of a centrifugal impeller with 20 full blades at 40° backward lean, an outlet diameter of 370 mm and rotating at 10 000 RPM corresponding to $S_R = 0.1$. The vaneless diffuser with radius ratio 1.67 is followed by a volute (Figure 7.5).

The circumferential pressure distributions, corresponding to minimum, medium, and maximum mass flow, shown in Figure 7.6, indicate that the volute is well adapted for minimum mass flow. The variation of the spanwise averaged values of V_{R2}, V_{u2}, and corresponding relative flow angle $\beta_{2,fl}$ are shown in Figure 7.7. The variation of $\widetilde{V}_{R2}$ at maximum mass flow is much larger than that of $\widetilde{V}_{u2}$ and not in line with the response model visualized in Figure 7.4 at constant $\beta_{2,fl}$. $\widetilde{V}_{u2}$ increases with increasing $\widetilde{V}_{R2}$. The value of $\tilde{\beta}_{2,fl}$ is almost constant at minimum mass flow, as expected for a circumferential uniform pressure, but shows a 15° variation at maximum $\dot{m}$.

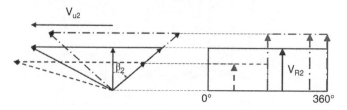

Figure 7.4 Impeller outlet velocity triangles at different mass flows and constant $\beta_{2,fl}$.

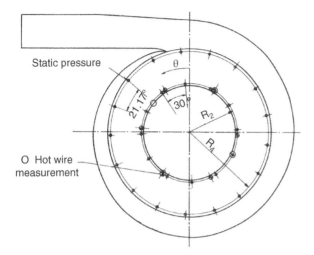

Figure 7.5 Cross sectional view of the centrifugal compressor impeller and volute (from Sideris et al. 1987a).

Figure 7.6 Circumferential variation of the static pressure at the diffuser inlet and outlet (from Sideris et al. 1987a).

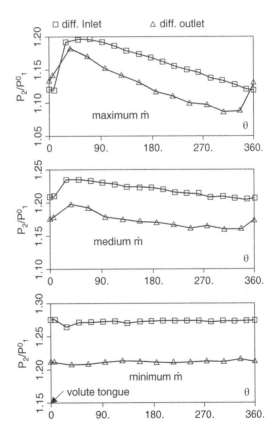

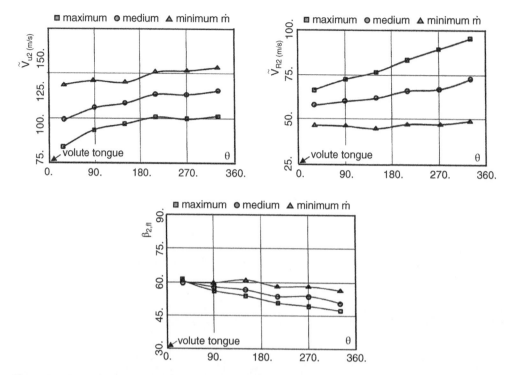

Figure 7.7 Circumferential variation of the spanwise averaged radial and tangential component of the velocity and relative flow angle at the diffuser inlet for different mass flows (from Sideris et al. 1987a).

Previous measurements allow the following conclusions:

- The unsteady impeller response results in a change in the relative outlet flow angle $\beta_{2,fl}$ such that V_{u2} is almost unaffected by the change in V_{R2}.
- The outlet pressure distortion has an impact on the impeller inlet flow by inducing two counter-rotating vortices. Impeller choking may block this upstream propagation.
- Contrary to what is assumed in most theoretical models, the energy input with a circumferentially distorted outlet pressure is larger than with uniform outlet conditions.

7.2 Theoretical Predictions

There are two main interaction mechanisms between the blade rows of a turbomachine:

- A wake interaction resulting from the convection of the wakes originating from the upstream blade row into the downstream one. If the distance is sufficiently small the wakes of the rotating impeller will be chopped by the downstream diffuser vanes before they are mixed out (Figure 7.8). This generates entropy waves affecting the viscous aspects of the flow (turbulence and losses) in the diffuser.
- A potential interaction by the static pressure distortion resulting from the blade loading in one blade row on the flow in the neighboring ones (upstream and/or downstream). The consequence is an unsteady flow in both components generated by the relative motion of the adjacent pressure fields. It can be described by the superposition of three flow components: a steady axisymmetric flow field, a non-uniform steady flow field according to the rotor frame

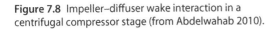
Figure 7.8 Impeller–diffuser wake interaction in a centrifugal compressor stage (from Abdelwahab 2010).

of reference, and a non-uniform steady flow field according to a stator frame of reference (Giles 1991). The non-uniform pressure field, which is steady in the relative frame of reference, is unsteady in the absolute one. The non-uniform steady pressure field in the absolute frame of reference is unsteady in the relative one. The main consequence is a circumferential and timewise variation of the velocity and loading, hence of the forces that are acting on the blades and vanes.

The potential interaction becomes more prominent at increasing pitch over gap ratio because:

- larger pitches allow larger circumferential pressure distortions at off-design operation that interact for a longer time on the neighboring blade rows
- smaller gaps between the blade rows do not allow much attenuation of the distortions.

Hence the interaction between a volute and an impeller can be very strong and generate large forces on the blades and shaft.

The calculation of this unsteady blade row interaction becomes more expensive when more blade passages are needed to obtain a periodic calculation domain. This is particularly true for volutes where the pitch of the stator spans the full circumference of the rotor. Hence all of the rotor blades must be analyzed together with the complete diffuser and volute. Furthermore, the potential interaction is the dominant effect for these machines because the wakes are rapidly mixed out in the diffuser.

Although the real flow is unsteady, most turbomachinery calculations are made with a steady flow solver to save on time and cost. They give an acceptable approximation at volute design point operation or with vaned diffusers. However, the error can be particularly large in the case of a vaneless diffuser with the downstream volute at off-design operation.

Unsteady compressible flow is different from an unsteady incompressible one. In compressible fluids, unsteadiness manifests itself by pressure waves propagating in the flow field at the speed of sound and reflecting on the boundaries. The degree of unsteadiness is characterized by the acoustic Strouhal number (Equation (7.2)). Unsteady flows of non-compressible fluids require a simultaneous change in the velocity over the whole flow field to satisfy continuity at every instant of time. Any change in the impeller outlet velocity imposes a proportional change in the inlet velocity because incompressible fluid cannot temporarily be accumulated in one location by a change in density. This is at the basis of the rigid column theory. The time change of momentum ($\rho V \Delta V$) can be very large, leading to large pressure pulses.

In what follows, different models to predict the effect of a circumferential pressure distortion on the flow in an impeller are presented. The emphasis is on the approximations and assumptions that are made and their impact on the accuracy of the results.

7.2.1   1D Model

A 1D model was presented by Frigne et al. (1985) to evaluate the unsteady impeller flow resulting from a rotating stall in the diffuser. A similar model by Lorett and Gopalakrishnan (1986) evaluates the response of a pump impeller to the pressure distortion by a volute at off-design operation. It assumes that the flow is 2D, incompressible, and inviscid. The whole mass of fluid in one blade channel accelerates or decelerates simultaneously. The model predicts the timewise variation of the radial velocity at the outlet of the rotating impeller for a prescribed circumferential static pressure distribution $P_2(\theta)$. The fluid accelerates when the instantaneous pressure rise in the impeller is larger than the locally imposed outlet pressure and decelerates when the outlet pressure is larger than the one corresponding to the impeller pressure rise:

$$\frac{\partial V_{R2}}{\partial t} = -\frac{U_2}{R_2}\frac{\partial V_{R2}}{\partial \theta} + \frac{\cos\beta_{bl}}{EL}\left(U_2 V_{u2} - U_1 V_{u1} + \frac{P_1^o - P_2(\theta)}{\rho} - \frac{V_{R2}^2 + V_{u2}^2}{2}\right) \tag{7.4}$$

EL is the effective length of the flow channel and is defined as

$$EL = \int_1^2 \frac{A_2}{A}dL \tag{7.5}$$

An estimation of the exit relative flow angle allows the calculation of V_{u2}:

$$V_{u2} = U_2 - V_{R2}\tan\beta_{2,fl} \tag{7.6}$$

The radial velocity variation can be obtained by a numerical integration in time, using an explicit one-step Lax–Wendroff discretization scheme:

$$V_{R2,j}^{k+1} = V_{R2,j}^k - \frac{CFL}{2}\left(V_{R2,j+1}^k - V_{R2,j-1}^k\right) + \frac{CFL^2}{2}\left(V_{R2,j+1}^k - 2V_{R2,j}^k + V_{R2,j-1}^k\right)$$
$$+ \frac{\cos\beta_{bl}}{EL}\Delta t\left[U_2 V_{u2,j}^k - U_1 V_{u1,j}^k + \frac{P_1^o}{\rho} - \frac{P_{2j}^k}{\rho} + \frac{1}{2}\left(V_{R2,j}^{2^k} + V_{u2,j}^{2^k}\right)\right] \tag{7.7}$$

where $\Delta t = \Delta\theta / \Omega$ to be time accurate, k is the index of the time step, $\Delta\theta = \frac{2\pi}{NUEL}$, the angular displacement of the impeller per time step, and j is the number of the NUEL angular positions of the impeller where the velocity is calculated.

Figure 7.9 shows a comparison between the experimental and calculated V_{R2} and V_{u2} distributions at the exit of the impeller operating with different outlet pressure distortions. The predictions are based on the assumption that the exit flow angle $\beta_{2,fl}$ is constant in Equation (7.6). The agreement is not very good. The largest discrepancy is at minimum mass flow, where the pressure distortion is largest. In reality, the measured values of V_{u2} are almost independent of the change in radial velocity, indicating that $\beta_2 \neq C^{te}$.

Calculations with the experimental $\beta_{2,fl}$ distribution (Figure 7.10a), corresponding to an almost constant V_{u2}, show very good agreement (Figure 7.10b). This proves that the change in outlet velocity is partly due to a change in relative velocity modulus and partly due to a change in the impeller outlet flow angle, for which an analytical prediction model is not yet available. The best approximations go in the direction of a constant tangential velocity for a varying radial velocity.

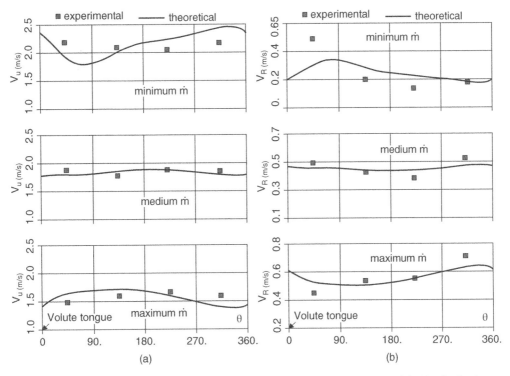

Figure 7.9 Comparison between the measured and calculated (a) V_{u2} distribution and (b) V_{R2} distribution at three different mass flows for constant β_2 (from Sideris et al. 1987b).

7.2.2 CFD: Mixing Plane Approach

Computational fluid dynamics (CFD) solves the flow equations on grid systems linked to the rotating and non-rotating blade rows. The simplest way to couple them is by the mixing plane approach, in which the circumferentially averaged flow quantities at the inlet/outlet of one stage (impeller, diffuser or volute) are imposed as outlet/inlet boundary conditions of the neighboring stages. The main advantages are that only steady flow calculations are required and that the periodicity of the flow allows the numerical domain to be limited to one pitch in each blade row. The main shortcomings are that the circumferential flow distortions, by the diffuser vanes or volute, are not accounted for and only steady blade forces can be predicted. The circumferential periodicity excludes any radial force on the shaft. This approach is well known and commonly used in design procedures, and will therefore not be discussed in detail.

However, attention must be given to inconsistencies that may result from the circumferential averaging of the flow quantities. The transfer of information at the inlet and outlet of the numerical domains is governed by the characteristic theory. The circumferentially averaged total flow conditions (P^o and T^o) and an indication of the flow direction or tangential momentum at the outlet of the upstream domain are imposed as inlet boundary conditions on the downstream domain. The calculated circumferentially averaged static pressure at the inlet of the downstream domain is imposed as outlet boundary conditions of the upstream one. This is physically correct as long as the velocity has a component in the downstream direction (i.e. $V_R > 0$ at the impeller-diffuser interface). This may not always be the case when the spanwise variation of the impeller outflow results in return flow at the hub or shroud side of the diffuser inlet. Imposing the static pressure on the downstream boundary in locations where the real flow re-enters

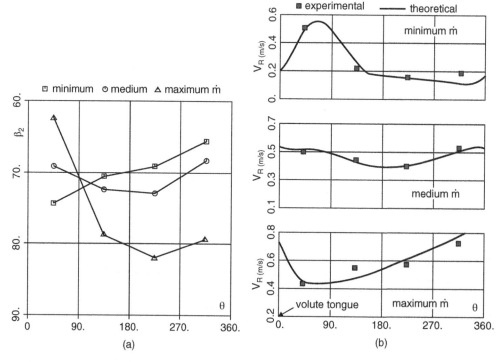

Figure 7.10 Comparison between the measured and calculated V_{R2} distribution for three different mass flows using the experimental β_2 distribution (from Sideris et al. 1987b).

the numerical domain is not in accordance with the characteristics theory, hence is physically incorrect. The consequence may be an erroneous prediction of the head and efficiency.

Liu et al. (2010) compared results of mixing plane calculations with those of time accurate unsteady calculations. They demonstrated that the mixing plane coupling at the impeller/vaned diffuser interface predicts the overall performance very well at design operation but that the error increases at off-design operation. Detailed Mach number distributions at the diffuser inlet and downstream divergent channel showed large discrepancies at all operating points. They attributed these local discrepancies to the absence of any upstream influence of the diffuser pressure distortion on the impeller flow because of the circumferential averaging. This upstream influence is not negligible because of the small vaneless space. It may be particularly large in transonic vaned diffusers by the upstream shock extension normal to the very tangential vane suction side. They estimated that the upstream influence of the diffuser on the mass averaged total pressure extends over 30% of the impeller blade chord.

Large discrepancies between the detailed Mach number distribution predicted by unsteady and mixing plane calculations at off-design operation, including erroneous location of the separated flow zone, are confirmed by Benichou and Trébinjac (2016). Such a difference may have a large impact on the predicted stability limit (Marsan et al. 2015).

Everitt and Spakovszky (2013) emphasize the importance of a physically correct averaging of the impeller outflow conditions. They advocate calculating the diffuser inlet conditions from the conservation of mass, momentum, and energy at the impeller outlet (Section 3.3.4) and imposing a diffuser inlet flow angle defined by $\alpha = \arctan(\widetilde{V_u}/\widetilde{V_R})$. The different ways of averaging flow quantities and minimizing the errors on mass, momentum, and energy transfer that may result from it have been discussed in detail by Holmes (2008).

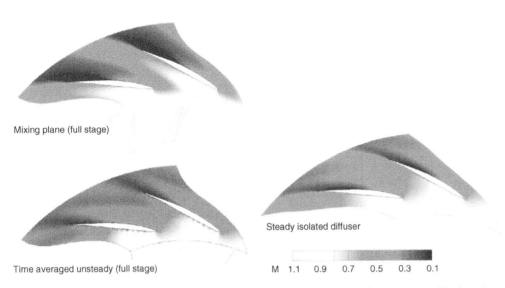

Mixing plane (full stage)

Steady isolated diffuser

Time averaged unsteady (full stage)

M   1.1   0.9   0.7   0.5   0.3   0.1

Figure 7.11 Vaned diffuser flow at midspan by different prediction methods (from Everitt and Spakovszky 2013).

Everitt and Spakovszky (2013) also claim that a diffuser-only calculation provides a more accurate prediction of the vaned diffuser flow than a mixing plane calculation (Figure 7.11). The inlet conditions of the isolated diffuser calculations are the ones obtained from a single channel impeller calculation with a pitch and spanwise uniform static pressure at the outlet. These results differ from those of a mixing plane calculation because the spanwise variations are also eliminated. The consequence is a better approximation of the suction side expansion and the absence of a too large separation zone on the vane pressure side predicted by the mixing plane calculations.

Experimental and numerical observations by Baghdadi (1977) and Dawes (1995) confirm that the spanwise variation of the circumferentially averaged impeller outlet conditions has a larger impact on the mixing plane calculations of the impeller flow than the high frequency unsteadiness resulting from the pitchwise non-uniformity of the impeller outflow and that this may be responsible for local flow reversal at the impeller hub or shroud side.

Imposing a circumferentially averaged static pressure at the outlet boundary of the upstream impeller means that the pressure may locally be different from the one calculated at the inlet of the next blade row and unphysical discontinuities will appear on the iso-pressure graphs. The same applies for the other averaged flow conditions that are imposed at the inlet or outlet boundaries of a neighboring numerical domain.

7.2.3 3D Unsteady Flow Calculations

Full 3D unsteady time accurate Navier–Stokes calculations are needed for a correct prediction of the rotor stator interaction. Special problems are the non-periodicity of the flow and the definition of the time step to assure accuracy. The time step must be the same in the whole numerical domain, to be time accurate, and related to the rotational speed of the rotor:

$$\Delta t = \frac{2\pi}{\Omega \, NUEL} \tag{7.8}$$

Since the downstream static pressure is not uniform around the circumference, the flow variables will be different on each periodic grid line so that the classical periodicity conditions can no longer be applied. An impeller blade will experience the same outlet flow conditions only after it has rotated over a period of the downstream pressure distortion.

There are three basic methods to handle this phenomenon numerically. The first one is to extend the numerical domain in the circumferential direction until periodic boundaries are obtained. In case of an impeller volute interaction, periodicity occurs only every 360° and the numerical domain must cover the whole impeller and volute. For 3D flows this requires sufficiently large computer storage because all flow variables at all grid points have to be stored in memory.

Phase-lagged periodicity conditions assume that the unsteadiness is only due to the rotation of the impeller (Erdos and Alzner 1977). This allows consideration of only one blade channel as computational domain. Storing in each impeller position the flow conditions close to the periodic grid boundaries makes them available as boundary conditions for the flow calculations in neighboring channels at the time the blade channel has moved over one pitch or one rotation minus one pitch. Convergence is improved when the numerical domain is extended to more than one pitch (Sideris et al. 1987a).

Local inclination in time is an alternative numerical technique, developed by Giles (1988, 1991), for rotor stator interaction calculations without the need to analyze the whole numerical domain at once (Zhou and Cai 2007). The effect of inclining time is that the relative speed of the characteristic waves is modified. This technique, originally developed for rotor stator interactions, can also be used to accelerate the convergence of numerical schemes.

Non-linear harmonic coupling is a way to drastically reduce the computational effort for unsteady multistage turbomachinery calculations. The method is based on solving the steady flow transport equations for the time mean flow and for the time harmonics. The periodic perturbation in a blade row is calculated in the frequency domain by time-marching to the pseudo steady state solution of the transport equations associated with the blade passing frequency and higher harmonics. The number of harmonics (up to 5) is a user input. It is claimed that this approach allows a two to three orders of magnitude gain in CPU compared to full unsteady calculations. However, this technique is limited to stage interactions with a well-defined periodicity and is not suited to rotating stall calculations (He and Ning 1998; Vilmin et al. 2006, 2013).

The following results are obtained by a 3D unsteady time-dependent Euler solver for compressible flows (Fatsis 1995). The present computer capacity allows these calculations to be done by means of a more accurate Navier–Stokes solver in a reasonable time scale. However, the simplification to inviscid flow allows a clearer illustration of the dynamics of the flow because it is not perturbed by viscous phenomena. Results are shown for two different impellers (Figure 7.12).

7.2.3.1 Impeller with 20 Full Blades

The first impeller has 20 full blades at 40° backward lean (Fatsis et al. 1995). The experimentally defined circumferential static pressure distribution corresponding to maximum mass flow (Figure 7.6) is imposed as an outflow boundary condition. The variation of the shroud Mach number distribution during one rotation at $S_R = 0.25$ is shown on Figure 7.13. The changes with circumferential positions are due to a pressure wave traveling twice upstream and downstream in the blade channel during one rotation.

The flow is choked at the shroud streamsurface for circumferential positions $0° < \theta_r < 90°$. In this case the pressure wave generated by the sudden pressure increase at the tongue propagates upstream and modifies the extent of the supersonic flow region by changing the position and strength of the shock. The latter has propagated up to the leading edge and the shock has disappeared at $\theta_r = 135°$, and the complete flow field is subsonic. The flow progressively reaccelerates at higher values of θ and choking appears again at $\theta_r = 270°$, but for

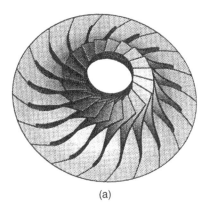
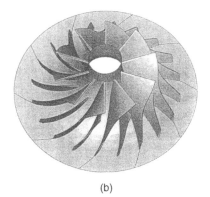

(a)   (b)

Figure 7.12 3D view of the centrifugal impeller with (a) full blades and (b) splitter blades.

a shorter time. Although the imposed pressure distortion has only one maximum per rotation, one observes two maxima per rotation in the Mach number distribution, one at 45° and one at 270°. This illustrates how the pressure wave, reflected at the leading and trailing edges, travels twice upstream and downstream during one rotation. This is in agreement with the value of the acoustic Strouhal number $S_r = 0.25$ expressing that a pressure wave can travel four times the length of the blade channel during one rotation.

The classical periodicity conditions (equal Mach number or pressure on suction and pressure side at the trailing edge) are no longer satisfied. This is because the calculations are made in a channel and the Mach numbers plotted are the ones on the pressure and suction side of two adjacent blades.

7.2.3.2 Impeller with Splitter Vanes

Figure 7.12 shows a centrifugal impeller with 10 full and 10 splitter blades with 30° backward lean. Shown in Figure 7.14 are the iso-Mach contours at one instant of time on the hub surface of the impeller–volute combination at a larger than optimal mass flow. One observes a large variation of the Mach number between the different circumferential positions. A maximum leading edge Mach number at $\theta_r = 0°$ and a second one with smaller amplitude at $\theta_r = 180°$ are the consequence of a pressure wave traveling twice upstream and downstream in the blade channel during one rotation and in line with an acoustic Strouhal number of 0.25.

These upstream and downstream traveling waves are somewhat less clearly visible in an impeller with splitter blades than in one with only full blades because they are perturbed by reflection at the leading edge of the splitter blades and by the phase difference between the waves traveling on each side of the splitters.

7.2.4 Inlet and Outlet Flow Distortion

The wavy reaction of the impeller flow to a circumferential variation of the outlet pressure is clearly reflected in the spanwise mass averaged relative flow angle (β_2), absolute flow angle (α_2), radial velocity (V_{R2}), and absolute tangential velocity (V_{u2}) at $R/R_2 = 1.028$ for the impeller with 20 full blades operating at maximum mass flow with $S_r = 0.25$ (Figure 7.15). The dashed line at 33° indicates the position of the volute tongue. In the same figure, stars (*) represent the quasi steady solution obtained by imposing the local value of the distorted pressure as outlet condition in a 3D steady flow calculation of one blade channel. One can conclude that quasi steady calculations cannot give reliable results, even at the relatively low acoustic Strouhal number of 0.25.

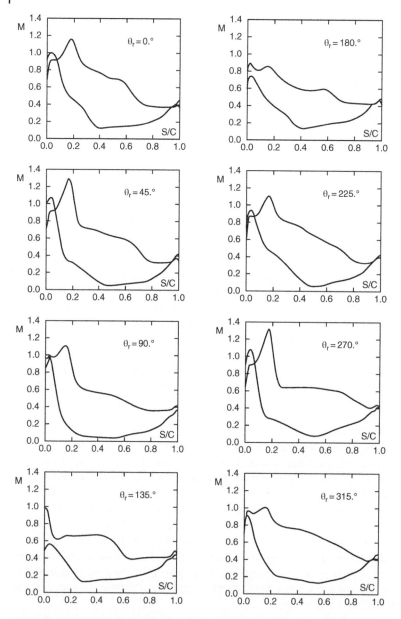

Figure 7.13 Variation of the relative Mach number versus the blade length at the shroud streamsurface of the impeller with 20 full blades operating at maximum mass flow and $S_r = 0.25$ (from Fatsis et al. 1995).

The β_2 distribution in Figure 7.15 shows a significant variation close to the tongue, due to a sudden increase in the static pressure at this position. A second peak occurs at $\theta = 200°$, a location where the imposed outlet static pressure changes smoothly. This second distortion must therefore be attributed to the reflection of the pressure waves at the inlet and outlet of the impeller channel. Its amplitude is smaller because of damping in the blade passage. The other flow parameters also show a bimodal variation per rotation with larger amplitude than the quasi steady results. These results make clear that the assumption of constant outlet relative

Figure 7.14 Iso-Mach lines on the hub contour of the splitter blade impeller at larger than optimum mass flow and $S_r = 0.25$ (from Hillewaert et al. 1999).

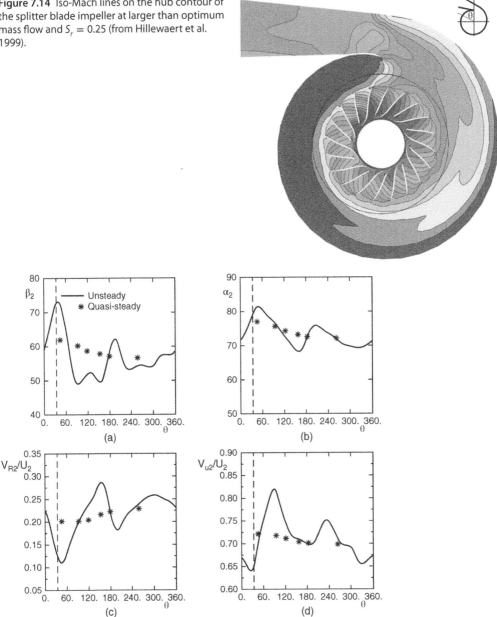

Figure 7.15 Circumferential distortion of the spanwise mass averaged (a) α_2, (b) β_2, (c) V_{R2}, and (d) V_{u2} at $R/R_2 = 1.028$ of the the impeller with 20 full blades operating at maximum mass flow and $S_r = 0.25$ (------ tongue position) (from Fatsis et al. 1995).

flow angle, generally used in 1D models, will not provide accurate predictions. A possible alternative at Strouhal numbers greater than 0.1 is a constant $V_{u,2}$.

A similar transformation of a one wave pressure distortion into a two wave impeller out-flow is also experimentally observed by Uchida et al. (1987) (Figure 7.3) and Yang et al. (2010), confirming that the unsteadiness is governed by pressure waves traveling up and down in the

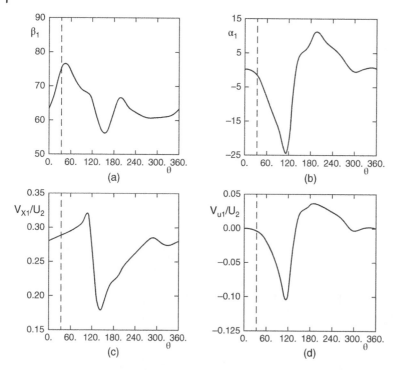

Figure 7.16 Circumferential distortion of the mass averaged (a) β_1, (b) α_1, (c) V_{x1}, and (d) V_{u_1} at the leading edge of the impeller with 20 full blades operating at maximum mass flow and $S_r = 0.25$ (- - - - tongue position) (from Fatsis et al. 1995).

flow channel. Although the predictions based on the 3D unsteady inviscid model may some-what overestimate the real unsteady flow changes, they are not unrealistic because the unsteady Navier–Stokes calculations by Liu et al. (2010) show variations of the absolute flow angle at the inlet of a vaned diffuser that are of the same order of magnitude.

Large circumferential variations of the mass averaged β_1, α_1 and of the non-dimensional V_{x1}, V_{u1} are observed at the leading edge of the impeller (Figure 7.16). Although the shroud section is choked, one observes a slight increase in the average inlet axial velocity for $0° < \theta < 120°$. This increase in mass flow is possible because of the simultaneous decrease in the relative flow angle (Figure 7.16a), resulting from an increased pre-rotation of the incoming flow (Figure 7.16d). One should also keep in mind that the values are averaged over the blade height from hub to shroud. The axial velocity can still increase at the hub even when the shroud section is choked.

The sudden decrease in axial velocity occurring at $\theta = 120°$ is when the impeller unchokes because the pressure wave that started when the trailing edge was at $\theta = 30°$ has reached the leading edge. This 90° phase shift between the point of minimum outlet radial velocity and the point of minimum axial velocity at the leading edge results from the impeller rotation during the time needed by a pressure wave to travel from the trailing edge to the leading edge. It is followed by a gradual increase in the axial velocity up to the position at $\theta = 270°$ where choking occurs again.

The alternating sign of the absolute tangential velocity corresponds to two counter-rotating vortices upstream of the impeller where the total circulation has not yet changed. These findings are in line with the observations of Pfau (1967).

7.2.4.1 Parametric Study

The variation of the impeller flow with the acoustic Strouhal number has been analyzed by imposing sinusoidal pressure waves around the periphery at the outlet of the impeller with 20 full blades. Calculations are for fully subsonic (10 000 RPM) and transonic (25 000 RPM) impeller flow. A one wave per circumference perturbation corresponds to respectively $S_r = 0.1$ and $S_r = 0.25$. Different Strouhal numbers are obtained by varying the number of waves around the periphery.

The variation with S_r of $\Delta\beta_2$ and ΔV_{R2} at radius $R/R_2 = 1.028$ (Figure 7.17) shows a gradual increase in $\Delta\beta_2$ up to the acoustic Strouhal number $S_r = 0.5$, i.e. when the time a pressure wave needs to travel back and forth in the impeller equals the time for one rotation. In this way the pressure wave is enforced at each rotation and its amplitude is only limited by viscous damping. At higher or lower Strouhal numbers, the expansion waves returning to the outlet counteract the waves generated by the circumferential pressure perturbation and the amplitude decreases.

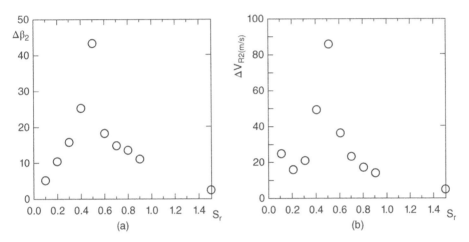

Figure 7.17 Variation of the (a) amplitude $\Delta\beta_2$ and (b) ΔV_{R2} as a function of the acoustic Strouhal number (S_r) for the impeller rotating at 10 000 RPM (from Fatsis et al. 1995).

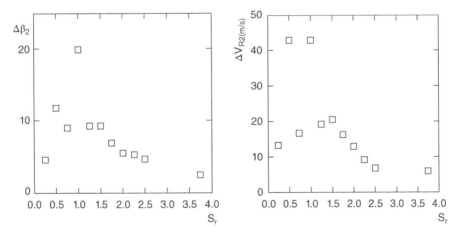

Figure 7.18 Variation of the (a) amplitude $\Delta\beta_2$ and (b) ΔV_{R2} as a function of the acoustic Strouhal number (S_r) for the impeller rotating at 25 000 RPM (from Fatsis et al. 1995).

At $S_r = 0.1$ the pressure wave results in a relatively large variation of the radial velocity with only a small variation of β_2. This quasi steady behavior has completely disappeared at higher values of S_r.

S_R values larger than 1 correspond to the perturbation by a diffuser with 10 vanes or more and, as shown in Figure 7.17, have a rapidly decreasing impact on the outlet flow angle. This is in line with the observations of Dawes (1995) and Baghdadi (1977) that the high frequency unsteadiness in the vaned diffuser resulting from the pitchwise flow perturbation at the impeller outlet has only a small influence on the diffuser flow.

A different variation of $\Delta\beta_2$ is observed at 25 000 RPM when the impeller flow is transonic (Figure 7.18). The acoustic Strouhal number for a one sinus wave per circumference is 0.25. One observes two peaks, one at $S_r = 0.5$ and one at 1.0, with an amplitude that is about half the one in the subsonic impeller flow. The pressure waves generated at the downstream boundary cannot propagate up to the leading edge because of the supersonic pocket. The small change in the outlet angle results from the limitation of the change in outlet radial velocity by inlet choking. Pressure waves reach the leading edge only in the subsonic part near the hub, which results in only a small peak at $S_R = 0.5$. The second peak at $S_R = 1.0$ is likely to be the consequence of a reflection of the pressure wave on the shock halfway in the channel but attenuated by a displacement of the shock.

7.2.5 Frozen Rotor Approach

The frozen rotor approach, implemented in most solvers for turbomachinery, has been advocated to approximate the impact of a circumferential flow distortion on the performance and forces. The method is mostly used at impeller volute off-design operation where the circumferential pressure distortion can be very large because it builds up over the full 360°. The frozen rotor coupling links the flow on the common boundary, with each blade row in a fixed azimuthal position. The inertial forces due to the time variation of the flow are neglected. Only the pressure gradients arising from the steady blade loading are considered.

Similar to unsteady flow calculations, this approach also requires a numerical domain defined by boundaries that are periodic to both the rotor and stator. Hence a much larger circumferential extension of the computational domain than with the mixing plane approach is required. The transfer of information at the interface is by a point by point link between the flow conditions on each side. This does not exempt the boundary conditions from respecting the physics, defined by the characteristic equations, and be adapted to the local flow direction at the interface (Holmes 2008).

The main argument for using the frozen rotor approach is a significant saving on computational cost in comparison to an unsteady calculation. It is sometimes argued that repeating such calculations for different relative positions of stator and rotor may provide an approximation of the time evolution of the flow and forces. However, frozen rotor calculations do not account for unsteady inertial effects and every result is different from the correct unsteady one. It is unclear how the juxtaposition of incorrect solutions could result in a correct one. Each calculation takes less time than a correct unsteady one but it is questionable if repeating the calculation at different azimuthal positions is still much more economical that one correct unsteady calculation.

A second source of errors related to frozen rotor approximations are the extra pressure losses resulting from the incorrect coupling between the fixed and rotating frame (Hillewaert et al. 2000; Yi and He 2015). This is illustrated by comparing frozen rotor results with those of unsteady flow calculations in an impeller–volute combination designed by Baun and Flack (1999). The same inlet total and outlet static conditions are imposed in both calculations.

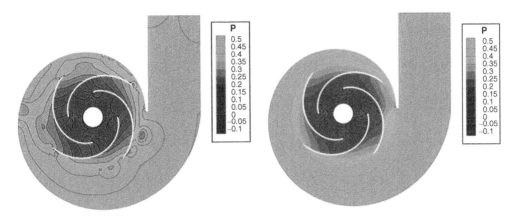

Figure 7.19 Static pressure distribution at 82% design mass flow by frozen rotor calculations (left) and unsteady calculations (right) (from Van den Braembussche et al. 2015).

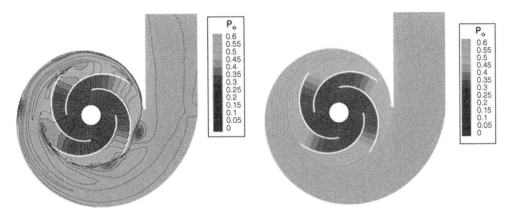

Figure 7.20 Total pressure distribution at 82% design mass flow by frozen rotor calculations (left) and unsteady calculations (right) (from Van den Braembussche et al. 2015).

Calculations are made with an inviscid solver and should provide a pure potential interaction without total pressure losses (Van den Braembussche et al. 2015).

Unsteady and frozen rotor calculation methods predict similar static and total pressure distributions at the volute design mass flow, i.e. in the absence of a circumferential pressure distortion. The results shown in Figures 7.19 and 7.20 are at 82% design mass flow. They are at the same angular position of the impeller because large variations in the flow fields by potential interaction and wakes impinging or not on the volute tongue can be expected considering the small gap between the impeller and volute.

The unsteady results show a rather smooth and periodic static pressure distribution in the impeller that is very similar to the one at design mass flow. The flow seems almost not influenced by the pressure distortion. The reason is that the flow in a rotating blade channel is exposed to the varying outlet pressure during the time the fluid needs to travel from the leading edge to the trailing edge. Inertia reduces the impact of a change in back pressure. The resulting flow is also in some way time averaged over the varying outlet conditions. Dick et al. (2001) claim that at 40% of nominal mass flow the transit time in the impeller shown on Figures 7.19 and 7.20 approaches the time of one rotation.

The frozen rotor calculations show an asymmetry in the rotor static pressure distribution. The reason is that the flow in every blade channel is permanently exposed to the same high or low outlet pressure. Neighboring channels influence each other only by a change in the inlet and outlet periodicity.

Although calculations are made with an inviscid Euler solver, important wake-like non-uniformities are observed in the total pressure distribution of the frozen rotor (Figure 7.20). They are at the origin of the lower exit total pressure than predicted by the unsteady calculations.

The wake-like distortions are the consequence of the inconsistency in the frozen rotor coupling system (Hillewaert et al. 2000). The relative velocity in the rotor $\vec{W}$ is related to the absolute velocity $\vec{V}$ by the constant rotor peripheral velocity $\vec{U}$. Substitution in the unsteady formulation of the momentum equation in the relative frame of reference:

$$\rho\left(\frac{\partial \vec{W}}{\partial t}\right)_r + \rho \vec{W}.\nabla \vec{W} + \nabla P = 0$$

results in the following relation for the absolute velocity at the interface:

$$\rho\left(\frac{\partial \vec{V}}{\partial t}\right)_r - \rho \vec{U}.\nabla \vec{V} + \rho \vec{V}.\nabla \vec{V} + \nabla P = 0 \tag{7.9}$$

This formulation should be equivalent to the one in the absolute frame of reference

$$\rho\left(\frac{\partial \vec{V}}{\partial t}\right)_a + \rho \vec{V}.\nabla \vec{V} + \nabla P = 0. \tag{7.10}$$

The subscripts a and r refer to a definition in the absolute and relative frame of reference. Comparing Equation (7.9) with Equation (7.10), it is clear that neglecting the unsteady terms, i.e. coupling steady flow in the moving frame to steady flow in the absolute frame of reference, introduces an inconsistency. The latter is due to the peripheral variation of the velocity component tangential to the rotor movement:

$$-\rho \vec{U}.\nabla \vec{V} = -\rho U \frac{\partial V_u}{R \partial \theta} \tag{7.11}$$

In an inviscid calculation, this variation relates to the blade loading, i.e. the potential interaction. In subsonic flows the impact of the blade loading extends upstream and downstream over a distance of approximately one blade pitch. The pitch of a volute being a full circumference the distortion propagates sufficiently far upstream to distort the pressure and velocity field in the impeller.

Considering Equation (1.47) the relative stagnation pressure P_r^o in the moving frame can also be written as:

$$P_r^o = P + \rho\frac{W^2}{2} = P + \rho\frac{V^2}{2} + \rho\frac{U^2}{2} - \rho U V_u = P_a^o + \rho\frac{U^2}{2} - \rho U V_u \tag{7.12}$$

relating P_r^o to the total pressure P_a^o in the stationary frame of reference. The unsteady Bernoulli equation in a general frame of reference reads

$$\rho\frac{\partial(W^2/2)}{\partial t} + \vec{W}.\nabla P_r^o = 0$$

where P_r^o is the stagnation pressure relative to the frame of reference under consideration. For steady relative flow in the rotor domain the equation states that P_r^o is convected along streamlines. Since P_r^o is constant over the inlet boundary it must also be constant over the outlet boundary of an inviscid calculation. Hence the pitchwise variation of the absolute stagnation pressure ∂P_a^o observed at the volute inlet results from the variation of the absolute tangential velocity at the interface (Equation 7.12)

$$\partial P_a^o = \rho U \partial V_u \tag{7.13}$$

This error is convected downstream by the steady flow and is responsible for the streamwise wake-like distortions in the volute.

The inconsistency of the frozen rotor coupling is also reflected in the unphysical variation of the static pressure at the boundary, i.e. a discontinuous change in slope of the iso-pressure lines at the interface on Figure 7.19. Further away from the interface the resulting distortion of the static pressure evens out.

Frozen rotor calculations may provide an acceptable approximation of the upstream propagation of the pressure distortion by the volute if the diffuser is sufficiently long to mix out the impeller wakes before reaching the volute in order not to perturb the flow around the tongue. They fail to predict correctly the unsteady impeller response to an outlet distortion.

Yang et al. (2010) have studied the circumferential shift of a pressure distortion in a compressor operating at $S_R = 0.3$. They experimentally observed $\approx 120°$ difference between the location of minimum outlet pressure and the one at the splitter blade leading edge whereas the geometrical distance is only $\approx 15°$ (Figure 7.21). This circumferential shift results from the rotation of the impeller during the time needed by the pressure wave to travel with a speed of $a - W$ from the outlet to the splitter blade leading edge. Frozen rotor calculations show a much smaller shift and a larger amplitude of the distortion. This incorrect estimation gives rise to an error on the radial force amplitude and direction (Section 7.3).

The impact of the coupling systems on performance has been evaluated by Dick et al. (2001) by means of a commercial Navier–Stokes solver applied to the same radial impeller. The results of correct unsteady calculations show an overprediction of the head that is increasing with decreasing mass flow (Figure 7.22). However, these calculations do not account for the shroud leakage flow, for which the impact on performance increases with decreasing mass flow. The frozen rotor calculations show a systematic underestimation of the pressure rise due to the numerical errors at the interface and additional losses resulting from the spurious wakes that are generated at the interface. The performance is underestimated and changes with rotor position, so no improvement can be expected from averaging the results obtained in different positions.

The mixing plane coupling system also results in an underestimation of the head at off-design operation. A first reason for this discrepancy is an incorrect prediction of the wake recovery by the sudden mixing. Further evaluations by Dick et al. (2001) showed an increasing amount of return flow at each blade passage when decreasing the mass flow. This illustrates the error

Figure 7.21 Phase shift between diffuser inlet and splitter vane leading edge pressure distribution (from Yang et al. 2010).

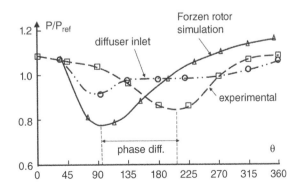

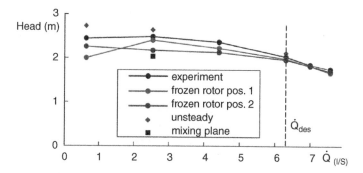

Figure 7.22 Head as a function of volume flow obtained with different coupling systems (from Dick et al. 2001).

resulting from not respecting the physically correct boundary conditions by imposing the static pressure as an outlet boundary condition in locations with return flow. Separate calculations of the impeller and diffuser respecting the physics at the interface, as proposed by Everitt and Spakovsky (2013), may be an outcome.

Furthermore, mixing plane and frozen rotor approximations assume steady impeller flow and are unable to predict the increase in energy input when operating with a circumferentially varying outlet pressure, experimentally verified by Bowerman and Acosta (1957).

7.3 Radial Forces

The circumferential variation of the impeller outlet pressure results in a different blade loading and hence different forces on each blade. They do not compensate each other and result in a non-zero radial force on the shaft. The amplitude increases with decreasing distance between impeller and volute. Force measurements by Jaatinen et al. (2009) indicate a drastic reduction in the radial forces when installing a vaned diffuser with a small dependency on solidity and almost no variation with mass flow.

7.3.1 Experimental Observations

Radial force measurement systems have been developed and used by Agostinelli et al. (1960), Okamura (1980), Chamieh et al. (1985), Kawata et al. (1983), Moore and Flathers (1998), Reunanen (2001), and many others. Most of them are for pumps, where the radial forces can be three orders of magnitude larger than in compressors because of the higher density of the fluid. However, there are also compressor applications where radial forces need to be controlled. Squeezed film damper bearings allow a better control of subsynchronous and synchronous vibrations in compressors operating at high pressure levels. The radial forces must be kept under control because the bearing is free to move in the housing with the oil as damping. The increasing interest in magnetic bearings is another incentive to control the radial forces because they have only a limited sustainable radial trust.

Compressors equipped with magnetic bearings allow a direct measurement of the forces by the bearings. Older systems measure the forces by means of strain gauge balances fixed to the bearings (Figure 7.23). Special attention is needed to avoid errors due to the rotor weight, the seals, and coupling forces. Meier-Grotrian (1972) also mentions measurement errors resulting from the circumferentially non-uniform distribution of the axial forces F_x on the impeller backplate resulting from the outlet pressure distortion.

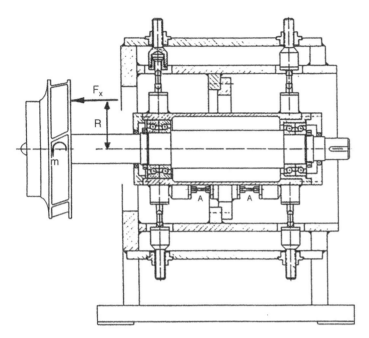

Figure 7.23 Example of a strain-gauge balance to measure the radial forces on the impeller (from Meier-Grotrian 1972).

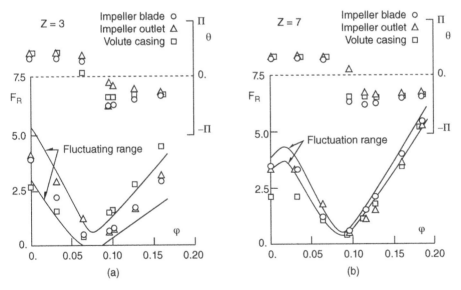

Figure 7.24 Radial trust and fluctuating range for (a) a three-bladed impeller and (b) a seven-bladed impeller (from Hasegawa et al. 1990).

Radial forces are presented in one of the following ways:

- Variation of the force amplitude and direction θ as a function of volume flow (Figure 7.24). A minimum radial force is observed near the design point, corresponding to the volute optimum mass flow, where there is no circumferential pressure distortion, and a linear increase in the forces at increasing and decreasing volume flow. For pumps the force levels off towards the shut-off mass flow. This has not been observed in radial compressors because such extremely small mass flows are normally not attainable.

- Variation of the amplitude and direction of the radial force vector, defined by the vector end-point at different volume flows, as shown in Figure 7.25. The forces change direction from $\theta \approx 80°$, when the volute operates at lower than optimum mass flow and the circumferentially increasing pressure pushes the impeller towards the volute inlet, to $\theta \approx 340°$ when the volute operates at higher than optimum mass flow and the circumferentially decreasing pressure sucks the impeller towards the volute outlet.

The direction and amplitude of the radial forces strongly depend on the impeller geometry (Figure 7.26). The force direction for a low NS_C impeller has a θ value nearly 90° larger than that of a high NS_C impeller. The variation of the forces with mass flow for a pump at high $NS_C = 112$ is similar to the one shown in Figure 7.25 for a compressor at $NS_C = 116$, indicating that the same mechanisms apply to both. Different from the pump is that the compressor operating range and hence the force are limited by compressor stall.

The effect of the number of blades on the magnitude of the radial force was studied experimentally by Hasegawa et al. (1990). Figure 7.24 compares the measured radial thrust for an impeller with seven blades and one with three blades. The point of zero force has shifted, as can be expected from the difference between the performance curve of a three- and seven-bladed impeller. The magnitude of the radial thrust of the seven-blade impeller is larger than that of the three-bladed impeller. However, the amplitude of the fluctuations is twice as large in the latter.

Stepanoff (1957) proposes following expression to predict the radial force:

$$F_R = K\rho\Delta H^i D_2 b_2 \tag{7.14}$$

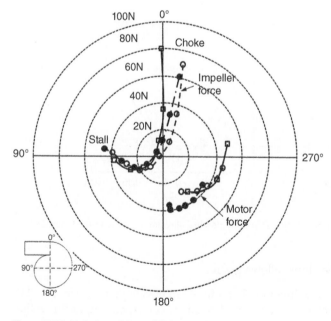

Figure 7.25 Variation of the radial force vector with volume flow for two different volutes (from Reunanen 2001).

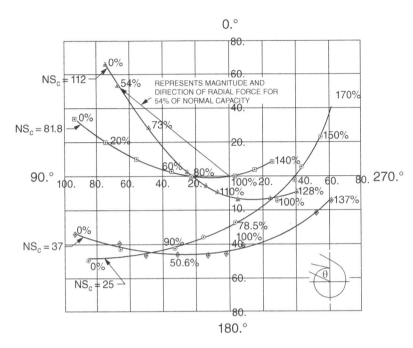

Figure 7.26 Polar plot of radial forces on impellers of different specific speed (from Agostinelli et al. 1960).

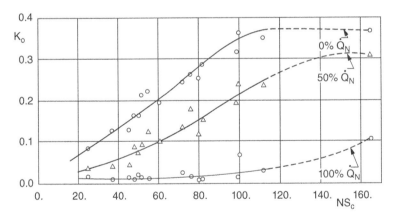

Figure 7.27 Variation of K_o coefficient as a function of specific speed NS_C for different values of the volume flow (from Agostinelli et al. 1960).

where K is an empirical factor defined by

$$K = K_o \left(1 - \frac{\dot{Q}}{\dot{Q}_N} \right)^2 \tag{7.15}$$

K_o is the value of K at zero mass flow and is equal to 0.36. Based on experimental observations Agostinelli et al. (1960) propose a value of K_o that depends on a specific speed (Figure 7.27). The value of K at 0% volume flow is therefore a better approximation of K_o for different values of NS_C.

A more detailed expression for the radial force on centrifugal impellers, which takes into account the volute geometry, is proposed by Biheller (1965):

$$F_R = 0.22061 U_2^2 \rho D_2 b_2 10^{-1.13 A_{th}/A_5}$$

$$* \left[1 + k^2 \left(\frac{\dot{Q}}{\dot{Q}_N} \right)^2 - 2k \frac{\dot{Q}}{\dot{Q}_N} \cos \left(\frac{3}{8} \theta_v - \frac{\pi}{2} + \left(\frac{\pi}{2} - \frac{\theta_v}{4} \right) \frac{\dot{Q}}{\dot{Q}_N} \right) \right] \qquad (7.16)$$

This expression is valid for spiral volutes ($\theta_v = 0$), fully concentric collectors ($\theta_v = 2\pi$) or combinations of a concentric collector ($0 < \theta < \theta_v$) with a spiral volute (from $\theta_v < \theta < 2\pi$). A_{th} is the minimum volute cross section at the tongue and A_6 is the volute outlet cross section.

The following values of k are recommended

$k = 1$	for $\theta_v = 0$
$k = 4.7$	for $\theta_v = 2\pi$
$k = 0.304 \times 10^{1.13 A_{th}/A_5}$	for $0 < \theta_v < 2\pi$

The direction of the radial force is defined by:

$$\alpha_F = \frac{\pi}{2} - \frac{\dot{Q}}{\dot{Q}_N} \qquad \text{if } 0 < \frac{\dot{Q}}{\dot{Q}_N} < 1$$

$$\alpha_F = -\frac{\pi}{2} - \frac{\dot{Q}}{\dot{Q}_N} \qquad \text{if } 1 > \frac{\dot{Q}}{\dot{Q}_N} < 1.75$$

It is sometimes proposed that the radial force can be approximated by integrating the circumferential pressure distortion over the full impeller exit area, together with the pressure on the hub and shroud disk. This is much easier than direct force measurements but inaccurate, both in terms of magnitude and direction of the force vector, because it neglects the reaction of the impeller flow to the outlet pressure distortion (Section 7.3.2). A large discrepancy between the measured radial bearing force and the force obtained from a pressure integration over the impeller outlet was observed by Aoki (1984) on a single blade impeller (Figure 7.28). This is a consequence of the large impeller flow distortion resulting in a large circumferential variation of the radial and tangential momentum at impeller outlet and inlet. This discrepancy decreases with increasing number of blades.

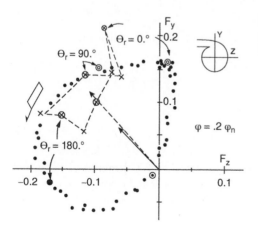

Figure 7.28 Radial trust obtained by circumferential pressure integration (x) versus measured bearing forces (o) on a single blade impeller (from Aoki 1984).

7.3.2 Computation of Radial Forces

In what follows a Cartesian coordinate system is used to define the forces acting on each individual blade and on the impeller shaft. The y axis is from the shaft center towards the volute tongue and the z axis is perpendicular to it (Figure 7.28).

There are two approaches to compute these forces. The first one integrates the pressure over the whole impeller surface, i.e. blades, back plate and hub for an unshrouded impeller and also on the inner and outer wall of the shroud of a shrouded impeller. The variation of the forces with the rotational position of the impeller versus volute or vaned diffuser, the weight of the fluid in the impeller, and the viscous forces are neglected.

$$\vec{F} = \iint_{\vec{S}_{i,j,k}} P_{i,j,k}\, d\vec{S}_{i,j,k} \tag{7.17}$$

The pressure in the shroud and backplate cavity is not always calculated in a CFD analysis. Suitable models are proposed by Childs (1986, 1992) and Will et al. (2011). Iverson et al. (1960) emphasizes the impact of the clearance between the diffuser and shroud cavity on the radial forces.

The x, y, and z components of the forces are obtained by integrating the pressure over the x, y, and z projections of the surface $d\vec{S}$. The radial force is then given by

$$\vec{F}_R = \vec{F}_y + \vec{F}_z \tag{7.18}$$

The contribution of the pressure on an almost radial backplate to the radial force is likely to be very small. Its contribution to the axial force can be quite large because a high pressure (nearly equal to the one at the impeller outlet) is applied over a large surface. It can be controlled by installing a labyrinth seal at a radial position between the shaft and the impeller outer radius to separate the zone of high pressure from a low pressure zone eventually connected to the inlet section (Figure 7.29). Another way to decrease the pressure difference between the hub and back-side-cavity is by drilling pressure balancing holes in the backplate. However, the flow leaking from the high pressure zone to the low pressure one perturbs the impeller flow. The axial force is further influenced by the blade forces resulting from the higher pressure on the pressure side than on the suction side.

The second method to calculate the axial and radial forces on the impeller is to apply the momentum and pressure balance on a control volume that contains the space between the impeller inlet and outlet, including the hub and shroud surface:

$$\frac{\partial}{\partial t}\left[\iiint_{Vol} \vec{V} \rho\, d(Vol)\right] + \iint_3 d\dot{m}\vec{V} + \iint_1 d\dot{m}\vec{V} + \iint_S P d\vec{S} + \iint_S \bar{\tau} dS = \vec{G} \tag{7.19}$$

Figure 7.29 Calculation of forces and control surface.

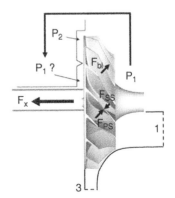

The first term is the rate of change in momentum of the fluids in the control volume due to the unsteady variation of the velocity. In the case of an impeller volute interaction the fluctuations are at the blade passing frequency and may be neglected. The second to fourth terms account for the change in momentum between the inlet and outlet and the pressure on all the walls of the control volume. The last two terms are the contribution of the shear forces and weight of the fluid contained in the control volume.

Neglecting the shear forces and weight of the fluid the total time-averaged force $\vec{F}$ on the impeller can be split into

$$\vec{F}_{\Delta Mx} + \vec{F}_{\Delta M_R} + \vec{F}_{\Delta M_u} + \vec{F}_P = \vec{F} \tag{7.20}$$

where $\vec{F}_{\Delta Mx}$, $\vec{F}_{\Delta M_R}$, and $\vec{F}_{\Delta M_u}$ are the forces due to the change of axial, radial, and tangential momentum between the inlet (1) and outlet (3) sections. $\vec{F}_P$ is the force due to the static pressure at the inlet and outlet sections and on the shroud and non-rotating parts of the hub and diffuser.

Written in summation form, the radial force (excluding the one on the backplate) reads:

$$\left(\sum_k \sum_j \rho \vec{V} d\vec{S} V_R \right)_3 + \left(\sum_k \sum_j \rho \vec{V} d\vec{S} V_R \right)_1$$
$$+ \left(\sum_k \sum_j \rho \vec{V} d\vec{S} V_u \right)_3 + \left(\sum_k \sum_j \rho \vec{V} d\vec{s} V_u \right)_1$$
$$+ \left(\sum_k \sum_j P d\vec{S}_R \right)_3 + \left(\sum_k \sum_j P d\vec{S}_R \right)_1$$
$$+ \left(\sum_i \sum_j P d\vec{S}_R \right)_S + \left(\sum_i \sum_j P d\vec{S}_R \right)_{H^{nr}} = \vec{F}_R \tag{7.21}$$

The indices j and k run in the circumferential and spanwise directions, respectively. i runs in the meridional direction. S is the shroud contour and H^{nr} is the non-rotating part of the hub contour, if there is one. The velocity and static pressure at each circumferential position are the instantaneous ones obtained at a given impeller position from the computation of the unsteady flow field.

The axial force (excluding that on the backplate) is given by:

$$\left(\sum_k \sum_j \rho \vec{V} d\vec{S} V_x \right)_3 + \left(\sum_k \sum_j \rho \vec{V} d\vec{S} V_x \right)_1$$
$$+ \left(\sum_k \sum_j P d\vec{S}_x \right)_3 + \left(\sum_k \sum_j P d\vec{S}_x \right)_1$$
$$+ \left(\sum_i \sum_j P d\vec{S}_x \right)_S + \left(\sum_i \sum_j P d\vec{S}_x \right)_{H^{nr}} = \vec{F}_x \tag{7.22}$$

The momentum balance method does not provide the forces on individual blades, required for a blade vibration analysis, but allows a better understanding of how the pressure distortion and momentum change contribute to the radial force as a function of the geometry.

Influence of Impeller Geometry

The influence of the impeller geometry on the radial force is evaluated by comparing the radial forces on the impeller with 20 full blades at 40° backward lean (Figure 7.12) with the ones on the Eckardt "O" impeller with 20 radial ending blades (Figure 7.30) (Schuster and Schmidt-Eisenlohr 1980). The first one is operating at 11 000 RPM and the latter at

Figure 7.30 Three-dimensional view of the Eckardt "O" impeller.

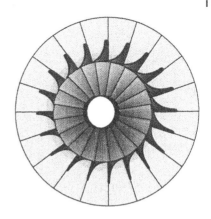

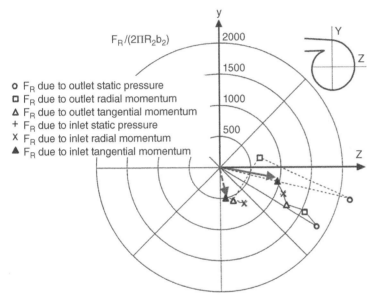

o F_R due to outlet static pressure
□ F_R due to outlet radial momentum
Δ F_R due to outlet tangential momentum
+ F_R due to inlet static pressure
X F_R due to inlet radial momentum
▲ F_R due to inlet tangential momentum

$F_R/(2\Pi R_2 b_2)$

Figure 7.31 Components of the radial force in the impeller with 20 full blades at 40° backward lean (———) and Eckardt "O" impeller with 20 radial ending blades (- - - -) at higher than optimal mass flow (from Fatsis et al. 1995).

14 000 RPM. Results are based on the unsteady impeller flow calculations with a circumferential pressure distortion corresponding to higher than optimal mass flow (Figure 7.6). The different momentum and pressure contributions to the radial force, divided by the respective impeller outlet section, are shown in Figure 7.31. The full lines are for the backward leaning impeller. The dashed lines are for the one with radial ending blades.

The pressure force on the radial ending impeller is to a large extent compensated for by the change in radial momentum, i.e. by a change of radial velocity at the outlet, and the total force is nearly half that for the impeller with backward leaning blades. The contribution of the variation in tangential momentum is also much larger than for the backward leaning impeller and perpendicular to the radial momentum force. This is not a consequence of a change in tangential velocity along the circumference because $V_u \approx U_2$, but results from the circumferential variation of the radial velocity.

The radial force is considerably larger in the impeller with backward leaning blades because of a much smaller compensation by the radial momentum. This can be explained using

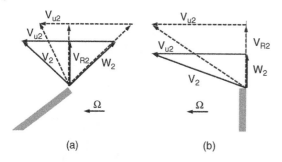

Figure 7.32 Difference in response to an outlet pressure distortion between (a) an impeller with backward leaning or (b) radial ending blades.

(a)   (b)

Figure 7.32. A local disequilibrium between the pressure rise in the impeller and the imposed outlet pressure results in a local change in the radial velocity. In the case of a backward leaning impeller operating at $S_R \approx 0.1$ the flow remains more or less tangent to the blade. Hence an increase in radial velocity, because of a decrease of the outlet pressure, provokes a decrease in the absolute tangential velocity and, as a consequence, a decrease in pressure rise in the impeller. The imbalance between the imposed outlet static pressure and impeller pressure rise is reduced. Such a corrective action results in a smaller change in radial velocity and hence a smaller contribution of the circumferential variation of the radial momentum to the radial force.

The pressure rise in radial ending impellers is almost independent of the radial velocity because a change in V_{R2} does not result in a significant variation in V_{u2}. There is no mechanism whereby the impeller can reduce the disequilibrium between the pressure rise in the impeller and the imposed outlet pressure. As a consequence the change in radial velocity and hence radial momentum is much larger.

The smaller contribution of the tangential momentum at the impeller outlet occurs for the same reasons. The increase in radial velocity, due to a local decrease in static pressure, results in a decrease of the tangential velocity at the outlet of an impeller with backward leaning blades. As a consequence, the product $\dot{m}V_{u2}$ in Equation (7.21) is almost constant along the periphery of a backward leaning impeller. The contribution of the tangential momentum to the radial force is thus negligible.

In the case of radial ending blades, an increase in mass flow does not alter the tangential component of the absolute velocity V_{u2} (dotted line velocity triangle in Figure 7.32b). However, the product $\dot{m}V_{u2}$ is not constant along the impeller periphery because of the large changes in V_{R2}. This change in outlet tangential momentum is almost perpendicular to the radial blade.

The larger the backward lean the smaller the non-pressure related contributions. This explains why the radial force in pump impellers, where very large backward lean angles up to 70° are not uncommon, may be approximated by integrating the outlet static pressure over the outlet area. This approximation further improves for shrouded impellers because the contribution of the pressure term becomes more important as it also acts on the axial part of the shroud outer wall.

Different results are obtained with frozen rotor calculations. The same pressure distortion applies continuously on the same flow channel, which results in an overreaction of the flow and hence an overprediction of the velocity changes. The consequence is an underprediction of the radial load as observed by Flathers and Bache (1999). Improvement in the numerical results was obtained by changing the inlet boundary conditions. It was assumed that imposing the mass flow at the inlet of the numerical domain closer to the leading edge limits the amplitude of the flow distortions and hence their impact on the radial forces.

Flathers and Bache (1999) also illustrated the large impact of the pressure distortion in the shroud and hub wall cavities. Considering the large surfaces involved, an error in the change in momentum will be less important in shrouded impellers than in unshrouded ones.

Figure 7.33 Comparison of the time-averaged unsteady and frozen rotor force at different mass flows (from Aksoy 2002).

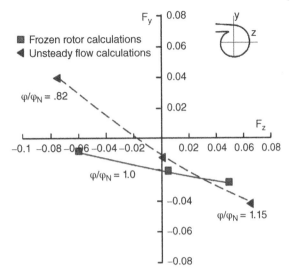

High specific speed impellers are likely to have more radial ending blades than low specific ones. This explains the similar variation of the force direction from large NS to low NS impellers (Figure 7.26) to that from radial to backward leaning impellers (Figure 7.31).

The non-dimensionalized time-averaged forces acting on the shaft of the low NS_c impeller discussed in Section 7.2.5, predicted by the unsteady and frozen rotor calculations, are compared in Figure 7.33. They differ in magnitude and direction at off-design operation. The error is rather small because the circumferential pressure distribution, which is almost identical for both calculations, is the dominant contribution in impellers with large backward lean. Much larger discrepancies can be expected for impellers with less backward lean.

Double volutes are commonly used in pumps to reduce the radial force on the shaft of large radial pumps. They are more complex to design and manufacture, and show higher losses.

7.4 Off-design Performance Prediction

CFD analysis of the impeller and volute performance at off-design operation requires a detailed geometry definition and expensive unsteady calculations. This section is about analytical methods suited for the analysis of a preliminary design or used in an optimization procedure. At the same time it intends to illustrate the interaction mechanisms between the impeller and volute flow.

An off-design performance model must include the following:

- A modeling of the impeller response to a circumferentially non-uniform pressure distribution at the outlet. The output is the circumferential variation of the total pressure, and tangential and radial velocity at the diffuser inlet.
- The calculation of the non-axisymmetric flow in a diffuser with a non-uniform inlet flow. The output is the velocity and total pressure distribution at the volute inlet.
- The prediction of the 3D flow in the volute taking into account the circumferential variation of the inlet flow. The output is a circumferential static pressure distribution at the volute inlet/diffuser exit.
- The calculation of the circumferential pressure distortion at the impeller exit for the pressure distribution imposed by the volute at the diffuser exit.

Each module is linked by the boundary conditions and will be described in more detail in what follows.

Simplified 1D volute flow and impeller interaction models have been proposed by Hübl (1975), Lorett and Gopalakrishnan (1986), and Chochua et al. (2005). They are extensions of the volute flow prediction method of Iverson et al. (1960) allowing for a circumferential variation of the volute inlet flow conditions. The latter ones are predicted by means of an impeller response model.

A similar but more detailed analytical calculation, taking into account the loss and flow mechanisms described in previous sections, has been proposed by Van den Braembussche et al. (1999). The method uses an iterative procedure to calculate the impeller–diffuser–volute interaction. It is composed of the following modules, which are repeated until the flow conditions at the interfaces remain unchanged.

7.4.1 Impeller Response Model

Lorett and Gopalakrishnan (1986) were the first to use the analytical impeller responds model explained in Section 7.2.1 to estimated the reaction of the impeller outlet velocity to a circumferential distortion of the outlet pressure. The model presented here assumes that the impeller flow at each circumferential position can be evaluated by a local linearization of the steady performance curve (Figure 7.34).

The local radial velocity and total pressure distribution at the impeller exit are defined as a function of the averaged values $\widetilde{P}_2$, $\widetilde{P}_2^o$, and $\widetilde{m}$ by means of the following expressions:

$$\dot{m}(\theta) = \widetilde{m} + (P_2(\theta) - \widetilde{P}_2)/\frac{dP_2}{d\dot{m}} \tag{7.23}$$

$$P_2^o(\theta) = \widetilde{P}_2^o + (\dot{m}(\theta) - \widetilde{m})\frac{dP_2^o}{d\dot{m}} \tag{7.24}$$

$\frac{dP_2}{d\dot{m}}$ and $\frac{dP_2^o}{d\dot{m}}$ are tangent to the impeller overall performance curve. Neglecting the change in density the local value of $\dot{m}(\theta)$ is a measure for the local radial velocity $V_R(\theta)$ at the impeller exit in position θ.

Based on experimental data (Sideris et al. 1987a) it is further assumed that the impeller outlet tangential velocity does not change as a function of the circumferential pressure distortion, hence the total temperature ratio is constant:

$$\frac{T_2^o}{T_1^o} = 1 + \frac{\widetilde{P_2^o}^{\left(\frac{\kappa-1}{\kappa}\right)} - 1}{\eta_{TT}} \tag{7.25}$$

In this way the diffuser inlet flow conditions are defined as a function of the local static pressure at NUEL equidistant points around the impeller periphery.

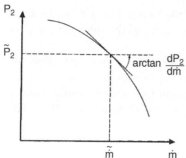

Figure 7.34 Local linearization of the impeller performance curve.

7.4.2 Diffuser Response Model

The calculation of the diffuser flow is based on the simplifying assumption that the different streamlines on the diffuser circumference (Figure 7.35) do not influence each other. The flow along each streamline is calculated separately by integrating the continuity, momentum, and energy equation in the radial direction by the 1D method described in Section 4.1.1. Starting from the inlet conditions, defined by the impeller response model, the velocity components and pressure in NUEL points at the inlet of the volute can be predicted. This provides an approximation of the circumferential shift of the total pressure and temperature between the diffuser inlet and outlet, and the velocity distortion at the volute inlet.

7.4.3 Volute Flow Calculation

The variation of the pressure and velocity over the cross section of a volute can be accounted for by correction terms for the swirl and radial variation of the pressure and through flow velocity (Hübl 1975). The following is based on the flow model described in Section 6.2.3. The volute channel is divided into NUEL segments defined by sections at equidistant angular locations around the circumference (Figure 7.36). Each cross section is covered with four identical quadrants of ellipses defined by the radii ($r_w(a)$ and $r_w(b)$).

Figure 7.35 Vaneless diffuser calculation.

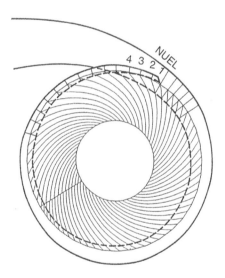

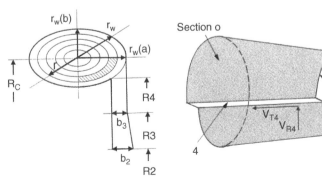

Figure 7.36 Volute calculation model.

The prediction of the crosswise variation of flow parameters is based on the observation that particles entering the volute close to the tongue stay at the center of the volute, and that particles entering the volute at larger values of θ swirl around the previous ones. In the case of inviscid flow, this assumption allows the determination of the velocity distribution in a cross section of position θ'' as a function of the radial velocity at the volute inlet at the positions θ'. The swirl velocity at each radius is related to the incoming radial velocity at the position θ' by the conservation of angular momentum (Figure 7.37):

$$V_S^i(r'(\theta'')) = \frac{V_{R4}(\theta')r_w(\theta')}{r'(\theta'')}$$

The relation between the position θ', where the fluid enters the volute, and the radial location r' at cross section θ'' is based on continuity. Suppose that the mass flow at $r < r'(\theta'')$ equals the mass flow that has entered the volute between $\theta = 0$ and θ':

$$\int_o^{\theta'} \rho_4(\theta)b_4R_4V_{R4}(\theta)d\theta = \int_o^{r'(\theta'')} \rho V_T 2\pi r dr \tag{7.26}$$

This relation between the incoming fluid and its location at a cross section allows the isentropic total pressure and temperature distribution over each cross section to be defined. A typical variation of $V_S^i(r'(\theta''), r')$ at a cross section is shown by the dashed line on Figure 7.38.

Experiments have shown that the real swirl distribution in a cross section has a forced vortex distribution. For a given swirl velocity on the wall $V_S(\theta'', r_w)$, this is defined by

$$V_S(\theta'', r) = V_S(\theta'', r_w)\frac{r(\theta'')}{r_w(\theta'')} \tag{7.27}$$

The real forced vortex swirl distribution is shown by the solid line on Figure 7.38.

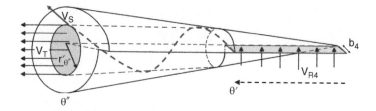

Figure 7.37 Schematic representation of the spiraling flow in a volute.

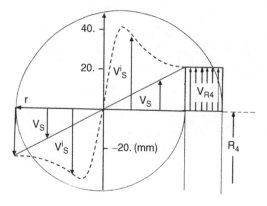

Figure 7.38 Swirl velocity distribution for inviscid flows (- - -) and real flows (—).

The difference between the inviscid and real swirl distribution defines the total pressure losses due to internal shear

$$P^o(\theta'', r) = P^{o,i}(\theta'', r) - 0.5\rho(\theta'', r)\left(V_S^{i,2}(\theta'', r) - V_S^2(\theta'', r)\right) \tag{7.28}$$

The radial equilibrium between the centrifugal forces (due to swirl) and the radial pressure gradient over a cross section

$$\frac{dP}{dr} = \rho\frac{V_S^2(\theta'', r)}{r(\theta'')} \tag{7.29}$$

defines the static pressure distribution over a cross section as a function of the wall static pressure. The last one is defined by an iterative procedure in which the wall static pressure is adjusted until continuity is satisfied. The throughflow velocity distribution is then defined by:

$$0.5\,\rho(\theta'', r)V_T^2(\theta'', r) = P^o(\theta'', r) - P(\theta'', r) - 0.5\,\rho(\theta'', r)V_S^2(\theta'', r) \tag{7.30}$$

The swirl velocity at the wall is iteratively adjusted until the throughflow velocity at the wall $(V_T(\theta'', r_w)$ and the diffuser exit tangentially velocity $(V_{T3}(\theta''))$ satisfy the relation

$$V_T(\theta'', r_w) = V_{T4}\frac{R_4}{R_c(\theta'')} \tag{7.31}$$

Once the throughflow velocity has been adjusted, the calculation of the inviscid total pressure and swirl velocity distribution is repeated.

Applying the tangential momentum equation on the NUEL elements of the volute (Figure 7.39) allows the calculation of a corrected static pressure at the exit of each volute element (θ_o) as a function of the flow parameters at the volute inlet (θ_i), outer wall (ow), and inner wall (iw) (Figure 7.39):

$$
\begin{aligned}
&\rho_4 V_{R4} 2\pi R_4^2 b_4 V_{T4} + R_C(\theta_i)A(\theta_i)\left(P_C(\theta_i) + \tilde{\rho}(\theta_i)\tilde{V}_T^2(\theta_i)\right) \\
&+ dA_{iw}\left(P_C(\theta_i) - \tilde{\rho}(\theta_i)\tilde{V}_T^2(\theta_i)\frac{R_C - R_{Ciw}}{R_{Ciw}}\right)R_{Ciw} \\
&- dA_{ow}\left(P_C(\theta_o) - \tilde{\rho}(\theta_o)\tilde{V}_T^2(\theta_o)\frac{R_C - R_{Cow}}{R_{Cow}}\right)R_{Cow} \\
&= R_C(\theta_o)A(\theta_o)\left(P_C(\theta_o) + \tilde{\rho}(\theta_o)\tilde{V}_T^2(\theta_o)\right)
\end{aligned}
\tag{7.32}
$$

Figure 7.39 Elements of the volute that define the terms in the tangential momentum equation.

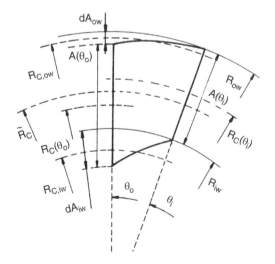

This calculation makes use of the average flow quantities calculated at the mean radial position of each cross section.

The main outcome of this calculation is the circumferential static pressure distribution and total pressure losses. The diffuser exit static pressure equals that at the center of the volute corrected for the pressure difference between R_C and R_4, due to the centrifugal forces:

$$P_4(\theta) = P_C(\theta) - 2\rho(\theta)\frac{V_T^2(\theta)}{R_C(\theta) + R_4}(R_4 - R_C(\theta))$$

7.4.4 Impeller Outlet Pressure Distribution

The diffuser exit static pressure distortion propagates radially upstream in the vaneless diffuser and results in a new circumferential distortion at the impeller outlet. In accordance with the experimental observations one supposes that the amplitude of the circumferential static pressure distortion at the impeller outlet equals that at the diffuser outlet pressure:

$$P_2(\theta) = \widetilde{P}_2 + (P_4(\theta) - \widetilde{P}_4) \tag{7.33}$$

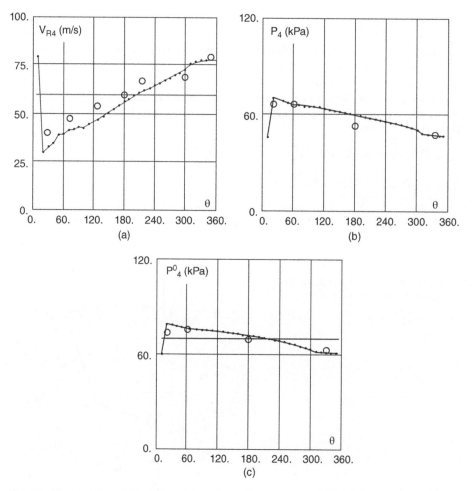

Figure 7.40 Circumferential variation of the calculated and measured (a) radial velocity, (b) static pressure, and (c) total pressure at the diffuser exit.

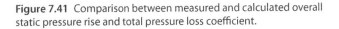

Figure 7.41 Comparison between measured and calculated overall static pressure rise and total pressure loss coefficient.

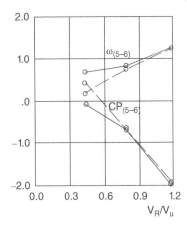

These four components of the model are repeated in an iterative system starting from the newly defined static pressure distribution at the impeller outlet.

7.4.5 Evaluation and Conclusion

The model has been verified by applying it to the elliptic volute for which the experimental data are presented in Section 6.2.4. The calculated circumferential variation of the radial velocity, and the static and total pressure at the diffuser exit are compared to the experimental one in Figure 7.40. Results are shown only at maximum mass flow because this is the operating point with maximum circumferential distortion.

Figure 7.41 shows a comparison between the measured and calculated static pressure rise and total pressure loss coefficients. One observes good agreement except at minimum mass flow, where larger losses and a lower static pressure rise are due to flow separation in the volute and exit diffuser, not accounted for by the method. The approach is not restricted to elliptic volutes but can easily be adapted to volutes with a rectangular cross section.

When evaluating this model one should keep in mind that the only empirical parameter is the friction coefficient in the vaneless diffuser and volute, and that the main geometrical parameters are analytically defined. The good agreement between experimental and predicted values allows it to be concluded that the main phenomena related to swirl and internal shear, and the strong interaction between the different components, are correctly captured by the model.

8

Stability and Range

A sufficiently large compressor range is a major design target for centrifugal compressors. In most cases it is limited by choking and surge. However, when operating at high pressure levels, the operating range may also be limited by flow induced vibrations resulting from rotating stall or other fluid dynamic excitations (Bonciani et al. 1980).

Most prediction methods are relatively accurate in terms of choking limit but often show large discrepancies in the prediction of the minimum mass flow. This is related to the complexity of the phenomena that are involved and the interaction between them.

Although surge and stall are very different phenomena, both names are commonly used for the same unsteady phenomena. One should therefore start with a clear definition.

The fluid in a decelerating flow experiences an adverse pressure gradient. As the kinetic energy is much smaller in the boundary layers than in the free stream, that part of the fluid may fail to overcome the pressure gradient, fixed by the free stream, and return flow or an inversion of the through flow velocity component occurs. From the macroscopic point of view, this reorientation of the flow may be steady (steady stall) or unsteady (rotating stall).

In steady stall, such as in the two-zone model flow, zones of separated flow rotate at the same speed as the impeller and do not constitute a dynamic load on the impeller blades (Figure 3.40). Rotating stall is when the axi-symmetric character of the flow is replaced by zones of high and low kinetic energy or separated flow, rotating at a different speed to the impeller. (Figure 8.1). The circumferentially averaged flow rate in the compressor remains constant but local flow rate changes with time.

Rotating stall is not limited to the region of positive or zero slope of the pressure rise curve, but is also frequently observed on the negative slope side. The unsteadiness of the flow not only results in a deterioration in the compressor performance, but also constitutes a source of mechanical excitation of the blades and shaft. Rotating stall can be tolerated as long as the blade loading variation is within acceptable limits or does not create any discontinuity in the performance curve. It is likely to give rise to an increase in the noise level. However, depending on the operating conditions and impeller geometry it may limit the operating range because of vibrational problems (Haupt et al. 1985a,b). Rotating stall becomes particularly important at high levels of inlet pressure because the forces corresponding to the velocity variations increase with the density of the working fluid (Bonciani et al. 1980), or when the frequency corresponds to the resonance frequency of a compressor component.

Surge is the flow condition in which the whole compressor system becomes unstable, resulting in violent changes in the inlet and outlet conditions with a typical low frequency noise whereby the overall mass flow changes with time (Figure 8.2). It can be described as a self sustained flow oscillation in which the impeller acts as an exciter and the other components (inlet and outlet volume, throttle valve, and pipes) as a resonator. The onset and frequency of these

Design and Analysis of Centrifugal Compressors, First Edition. René Van den Braembussche.
© 2019, The American Society of Mechanical Engineers (ASME), 2 Park Avenue, New York, NY, 10016, USA (www.asme.org).
Published 2019 by John Wiley & Sons Ltd.

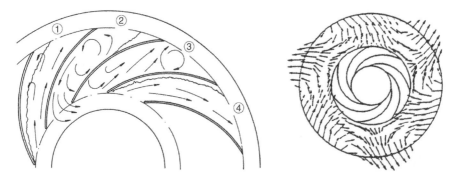

Figure 8.1 Circumferential distortion of impeller and diffuser flow (from Lennemann and Howard 1970; Yoshida et al. 1993).

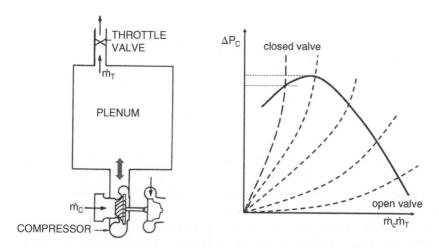

Figure 8.2 Compressor system and flow throttling.

oscillations depend on the shape of the pressure rise curve, the length of the inlet and outlet duct, the volume of the reservoirs, and the characteristics of the throttling device. Surge normally occurs at zero or positive slope of the performance curve ($\frac{d\Delta P_C}{d\dot{m}_C} \geq 0$).

Surge is a system instability whereas stall is a local phenomenon. This does not exclude that stall can be the triggering mechanism for surge but it is not a necessary condition. A more detailed definition of surge and stall, based on a theoretical description and the main characteristics, are the main topic of the present chapter. It is an attempt to classify the different phenomena in a logical way. This is complemented by prediction methods and correlations, which although they do not always fit into the theoretical models, also give good results.

The interventions that allow avoiding or postponing the occurrence of these instabilities or reduction of their impact are discussed in Chapter 9.

8.1 Distinction Between Different Types of Rotating Stall

Rotating stall can be generated by the destabilization of the impeller or diffuser flow, or by an unsteady interaction between impeller and diffuser. Different origins will result in different flow patterns and can best be illustrated by a series of experiments performed by Frigne (1982) and

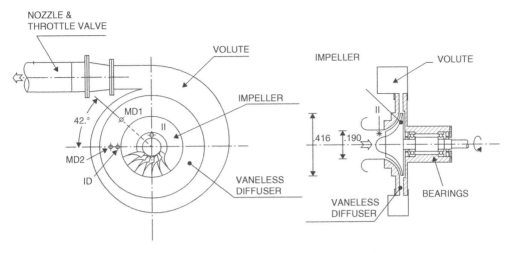

Figure 8.3 Configuration A of the centrifugal test compressor (from Frigne et al. 1984).

described in Frigne et al. (1984) on an impeller operated with different stator configurations. The three configurations are shown in Figures 8.3 and 8.7.

A centrifugal impeller with 20 radial ending blades and $b_2/R_2 = 0.077$ was first operated with a vaneless diffuser of radius ratio (R_4/R_2) 1.62 (Configuration A). The diffuser is followed by a volute of rectangular cross section, an orifice plate to measure the mass flow, and a throttle valve. The compressor inlet is connected to the ambient pressure by a short cylindrical section. Besides the conventional pressure and temperature probes, required for the overall performance measurements, four hot-wire probes are also installed: one at the impeller inlet (II) to detect the upstream propagation of the perturbations, one at the diffuser inlet (ID), and two at $R/R_2 = 1.3$ at 42° shift in angular position inside the diffuser (MD) to determine the propagation speed (ω_σ) and the number of the stall cells (λ).

Assuming that the stall cells rotate in the same direction as the impeller, the number of cells λ can be defined as a function of the period τ_2 and phase shift τ_1 between the two signals MD1 and MD2 (Figure 8.4):

$$\lambda = \frac{360}{42}\frac{\tau_2 - \tau_1}{\tau_2} \tag{8.1}$$

The angular speed ω_σ is defined by

$$\omega_\sigma = \frac{2\pi}{\lambda\,\tau_2} \tag{8.2}$$

The impeller, diffuser, and overall compressor static pressure rise coefficient *CP* for Configuration A are shown in Figure 8.5 for three values of the rotational speed. They are plotted versus α_2 because, as will be shown later, the absolute inlet flow angle is the critical parameter

Figure 8.4 Hot wire signal in MD1 and MD2 to calculate the number of cells and rotational speed.

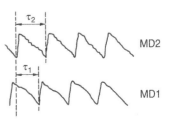

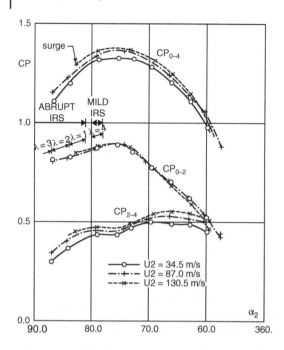

for diffuser stability. The angle α_2 is defined from slip factor correlations and verified against the measured total temperature rise. Due to the close coupling between the impeller and the throttling valve, the compressor can be operated without surge down to very small mass flows where the pressure rise curve has a positive slope.

A first type of rotating stall with small amplitude pressure oscillations is observed at $\alpha_2 = 78°$, both at impeller inlet and diffuser inlet, and therefore is called the mild impeller rotating stall (MIRS on Figure 8.6). A large number of stall cells ($\lambda = 4$ to 5) are rotating at $\omega_\sigma/\Omega = 0.14$. At decreasing mass flow, the peak-to-peak velocity variation grows to a maximum of 6.5% of

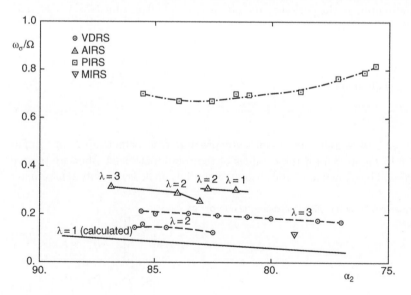

Figure 8.6 Variation of stall cell relative rotational speed.

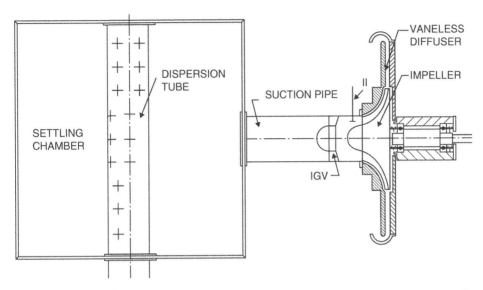

Figure 8.7 Modified test facility for zero outlet distortion (Configurations B and C) (from Frigne et al. 1984).

the mean velocity. The velocity variations are almost sinusoidal (no higher harmonics). This MIRS has also been observed in other compressors but its impact is very small and it will not be discussed further. No clear theoretical explanation for it has been found but rotating waves preceding other types of stall are not uncommon.

Reducing the mass flow further to $\alpha_2 = 81.5$, a second type of rotating stall appears suddenly. The number of stall cells varies between 1 and 3, and the rotational speed is between 20% and 30% of the impeller rotational speed. The peak-to-peak velocity variations are about 30% of the mean velocity and do not change with mass flow. Because of its sudden appearance at high amplitude upstream and downstream of the impeller, this type of rotating stall is called abrupt impeller rotating stall (AIRS on Figure 8.6). It was also observed by Abdelhamid and Bertrand (1979) and in combination with a low frequency rotating stall pattern (Abdelhamid 1981).

In order to eliminate the possible influence of the circumferential pressure perturbation by the volute, and to investigate the effect of the diffuser outlet to inlet radius ratio, a second series of tests was performed after the following modifications of the geometry were made (Figure 8.7). The throttle valve and orifice plate have been installed at the compressor inlet in combination with a large inlet settling chamber. The diffuser radius ratio was increased to $R_4/R_2 = 1.92$ with a circumferentially uniform outlet at atmospheric pressure.

The overall performance curves (Figure 8.8a, Configuration B) are very similar to the previous ones except that the impeller characteristics exhibit a minimum at $\alpha_2 = 81°$ and surge occurs near the maximum pressure rise for the highest values of U_2.

At the lowest speed line, the rotating stall is observed at $\alpha_2 = 76°$, at both impeller inlet and diffuser inlet. One to three stall cells are rotating at $\omega_\sigma/\Omega \approx 0.7$ to 0.8. The peak-to-peak velocity variations increase from 5% of the mean velocity to a maximum of 10% when the impeller static pressure rise coefficient has a minimum. Because of the progressive increase in amplitude, this instability is called progressive impeller rotating stall (PIRS on Figure 8.6). It is due to an unsteady propagation of separated flow zones in the impeller, similar to rotating stall in axial compressors.

Vaneless diffuser rotating stall (VDRS) is observed in a third configuration after the impeller is unloaded by means of 40° prerotation, allowing a lower mass flow for the same impeller relative

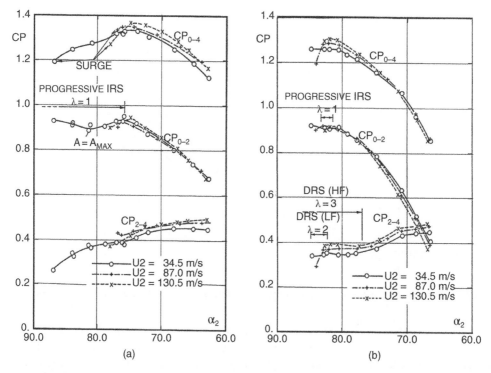

Figure 8.8 Progressive impeller rotating stall: (a) Configuration B and (b) Configuration C (from Frigne et al. 1984).

inlet flow angle. The pressure rise curves are shifted to larger diffuser inlet flow angles without destabilizing the impeller flow (Figure 8.8b, Configuration C).

Rotating stall is observed only in the diffuser starting at $\alpha_2 = 77°$ (which corresponds to a minimum in the diffuser pressure recovery). Three stall cells rotate at a relative speed $0.17 < \omega_\sigma < 0.21$. The relative amplitude of the velocity variations amounts to 10% and is almost independent of flow rate. The spectral analysis of the velocity signal shows no higher harmonics and no unsteadiness at the impeller inlet. A similar type of stall is observed at $\alpha_2 = 83.6°$, but with two stall cells, rotating at $\omega_\sigma/\Omega = 0.13$ to 0.16. This is also diffuser rotating stall and, as will be demonstrated later, has the same origin as the previous type.

The different types of rotating stall, observed in these experimental investigations, are summarized in Table 8.1. Except for vaned diffuser and return channel rotating stall, it contains the major types of rotating stall in centrifugal compressors. Each of them will be discussed in more detail in the following paragraphs from both the theoretical and experimental point of view.

8.2 Vaneless Diffuser Rotating Stall

Vaneless diffuser rotating stall (VDRS) is a phenomenon whereby a circumferentially periodic flow pattern rotates in the diffuser at sub-synchronous speed. It is the consequence of a rearrangement of the flow into areas of larger outflow and areas of stalled flow. This is illustrated by the experimental and numerical results of Tsujimoto et al. (1994) (Figure 8.9) showing the velocity and pressure fluctuations during rotating stall. The circumferential variation of the velocity clearly shows areas of reduced trough flow, limited by two counter-rotating vortices.

Table 8.1 Characteristics of the different types of rotating stall (from Frigne et al. 1984).

Type	Impeller rotating stall			Diffuser rotating stall	
Characteristic	Mild	Abrupt	Progressive	High frequency	Low frequency
Configuration	A	B	B and C	C	C
Amplitude A in diffuser	$\alpha_2 \nearrow \gg A \nearrow$ $A_{max} \searrow$	$\alpha_2 \nearrow \gg A =$ Cte	$\alpha_2 \nearrow \gg A \nearrow$ $A_{max} \searrow$	$\alpha_2 \nearrow \gg A =$ Cte	$\alpha_2 \nearrow \gg A \nearrow$ $A_{max} \searrow$
	$A_{max} = 0.065$ $U_2 = 43.5$ m/s	$A_{max} = 0.30$ $U_2 = 43.5$ m/s	$A_{max} = 0.10$ $U_2 = 43.5$ m/s		$A_{max} = 0.10$ $U_2 = 43.5$ m/s
Number of cells (λ)	$\lambda = 5$ ($U_2 = 43.5$ m/s)	$81° < \alpha_2 < 84°$: $\lambda = 1$	$\lambda = 1$ (always dominant)	$\lambda = 3$ (no harmonics)	$\lambda = 2$ (no harmonics)
	$\lambda = 4$ ($U^2 = 87.0$ m/s)	$81° < \alpha_2 < 84°$: $\lambda = 2$	$\lambda = 2$ (higher harmonics)		
	$\lambda = 4$ ($U^2 = 130.5$ m/s)	$81° < \alpha_2 < 84°$: $\lambda = 3$	$\lambda = 3$ (higher harmonics)		
	No harmonics	$\lambda = 1$ at $U_2 = 130.5$ m/s			
ω_σ/Ω	0.14	$0.26 \rightarrow 0.31$	$0.82 \rightarrow 0.67$	$0.17 \rightarrow 0.21$	$0.13 \rightarrow 0.16$
when $\alpha_2 \nearrow$	Constant	$\omega_\sigma/\Omega \nearrow$	$\omega_\sigma/\Omega \searrow$ minimum $\nearrow$	$\omega_\sigma/\Omega \nearrow$	$\omega_\sigma/\Omega \nearrow$

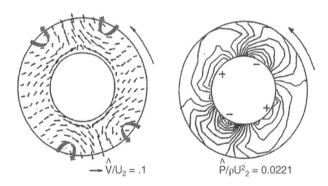

$\longrightarrow \hat{V}/U_2 = .1$ $\hat{P}/\rho U^2_2 = 0.0221$

Figure 8.9 Experimental velocity and calculated pressure distribution in a vaneless diffuser during rotating stall (from Tsujimoto et al. 1994).

This requires an increase in flow in the remaining areas, resulting in a more radial velocity and locally enhanced stability. This flow pattern is in agreement with the results of Wachter and Rieder (1985), who measured a high static pressure at low velocity (stalled flow) and a lower pressure at higher (outgoing) velocity.

VDRS has been observed and commented by Jansen (1964b), Senoo and Kinoshita (1978), Abdelhamid et al. (1979), Abdelhamid (1981), Imaichi and Tsurusaki (1979), Van den Braembussche et al. (1980), Kinoshita and Senoo (1985), Nishida et al. (1988, 1991), and Kobayashi et al. (1990).

Jansen (1964b) was the first to demonstrate that diffuser rotating stall can be triggered by a local inversion of the radial velocity component in the vaneless diffuser without interaction with the impeller. His theoretical approach starts from the continuity and momentum equations in the radial and tangential directions, for unsteady, inviscid, and incompressible flow. The flow field is described by a steady axisymmetric free vortex flow on which an unsteady perturbation is superposed. The following linearized wave equation describes the unsteady component of the streamfunction in the circumferential plane of the vaneless diffuser:

$$\frac{R}{V_u}\frac{\partial \nabla^2 \Psi}{\partial t} + \frac{1}{R\tan\alpha}\frac{\partial \nabla^2 \Psi}{\partial R} + \frac{1}{R^2}\frac{\partial \nabla^2 \Psi}{\partial \theta} = 0 \tag{8.3}$$

Continuity is automatically satisfied by specifying the velocity perturbations $\hat{V}_R$ and $\hat{V}_\theta$ in the radial and tangential directions as a function of a streamfunction Ψ:

$$\hat{V}_R = \frac{1}{R}\frac{\partial \Psi}{\partial \theta} \qquad\qquad \hat{V}_u = -\frac{\partial \Psi}{\partial R}$$

The boundary conditions are a uniform pressure at the diffuser outlet (Figure 8.10):

$$\frac{\partial P}{\partial \theta} = 0 \qquad\qquad \frac{\partial \hat{P}}{\partial R} = 0$$

where $\hat{P}$ is the pressure perturbation, and steady and uniform inlet flow:

$$\hat{V}_R = 0 \qquad\qquad \hat{V}_u = 0$$

The disturbances can be described as a periodic wave:

$$\Psi = A_\Psi(R)e^{i\lambda\theta}e^{St} \tag{8.4}$$

where $A_\Psi(R)$ defines the radial variation of the amplitude of the waves and λ is the number of waves around the circumference. After substitution of Equation (8.4) in (8.3) one obtains the following solution for the exponential growth rate S of a perturbation:

$$S = -\left(\frac{A_\sigma}{\tan\alpha} - i\,F(\tan\alpha)\right)\frac{\lambda V_{u,2}R_2}{R_4^2} \tag{8.5}$$

The system will be unsteady (increasing amplitude of the velocity perturbations) at the operating conditions where the real part of S becomes positive, i.e. for $A_\sigma/\tan\alpha < 0$ or $\alpha > 90°$.

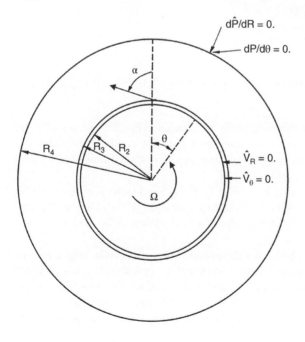

dP̂/dR = 0.

dP/dθ = 0.

Figure 8.10 Vaneless diffuser stability calculation model.

$\hat{V}_R = 0.$

$\hat{V}_\theta = 0.$

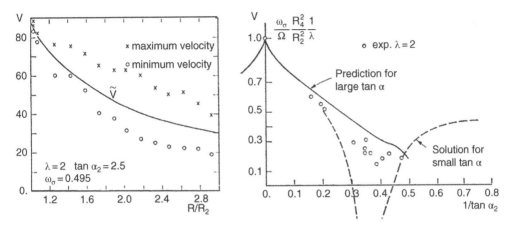

Figure 8.11 Experimental velocity fluctuation (left) and comparison between predicted and experimental rotational speed of stall cells (right) (from Jansen 1964b).

This analysis attributes the onset of an unsteady periodic motion in the vaneless diffuser to the presence of local return flow. Experimental data (Jansen 1964b) also show that the amplitude of the velocity fluctuations is at a maximum in the center of the diffuser and decreases towards the exit (Figure 8.11).

The rotational speed of the perturbation ω_σ is defined by the imaginary part of S:

$$\frac{\omega_\sigma}{\Omega} = iF(\tan \alpha)\frac{\lambda R_2^2}{R_4^2} \tag{8.6}$$

indicating a dependence of the relative rotational speed ω_σ/Ω on the wave number (λ), radius ratio (R_4/R_2), and flow angle (α). The function $F(\tan \alpha)$ for $\lambda = 2$ is compared with experimental values in Figure 8.11.

The main conclusions from this theory are:

- rotating stall is triggered by local return flow in the diffuser
- rotating stall can exist with circumferentially uniform inlet velocity and outlet pressure
- the rotational speed ω_σ is between 5% and 22% of Ω and proportional to the number of stall cells.

It is likely that return flow will first occur in the boundary layer, where, under the influence of the radial pressure gradient, the streamlines are more inward curved than in the middle of the diffuser passage. Although the triggering of VDRS is by the boundary layer (local return flow and increased blockage), the inviscid calculations of Jansen (1964b) made clear that the dynamic characteristics (number of stall cells and propagation speed) depend on the inviscid part of the flow.

This destabilizing effect of the boundary layer is confirmed by Frigne et al. (1985) by means of a time-evolving calculation of the strong interaction between the inviscid core flow in the center of the diffuser and the unsteady boundary layers along the walls. Depending on the diffuser inlet flow angle α_2 and wave number λ of the initial perturbation, the calculations predict either the return to a steady axisymmetric flow ($\lambda = 1$, $\alpha_2 = 80°$) or a growing amplitude of the rotating non-uniform flow pattern of ($\lambda = 3$, $\alpha_2 = 80°$) (Figure 8.12).

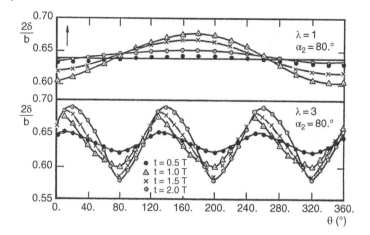

Figure 8.12 Increase or decrease in the perturbation amplitude depending on the wave number λ and inlet flow angle (from Frigne et al. 1985).

8.2.1 Theoretical Stability Calculation

Jansen's theory leads to a practical stability criterion when combined with a viscous flow calculation method to predict return flow in the boundary layers. Such a calculation using an inviscid flow core in the center and 3D boundary layers near the walls was presented by Jansen (1964a). However, an unfortunate choice of the boundary layer profile resulted in a prediction of return flow near the walls as soon as the two boundary layers filled the whole channel width. The resulting conditions for diffuser stability, predicted in this way, were therefore too restrictive for narrow diffusers.

The boundary layer calculations reported by Senoo and Kinoshita (1977) (Section 4.1.3) predict a limiting inlet flow angle $\alpha_{2,ret}$, a function of the diffuser width, above which return flow occurs in the diffuser. Typical results of such a calculation for a spanwise symmetric flow are shown in Figure 8.13.

The freestream angle, α, continuously decreases with radius ratio. The angle of the wall streamline, $\alpha + \varepsilon$, first increases and then starts to decrease once the boundary layers fill the whole diffuser width ($\delta_S + \delta_H = b$). Return flow occurs when $\alpha + \varepsilon$ exceeds 90°. This is more likely to occur at larger inlet flow angle, as illustrated by the dotted lines for $\alpha_2 = 82°$ on Figure 8.13.

Based on comparisons with experimental data, Senoo and Kinoshita (1978) concluded that VDRS will occur when the diffuser inlet flow angle exceeds the critical value α_{2c}, defined by

$$\frac{90° - \alpha_{2c}}{90° - \alpha_{2,ret}} = 0.88 \tag{8.7}$$

where α_{ret} is the diffuser inlet flow angle at which return flow occurs downstream in the diffuser.

3D boundary layer calculations also allow the influence of aerodynamic and geometric parameters on the vaneless diffuser stability to be evaluated. As shown in Figure 8.14, narrow diffusers are more stable than wide ones, i.e. larger values of α_2 are allowed before rotating stall occurs.

Calculations at different Reynolds numbers show that wide diffusers become more stable with increasing Reynolds number, but that the latter has no effect on narrow diffusers (Figure 8.15). The increasing stability of wide diffusers with Reynolds number is because the more turbulent wall boundary layers are more resistant to the radial pressure gradient.

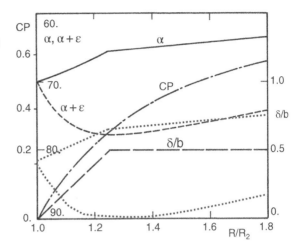

Figure 8.13 Variation of flow angle and boundary layer based on flow calculations for two different inlet flow angles (from Senoo and Kinoshita 1977).

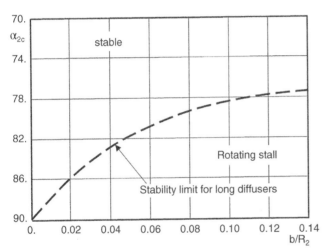

Figure 8.14 Variation of critical inlet flow angle based on flow calculations (from Kinoshita and Senoo 1985).

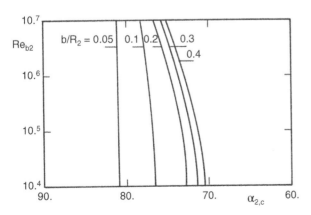

Figure 8.15 Variation of critical inlet flow angle with diffuser width and Reynolds number (from Senoo and Kinoshita 1977).

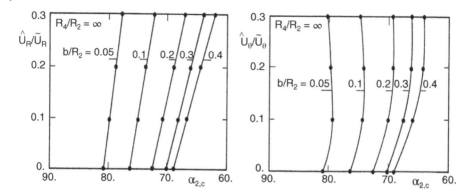

Figure 8.16 Impact of inlet distortion on the inlet flow angle at return flow (from Senoo and Kinoshita 1977).

The non-uniformity of the radial velocity component at the inlet has an important influence on both narrow and wide diffusers (Figure 8.16). The influence of a tangential velocity perturbation is smaller, especially for narrow diffusers. The main problem, however, is the difficulty in estimating the diffuser inlet flow conditions. The inlet flow can be distorted because of the non-uniform impeller exit velocity (jet-wake), a circumferential pressure distortion by the volute or a sudden change in passage width at the transition from impeller to diffuser. This is not accounted for by the theory and will be discussed in more detail in Section 8.2.3.

The variation of $\alpha_{2,c}$ with inlet Mach number is shown in Figure 8.17. $\alpha_{2,c}$ decreases with increasing Mach number for both narrow and wide diffusers because of the additional pressure rise due to compressibility.

The dependence of stability on the diffuser radius ratio R_4/R_2 is summarized in Figure 8.18 for incompressible uniform inlet flow. This graph allows an estimation of the extent of the separation zone for given values of b_2/R_2 and inlet flow angle. For narrow diffusers, return flow starts close to the diffuser inlet but disappears further downstream. The stabilizing effect of fully developed flow ($\delta H + \delta S = b$) starts close to the inlet of narrow diffusers and further downstream in wide diffusers. The vertical dashed line on Figure 8.18 shows the radial extent of the return flow zone for $b_2/R_2 = 0.05$ and inlet flow angle of 82°. In wider diffusers the return flow starts at a larger radius ratio. The figure also shows that return flow may be postponed to higher inlet flow angles, by reducing the radius ratio. This explains why shorter diffusers are

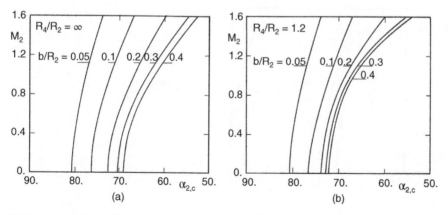

Figure 8.17 Variation of return flow angle with Mach number for (a) a long diffuser and (b) a short diffuser (from Senoo and Kinoshita 1977).

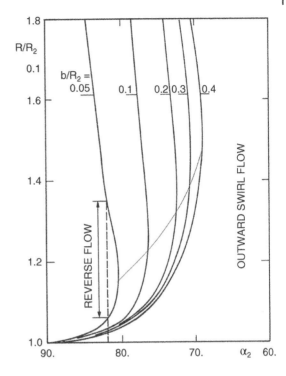

Figure 8.18 Influence of diffuser width and length on the location of the reverse flow (from Senoo and Kinoshita 1978).

more stable than longer ones and why very high inlet flow angles can be tolerated in the short vaneless space upstream of vaned diffusers without triggering rotating stall. This effect is less pronounced for narrower diffusers, which have their critical area nearer to the diffuser inlet.

Watanabe et al. (1994) make a distinction between intermittent rotating stall (which are small zones of return flow having a small influence on diffuser performances) and permanent rotating stall, where large zones, extending up to the impeller exit, result in larger amplitude variations and a discontinuous drop of the diffuser pressure recovery.

8.2.2 Comparison with Experiments

When comparing experimental data with theoretical predictions by the Senoo theory, one must make sure that the measured unsteadiness is VDRS and not AIRS. Both phenomena have a low rotational speed and are easily confused. The published data do not always contain all the necessary information to make this distinction. Criteria to assume VDRS are a smaller amplitude of the velocity and pressure variations than for AIRS and there is no discontinuity of the pressure rise curve at the onset of rotating stall. VDRS has been observed also on the negative slope side of the compressor pressure rise curve.

A first comparison (Figure 8.19) concerns the diffuser critical inlet flow angle, α_{2c}, separating the zone of steady flow ($\alpha_2 < \alpha_{2c}$) from the zone of VDRS ($\alpha_2 > \alpha_{2c}$). The dashed line shows the diffuser critical inlet flow angle calculated by Kinoshita and Senoo (1985) based on Equation (8.7) for an infinitely long diffuser ($R_4/R_2 > 2$) with uniform inlet flow at low Mach number. For each experiment Table 8.2 lists the corresponding reference where more details can be found.

The flow angles have been defined in different ways: the circles are critical flow angles calculated from the measured mass flow and slip factor correlation, the squares are values calculated from measured temperature rise, and the triangles are values based on direct flow

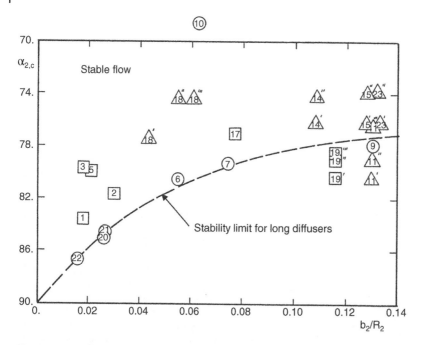

Figure 8.19 Diffuser critical inlet flow angle (comparison between experimental data and theoretical predictions of VDRS).

Table 8.2 Source of the experimental data shown in Figure 8.19.

Test point	Reference	Test point	Reference
0	Jansen (1964b)	1-2-3-5	Van den Braembussche et al. (1980)
6-7-9	Senoo and Kinoshita (1977, 1978)	10-11	Abdelhamid et al. (1979)
14-15	Abdelhamid and Bertrand (1979)	17	Frigne et al. (1984)
18	Wachter and Rieder (1985)	19	Abdelhamid (1981)
20-21-22	Kinoshita and Senoo (1985)	23	Kämmer and Rautenberg (1986)
24	Imaichi and Tsurusaki (1979)	.	.

angle measurements. This difference in data reduction methods is a first explanation for some of the discrepancies between experiments and predictions.

The points 6, 7, 9, 20, 21 and 22 have been used to define the constant in Equation (8.7) and therefore show a very good agreement. They are obtained with lightly loaded impellers $(W_2/W_1 \approx 1)$ at low Mach number, with a smooth transition from impeller exit to diffuser inlet and in the absence of a return channel.

Points 1 to 5 are obtained on industrial compressors with very narrow impellers. The flow perturbation resulting from the relatively large discontinuity at the diffuser inlet has a destabilizing effect and will be discussed in Section 8.2.3. Test points 2, 3, and 5 are obtained at low Reynolds number. Test point 1 is obtained on the same compressor as point 3 but at a much higher Reynolds number (1.7×10^6 compared to 1.1×10^4). Although the theory does not show a direct Reynolds number influence for narrow diffusers, it may affect the stability because it influences the diffuser inlet flow distortion $\hat{V}_R/\tilde{V}_R \neq 0$ created by the impeller blade loading.

The largest discrepancy is observed for test point 10. It is suspected that in this case a poorly adapted return channel is responsible for the decrease in diffuser stability.

Point 17 corresponds to the VDRS discussed in Section 8.1. Taking into account the impeller loading and the fact that a different method is used to define α_{2c}, the agreement is considered satisfactory.

Points 19′ to 19‴ illustrate the influence of a change in diffuser radius ratio for wide diffusers. The short diffuser 19′ ($R_5/R_2 = 1.6$) is more stable than the longer one 19‴ ($R_5/R_2 = 1.93$), where the stability limit agrees well with that of an infinitely long diffuser.

A similar shift in the stability limit is also observed between test points 14′ and 15′ ($R_4/R_2 = 1.55$) and test points 14″ and 15″ ($R_4/R_2 = 1.83$). Both test series are obtained with the same impeller but the somewhat larger discrepancy for 14′ must be attributed due to a velocity perturbation at the diffuser inlet because of the sudden spanwise reduction in the flow channel. Further reduction in diffuser width resulted in AIRS.

The data for test points 18″ and 18‴ are obtained with the same impeller as 18′ but with a sudden enlargement at the diffuser inlet, which explains the large discrepancy with the predictions. Furthermore, the flow angles of test points 18′ to 18‴ are not defined at the diffuser inlet but at a radius ratio of 1.3 where the flow angle has already decreased because of boundary layer growth.

Test points 11′ to 11‴ illustrate the influence of the diffuser inlet Mach number on stability. Geometry 11′ has a relatively wide but short diffuser ($R_4/R_2 = 1.52$) and is therefore more stable than predicted for infinitely long diffusers. An increase in inlet Mach number from 0.5 to 1.1 shows in a decrease in α_{2c} by 4°, which is exactly as predicted in Figure 8.17a.

Point 23′ corresponds to diffuser rotating stall at high inlet Mach number but the diffuser radius ratio is limited to 1.5 by means of a throttle ring. The decrease in the critical inlet flow angle with Mach number is compensated for by an enhanced stability because of the small radius ratio. Point 23″ shows a lower critical flow angle for a diffuser radius ratio of 2 after the throttling ring has been removed. The measured velocity perturbations show a large amplitude, suggesting that this could also be AIRS created by the volute flow perturbations propagating upstream to the impeller.

One can conclude from previous comparisons that the stability limit shown in Figure 8.19 allows a reliable prediction of VDRS if the corrections for Reynolds number, Mach number, diffuser radius ratio, and inlet flow distortion are accounted for.

Jansen (1964b) predicted that the rotational speed ω_σ of the stall cells is a function of the radius ratio, inlet flow angle, and number of stall cells (Equation (8.6)). This relation is only valid for large values of α_2, i.e. the flow conditions at which VDRS has been observed.

Figure 8.20 shows a comparison between the experimental data and theoretical predictions. The points (o) are experimental data obtained by Jansen (1964b) with the swirl created by a rotating screen at the diffuser inlet. The other data are obtained with real impellers and the numbering of the test data is the same as in Figure 8.19 and Table 8.2. The agreement is not perfect, but theory and experiments clearly show an increase in rotational speed with increasing values of $\tan(\alpha_2)$. This figure illustrates how all experimental data coincide if the relative rotational speed is divided by the number of cells.

8.2.3 Influence of the Diffuser Inlet Shape and Pinching

Shaft vibrations at low subsynchronous frequencies have occasionally been observed in compressors operating at very high pressure levels. The frequency of the vibrations (between 2% and 8% of the rotor speed) suggests that they might be related to vaneless diffuser rotating stall with one stall cell ($\lambda = 1$). This is the most critical VDRS pattern because there is no compensation

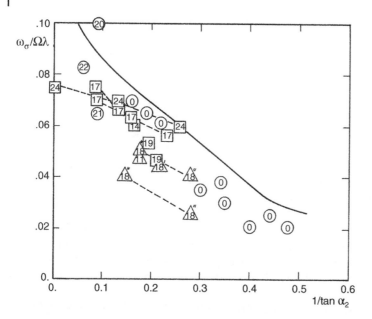

Figure 8.20 Rotational speed of stall cells: a comparison between experimental data and theoretical prediction.

of the radial force generated by the stall cell possible. However, in the case of low specific speed impellers, the onset of these instabilities could not be predicted by Senoo's stability criterion, which has triggered additional research on this type of instability.

It was assumed that the change in passage width between the impeller outlet and diffuser inlet (radial gap and diffuser pinching) and the way this transition is realized (round-off radius) could influence the flow stability. The impact of these geometrical features could be particularly large in the case of narrow diffusers because:

- even a small change in width causes a relatively large change in cross section
- the onset of instability for narrow diffusers is near the inlet (Figure 8.18), hence is likely to be influenced by the gap and round-off.

This problem has been studied in detail by Nishida et al. (1988, 1989, 1991) and Kobayashi et al. (1990). Similar research has been done by Cellai et al. (2003a,b), Ferrara et al. (2002a,b, 2006), and Carnevale et al. (2006) but only the first research will be discussed here in detail. Both unsteady pressure transducer and hot-wire measurements of the stall characteristics were made on the test configuration shown in Figure 8.21 for four different impellers, named A, B, C, and D. They have outlet widths of 5.5, 7.3, 10, and 3.5 mm, respectively, corresponding to b_2/R_2 values of 0.0366, 0.0486, 0.0666, and 0.0233. Impeller B is not a standard type of impeller (1 cm thick blades at the trailing edge). As the results are in line with the others they will not be discussed here.

The impellers have been combined with diffusers of different width and inlet shape:

- VL2.1T, VL2.7T, VL3.5T, VL3.5A, VL3.5B for impeller D
- VL5.5T, VL5.5A, VL5.5B, VL2.8T, VL3.8T for impeller A
- VL10T, VL10A, VL6.7T, 4.7T for impeller C.

The letters VL stand for vaneless diffuser, the number refers to the diffuser width in millimetres and the last letter characterizes the inlet shape defined in Figure 8.22 (i.e. VL3.8T is a vaneless diffuser of 3.8 mm width and T-shaped inlet).

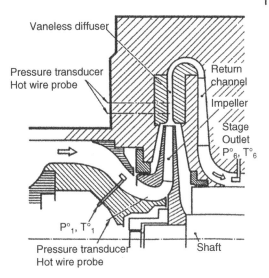

Figure 8.21 Cross section view of test compressor (from Nishida et al. 1988).

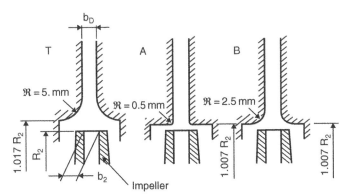

Figure 8.22 Inlet shapes of diffusers (from Kobayashi et al. 1990).

The impact of diffuser pinching on the operating range is illustrated in Figure 8.23. All data are for a T-type inlet. A decrease in the diffuser width from the initial 5.5 mm, which is equal to the A impeller exit width, to almost half of its original value results in a considerable extension of the stable operating range. A similar shift is observed when pinching the diffuser of the wider C impeller. Pinching the diffuser of the narrow D impeller with a T-type inlet from 3.5 to 2.1 mm is less effective.

There is no diffuser geometry for which the stable operating range extends to the surge limit. This does not necessarily restrict the compressor operating range. Operating compressors during rotating stall will probably increase the noise level but the impact on blade and shaft vibrations is likely to be small when the pressure level is low. Robust industrial compressors may then operate safely in rotating stall. This will not be the case, however, for advanced high specific speed compressors with thin blades or when the compressor is operated at very high pressure levels. In the latter case, the same relative pressure fluctuations in the diffuser result in large rotating radial forces acting on the impeller and shaft, and the operating range will have to be limited to the stall free zone because of too high vibration levels.

The results, summarized in Figure 8.24, are for the impellers D, A, and C with diffusers having the same width as the impeller but with a different inlet shape. The main trend when reducing the diffuser inlet round-off radius is a shift of the stall limit to smaller mass flows. There is almost

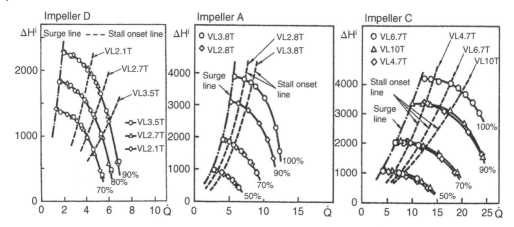

Figure 8.23 Effect of the diffuser width on the performance and stall limit (from Kobayashi et al. 1990; Nishida et al. 1988).

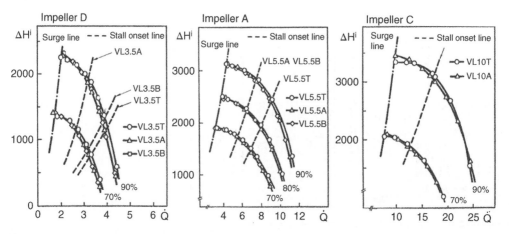

Figure 8.24 Effect of the inlet shape of the diffuser on the performance and stall limit (from Kobayashi et al. 1990).

no change in the stall limit when changing from a T- to a B-type diffuser inlet for the narrow impeller D ($b_2/R_2 = 0.0233$). Only a drastic reduction in the round-off radius to 0.5 mm (A-type diffuser) results in a considerable extension of the stable operation range but the stall limit is still far from the surge limit. Combining the A- or B- type diffusers with the wider A impeller ($b_2/R_2 = 0.0366$) is much more effective. The stall limit shifts in both cases to a considerably lower mass flow than the one obtained with the T-type diffuser. It makes no difference if an A- or T-type diffuser is combined with the widest impeller C.

It can be concluded from this that the diffuser inlet round-off radius has a large impact on the stability of narrow diffusers, but little or no effect on the stability of wider impeller/diffuser combinations. Changing the round-off radius has about the same impact on the stall limit of the A impeller as a change in the diffuser width (Figure 8.24 versus Figure 8.23). The change in diffuser width has no noticeable impact on performance. It is unexplained why, for all three impellers, the T-type diffuser on Figure 8.24 results in a higher pressure rise than the other ones. Reduced friction because of flow separation in the diffuser inlet area, where the velocities and hence friction are highest, is one possible reason.

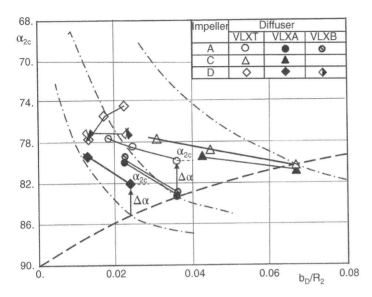

Figure 8.25 Experimental values of critical flow angle: influence of diffuser inlet shape (from Kobayashi et al. 1990).

Large discrepancies are observed when comparing previous experimental data with Senoo's criterion for diffuser rotating stall (Figure 8.25). The full lines connect the critical inlet angles for a given diffuser inlet shape at different values of b_D/b_2. Consider first the values on the right side of these lines where the diffuser width equals the impeller exit width ($b_D/b_2 = 1.0$). In the absence of inlet flow distortion, the rotating stall starting point should be on the line predicted by Senoo. A very good agreement, independent of the diffuser inlet shape, is observed for impeller C. The same good agreement is also observed for impeller A when combined with diffusers with a sufficiently small round-off radius (diffusers A and B). However, the discrepancy is nearly 4° when combined with a T-type diffuser.

None of the diffusers combined with impeller D have their stability limit in agreement with the Senoo correlation. In spite of the almost smooth transition from the impeller exit to diffuser A (only a small round-off radius and only a 1.05 mm radial gap between the impeller exit and diffuser inlet), rotating stall occurs at a value of α_{2c} that is 3.5° lower than predicted.

A smooth transition without inlet distortion is one of the conditions for Senoo's correlation (Kinoshita and Senoo 1985). The discrepancies between the experimental and predicted stability limit clearly depend on the diffuser inlet shape. The large round-off radius of diffusers B and T gives rise to cavities near the diffuser inlet which constitute a sudden increase in the diffuser width. This results in a local increase in the absolute flow angle α_D and large flow distortions. The flow in narrow diffusers is very sensitive to such inlet flow perturbations because the critical area for reverse flow is close to the inlet (Figure 8.18). This explains the increasing discrepancy at decreasing impeller width (from impeller C to impeller D). The discrepancy increases further with increasing round-off radius (from diffuser A to diffuser T) to reach more than 10° for diffuser T in combination with impeller D. The discrepancy increases further in all cases when reducing the diffuser width to values below the impeller exit width.

Wider impellers and diffusers are less sensitive to the diffuser inlet shape because the critical zone for return flow is at larger values of R/R_2. Hence the flow has time to recover from the inlet distortion before reaching the zone where reverse flow may trigger rotating stall. This explains the good agreement for impeller C with diffusers T and A, and impeller A with diffusers A and B.

When reducing the diffuser width below the impeller exit width ($b_D < b_2$) the critical inlet flow angle (full lines on Figure 8.25) decreases for all geometries. The dash-dot lines represent the change in diffuser flow angle at unchanged mass flow when reducing the diffuser width below the impeller outlet width. Because the stability limit of impeller C combined with a pinched diffuser is always below the dash-dot line, the diffuser will also be stable for mass flows smaller than the one at $b_D = b_2$.

Such an increase in the stall free operating range by pinching the diffuser is more difficult for impeller A. The critical flow angle first changes in the same way as the flow angle (the full lines for diffusers A and B almost coincide with the dash-dot line) and it is only after a reduction in the diffuser width to $b_D/R_2 < 0.03$ that any range extension can be expected.

Impeller A with a T-type diffuser and $b_D = b_2$ experiences rotating stall at a mass flow that is considerably larger than the one predicted by the Senoo correlation ($\alpha_2 < \alpha_{2c}$). For this combination a reduction in the diffuser width to less than 60% of its original value is required before stable operation can be achieved at the mass flow corresponding to the dash-dot line.

It turns out that diffuser pinching with a small round-off radius at the inlet is of decreasing effectiveness for impellers with a small outlet width (impeller D on Figure 8.25). The discrepancy with Senoo's correlation increases rapidly and more than 50% reduction of the diffuser width is needed to stabilize the flow in diffuser A with impeller D. The reason is that pinching with a small round-off radius results in a discontinuous variation of the cross section between the impeller and diffuser. The radial flow leaving the impeller is first turned to axial and then back to radial before entering the diffuser (Figure 8.26b). Such a sharp turn over large angles gives rise to local velocity peaks and flow separation is more likely to occur at the diffuser inlet. The impact increases with decreasing diffuser width.

A similar phenomenon to diffuser pinching occurs in low specific impellers when, because of assembly errors or thermal dilatation of a long shaft, the impeller is not well aligned with the diffuser inlet (Figure 8.26c). This also leads to reduced stability.

Kobayashi et al. (1990) claim that Senoo's correlation remains valid when $\Delta R/R_2 > 0.1$. ΔR is the distance between the start of the parallel diffuser after the round-off and the reverse flow starting point. As the radius ratio at which reverse flow occurs is defined only for $b_D/R_2 \geq 0.05$ on Figure 8.18, an extension to narrower diffusers was required to define the value of ΔR also for those geometries where the round-off is playing a role (Figure 8.27). They also claim that for values of $\Delta R/R_2 < 0.1$, Senoo's critical angle α_c should be reduced by the value $\Delta \alpha_c$ specified in Figure 8.28.

A corrected critical angle α_{2c}^{corr} can be defined as a function of b_2/R_2 for $b_D/b_2 < 1$

$$\alpha_{2c}^{corr} = 76 + (\alpha_{2c} - \Delta\alpha - 76)\frac{b_D}{b_2} \tag{8.8}$$

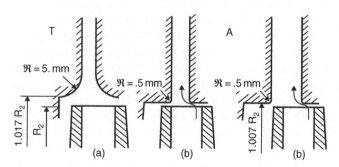

Figure 8.26 Flow at the inlet of narrow diffusers.

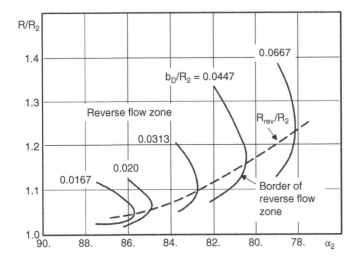

Figure 8.27 Vaneless diffuser reverse flow starting point for narrow vaneless diffusers (from Nishida et al. 1991).

Figure 8.28 Effect of diffuser reverse flow starting point on the onset of rotating stall (from Kobayashi et al. 1990).

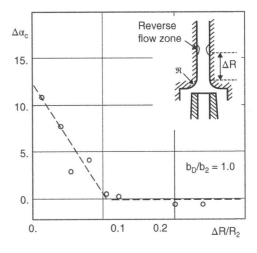

where α_{2c} is the diffuser critical flow angle according to Senoo's correlations (Equation (8.7)) for diffusers with $b_D/b_2 = 1$ and $\Delta\alpha$ the correction according to Figure 8.28. This correlation is illustrated on Figure 8.29 by the dashed lines converging at $\alpha_{2c} = 76°$.

A very different change in the stability limit is observed when combining the narrow impeller D with T-shaped diffusers of decreasing width. Results from Ferrara et al. (2002a), superposed on Figure 8.29, show a similar improvement in stability when pinching narrow diffusers with sufficiently large inlet round-off radius. Nishida et al. (1989) concluded from the unsteady pressure signals that in these cases, one or two stall cells rotate at $\omega_\sigma \approx 0.07\,\Omega$, similar to the classical vaneless diffuser rotating stall (VDRS). The increasing values of α_{2c} when pinching the diffuser are likely to be due to the smooth convergence that is created by the large round-off radius (T) and the decreasing size of the cavity, as illustrated in Figure 8.26a. The corresponding reacceleration of the flow improves uniformity of the flow and results in a decreasing but still large discrepancy with Senoo's correlation. This does not happen when the round-off radius is too small (inlet shape A) because it creates inlet flow distortion (Figure 8.26b).

The rotational speed of the stall cells (ω_σ/Ω) measured by Nishida et al. (1989), Frigne et al. (1984), and Ferrara et al. (2002a,b) slightly increases with diffuser inlet flow angle and is

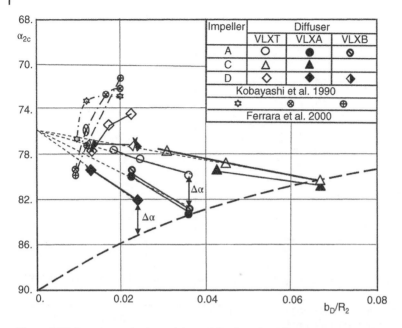

Figure 8.29 Experimental values of the stability limit for different inlet shapes and pinching (From Kobayashi et al. 1990).

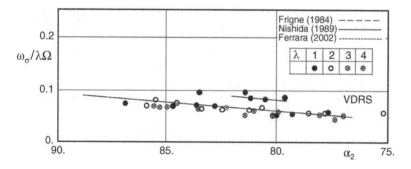

Figure 8.30 Vaneless diffuser stall rotational speed.

proportional with the number of stall cells (λ), as predicted by the model of Jansen (1964b) (Figure 8.30). Most violent subsynchronous vibrations have been observed in a frequency range of 5–9% of the impeller speed, which corresponds to the excitation by one stall cell. This is not surprising since the radial forces resulting from two or more rotating stall cells would in some way compensate each other and result in smaller amplitude vibrations at a two or three times higher frequency. Hence high amplitude vibrations are likely due to diffuser rotating stall with one stall cell. However, up to now there has been no theory known that can reliably predict the number of rotating stall cells in VDRS.

8.3 Abrupt Impeller Rotating Stall

Abrupt impeller rotating stall is characterized by large amplitude velocity and pressure fluctuations in the impeller and vaneless diffuser. It occurs on the negative as well as on the positive slope side of the pressure rise curve and because of their sudden increase in amplitude it often results in a discontinuity of that curve. The rotational speed is between 20% and 30% of the impeller speed, and is almost independent of the diffuser inlet flow angle. The frequency spectrum of the velocity and pressure variations show higher harmonics.

8.3.1 Theoretical Prediction Models

The stability of the flow in a vaneless diffuser has been studied theoretically by Abdelhamid (1980) by a method similar to that of Jansen (1964b) but considering also a possible circumferential distortion of the diffuser inlet flow. The variations of radial and tangential velocity and the static pressure perturbations in the diffuser are assumed to be small, compared to the mean values, and are described by the continuity and momentum equations in the tangential and radial directions. A constant static pressure distribution is imposed at the diffuser exit. The main difference with the theory of Jansen is the interaction between the diffuser and impeller, accounted for by following characterisitcs of the diffuser inlet flow:

$$Z_p = \frac{dP_2}{dV_{R2}} \tag{8.9}$$

$$Z_u = \frac{dV_{u2}}{dV_{R2}} \tag{8.10}$$

Both parameters are complex numbers relating the pressure rise and energy input to the local change in the radial velocity. For quasi steady flow, $Z_p < 0$ on the negative slope side of the impeller pressure rise curve ($d\psi_2/d\phi < 0$) (Figure 8.31a) and $Z_p > 0$ on the positive slope side ($d\psi_2/d\phi > 0$) (Figure 8.31b). Positive Z_p values can also occur on the negative slope side when the flow is strongly perturbed and instantaneous pressure changes are no longer along the steady performance curve.

The energy transmitted to the fluid by the impeller is the time integral of $\phi\psi$. In case of unsteady flow it is defined by

$$\int (\tilde{\phi} + d\phi)(\tilde{\psi} + d\psi)dt = \int (\tilde{\phi}\tilde{\psi} + \tilde{\phi}d\psi + \tilde{\psi}d\phi + d\psi d\phi)dt \tag{8.11}$$

The integral of the second and third terms of the RHS is zero. An extra unsteady energy addition will occur when $d\psi$ and $d\phi$ have the same sign, i.e. instabilities will grow in amplitude when $Z_p > 0$. The unsteady energy addition will generate flow oscillations with increasing amplitude until the extra energy is in equilibrium with the damping.

For quasi steady flow, Z_u can be approximated by (Figure 8.32)

$$Z_u = -\tan \beta_{2,fl} \tag{8.12}$$

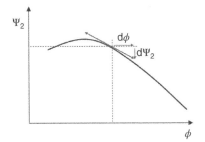
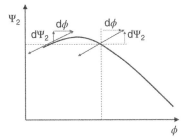

Figure 8.31 Parameters influencing Z_p.

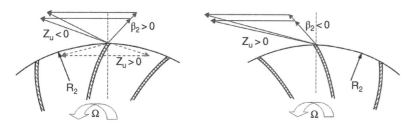

Figure 8.32 Parameters influencing Z_u.

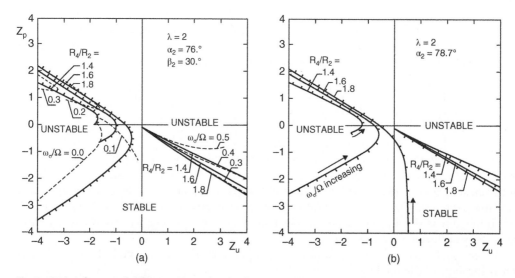

Figure 8.33 Influence of diffuser radius ratio, inlet flow angle, number of stall cells on vaneless diffuser flow stability, and rotational speed (from Abdelhamid 1980).

Negative Z_u values (Figure 8.32) correspond to quasi steady flow in impellers with radial ending or backward leaning blades ($\beta_{2,fl} > 0$). Positive Z_u values occur in impellers with forward leaning blades. They can also occur in backward leaning impellers (Figure 8.32a) when the flow is no longer tangent to the blade trailing edge because of strong perturbations or local return flow, due to volute impeller interaction or a partial blockage of the impeller outlet section.

The stability limits and characteristics predicted by Abdelhamid are plotted on Figure 8.33 as a function of Z_p and Z_u for different combinations of λ, R_4/R_2, ω_σ/Ω, and two values of α_2. Instabilities occur for positive as well as negative values of Z_u but their characteristics, summarized in Table 8.3, are different.

The characteristics for $Z_u < 0$ are very similar to those of VDRS described by the theory of Senoo and Kinoshita (1978). Negative values of Z_u are typical for impellers with backward leaning blades with the flow tangent to the impeller trailing edge, i.e. with steady impeller flow as assumed in the model of Jansen (1964b). The governing equations of the Abdelhamid model are for inviscid flows, hence are representative for wider diffusers where the boundary layers are less important. They are also unable to predict the onset of instabilities that are triggered by viscous effects but can predict the characteristics of VDRS because they are governed by the dynamics of the inviscid flow.

Table 8.3 Stability limits as a function of Z_u.

$Z_u < 0$	$Z_u > 0$
Shorter diffusers are more stable than long ones	Stability is almost independent of diffuser radius ratio
Rotating stall occurs at positive and negative values of Z_p	The flow is always unstable at $Z_p > 0$
	Rotating stall occurs at negative values of Z_p
Stability decreases with increasing α_2	Stability is almost independent of α_2
ω_σ/Ω varies from 0.1 to 0.3	ω_σ/Ω varies from 0.3 to 0.5 and
and increases with increasing Z_p	increases slightly with decreasing R_4/R_2
The number of stall cells varies between 1 and 3	The number of stall cells varies between 1 and 3

The characteristics at $Z_u > 0$ listed in Table 8.3 show similarities with the AIRS described in Section 8.2. The rotational speed is high and decreases slightly with diffuser radius ratio. This is in line with Equation (8.20) indicating that the rotational speed due to an impeller/vaned diffuser interaction is proportional to the ratio of impeller volume over total compressor volume. Although vaneless diffusers do not satisfy all the conditions required to use that equation, they show the same trend of increasing rotational speed with decreasing radial extent of the diffuser.

$Z_u > 0$ is the normal situation in impellers with forward leaning blades, as shown in Figure 8.32b. However, this type of instability can also occur in impellers with backward leaning or radial ending blades when the outlet flow is no longer tangent to the blades because of outlet flow distortions or local return flow. In the latter case the flow is no longer guided by the blades and the value of Z_u is even independent of β_2 (Figure 8.32a).

Strong impeller outlet flow perturbations are likely to occur when the downstream volute is operating at off-design and imposes a circumferential pressure distortion at the impeller outlet. It has been shown by Mizuki et al. (1978) that this can result in zones of return flow, even observable at the impeller inlet (Figure 8.34). The AIRS, observed in Configuration A of Frigne et al. (1984), disappeared after eliminating the circumferential pressure distortion by replacing the volute by an axisymmetric outlet geometry. AIRS due to the flow distortion by a discontinuous decrease of the spanwise width between the impeller and diffuser has been reported by Abdelhanid and Bertrand (1979).

The nearly constant rotational speed, the large amplitude of the perturbations, the unsteadiness of the impeller inlet flow, and the presence of higher harmonics all emphasize the important role the impeller blades play in this type of flow excitation. Positive values of Z_u due to the strong interaction between the diffuser and impeller are more likely to occur at high blade loading or when β_{2bl} is small or zero (Figure 8.32a). This is confirmed by Tsujimoto et al. (1994),

Figure 8.34 Variation of the impeller inlet and outlet velocity at volute off-design operation (from Mizuki et al. 1978).

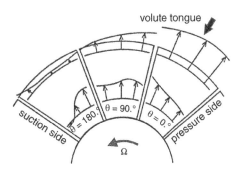

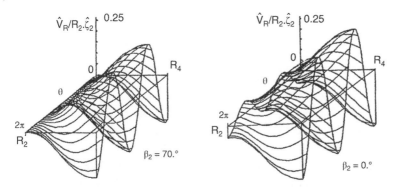

Figure 8.35 Variation of the impeller outlet velocity due to a circumferential perturbation of the diffuser flow ($\hat{\zeta}_2$ = unsteady vorticity at the impeller exit) (from Tsujimoto et al. 1994).

who calculated the impact of an outlet pressure distortion on the flow in a radial ending and a 70° backward leaning impeller. The results (Figure 8.35) show a large variation in the impeller outflow in the case of radial ending impellers and almost no variation in a 70° backward leaning impeller. One can conclude from this that AIRS is more likely to occur at small values of β_2 because it facilitates a distortion of the flow at the impeller outlet.

Figure 8.33 suggests that the AIRS stability limit is almost independent of the diffuser radius ratio. However, the volute off-design pressure distortion and hence also the value of Z_u is likely to increase with decreasing radius ratio.

The model of Abdelhamid predicts also a VDRS type of instability at positive values of Z_p. It is questionable if VDRS can exist under these conditions or if any initial instability will develop into a large amplitude AIRS because of the increase in vibrational energy expressed in Equation (8.11). The VDRS measurements shown in Figure 8.8b are only on the negative slope side of the impeller pressure rise curve ($Z_p < 0$) and change into surge when the slope of the impeller pressure rise curve is zero or positive.

This impeller–vaneless diffuser interaction leading to instabilities has also been theoretically studied by Moore (1989, 1991), who analyzed the dynamic behavior of radial impeller–vaneless diffuser combinations. The boundary conditions at the impeller–diffuser interface are a perturbation of the radial velocity in combination with a circumferentially uniform tangential velocity, i.e. the relative flow is no longer tangent to the blades. This is typical for the unsteady impeller response to an outlet distortion and corresponds to a vortical flow at the diffuser inlet. The system of equations is similar to that of Abdelhamid but is solved analytically. Unsteadiness is due to the extra energy addition resulting from a phase shift between the pressure and velocity variation. Results are presented in graphs indicating rotational speeds that at moderate radius ratios and low flow coefficients ($\phi_2 = V_{R2}/U_2$) are similar to those of AIRS. Moore describes the unsteadiness as rotating waves because at no moment is impeller or diffuser stall required. The theories of Abdelhamid and Moore allow the experimentally observed phenomena to be described, but practical predictions are limited by the difficulty of characterizing the impeller–diffuser interaction, i.e. estimating the flow conditions at the interface.

8.3.2 Comparison with Experimental Results

Distinction between VDRS and AIRS is not always very clear. Abdelhamid and Bertrand (1979) show the impact of a geometrical discontinuity at the exit of an impeller with 27° backward lean on the rotating stall pattern. A zero or small decrease in the diffuser width shows VDRS starting on the negative slope side of the continuous pressure rise curve (points 14 and 15 on Figure 8.19). Larger abrupt contractions at the diffuser inlet result in a discontinuity in the performance curve at the onset of the instabilities with high rotational speed. The rotational speed decreases with increasing radial extent of the diffuser. All these phenomena suggest that the larger flow perturbations at the diffuser inlet have triggered AIRS.

AIRS with perturbations at both low and high rotational speeds is also observed by Abdelhamid (1981) when decreasing the mass flow. As the observations were based on a power spectrum obtained from an average of 128 estimates, it was not possible to verify if both flow patterns exist simultaneously or alternatively. The measurements were made in diffusers with radius ratios of 1.94, 1.6, and 1.4. For the diffuser radius ratio of 1.4, the rotational speed was definitely higher than that predicted by the theory of Jansen (1964b). However, one has to admit that, except for the large rotational speed and higher harmonics in the unsteady pressure signals, there are also indications that some of the instabilities could be VDRS because the variation of the lowest rotational speed and critical diffuser inlet angle with radius ratio follow the predictions by Senoo and Kinoshita (1978) for wide diffusers.

8.4 Progressive Impeller Rotating Stall

A centrifugal impeller has a very complex 3D geometry. Curvature gives rise to important secondary flows further amplified by Coriolis forces and clearance flow (Fowler 1968). An accurate prediction of the stability limits is therefore much more difficult than for vaneless diffusers. This explains why most predictions of impeller rotating stall are more qualitative than quantitative and less accurate.

8.4.1 Experimental Observations

Two models of impeller rotating stall exist in the literature (Day 1993). Although they are based on studies about rotating stall in axial compressors, they do provide some insight into the onset and mechanisms involved in impeller rotating stall. Much less is known about rotating stall in centrifugal impellers.

The first model is based on the work of Emmons et al. (1955) and has been further investigated in the NACA model for 2D cascades by Graham and Guentert (1965). It predicts rotating stall when the incidence angle is at its maximum and a small local perturbation of the upstream flow creates a momentary overload and initiates flow separation in one blade passage (Figure 8.36a).

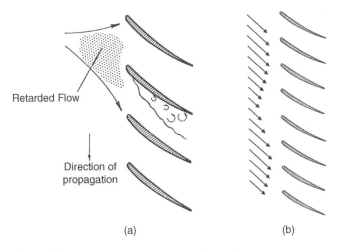

Retarded Flow

Direction of propagation

(a)   (b)

Figure 8.36 Axial compressor rotating stall models.

The flow is diverted to the next blade where separation will occur because the maximum incidence is exceeded. Part of the flow will also divert to the previous blade passage where the incidence decreases and an eventual separation disappears. In this way zones of stalled flow propagate in a direction opposite to the direction of rotation at a speed that is lower than the rotor speed. In a stator it moves in the direction of rotation. The result is a periodic variation of the blade loading. The cell of stalled flow can extend over one or several blade passages and more than one cell can exist on the circumference. This phenomenon is also observed on the negative slope side of the pressure rise curve and is known as spike stall.

The second model presented by Emmons et al. (1955) describes rotating stall by small wave-like flow perturbations rotating at steady subsynchronous speeds (Figure 8.36b). When decreasing the mass flow, the waves grow in amplitude and show small changes in rotational speed. They may occur for a longer period, cover a larger number of blades than the spike stall, do not noticeable change the pressure rise, and do not show a hysteresis when opening the throttle valve. Day (1993) has shown that the modal waves affect the formation of larger stall cells. In axial compressors this kind of stall usually starts near the point of maximum pressure rise. The phenomena are similar to the "fast waves" predicted by Moore (1989) but the onset is linked to the impeller inlet flow distortions. As they are triggered by the impeller, they even exist in combination with very short diffusers and have an almost constant rotational speed that can be very different from the one of spike stall.

Rotating stall in a radial impeller has been observed by Lenneman and Howard (1970). They used hydrogen bubbles to visualize the successive flow pattern in an impeller passage during impeller rotating stall. Figure 8.37a shows how the phenomenon is initiated by flow separation on the blade suction side (channel 1), blocking off the discharge section and reversing the flow direction. The fluid then starts to move out of the passage towards the inducer (channel 2), creating flow separation on the downstream pressure side. This backflow in combination with the Coriolis force creates a passage vortex that is in the same direction as the impeller rotation. Once the positive inflow is reestablished, the passage vortex changes sign and is again opposite to the impeller rotation (channel 3). Together with the more favorable inlet conditions, this helps to reestablish normal flow conditions with return flow on the suction side. The suction side boundary layer then develops again (channel 4), and the whole cycle is repeated. Measurements of the stall show an increase in the rotational speed from 50% to 70% of the rotor speed when decreasing the mass flow.

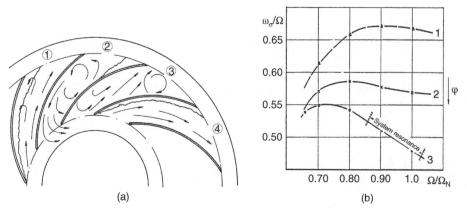

Figure 8.37 (a) Successive flow patterns during progressive impeller rotating stall and (b) variation of the rotational speed (from Lennemann and Howard 1970).

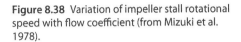

Figure 8.38 Variation of impeller stall rotational speed with flow coefficient (from Mizuki et al. 1978).

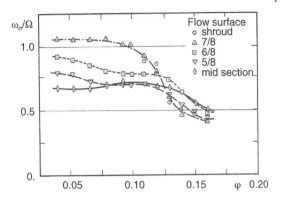

Measurements by Mizuki et al. (1978) on a radial compressor revealed similar values of the rotational speed (Figure 8.38). The value of ω_σ/Ω may even increase from 50% at onset to 100% at minimum mass flow. The stall cells grow from shroud to hub and progressively fill the entire flow passage. This explains the progressive increase in amplitude of the velocity perturbations measured during PIRS (Section 8.1). Impeller rotating stall generated in the inducer is also described by Kämmer and Rautenberg (1982), who measured two stall cells rotating at 44% of the rotor speed.

Interesting studies of the unsteady pressure fields in a centrifugal impeller during rotating stall are presented by Chen et al. (1993, 1994) for a 30° backward leaning impeller and Haseman et al. (1993) for an impeller with radial ending blades. Figure 8.39 displays a set of instantaneous pressure measurements taken along the shroud of the backswept impeller at an operating point just into rotating stall. The eight schematics represent the pressure at different instants of time during one complete stall cycle. The light shaded areas are at low pressure, the darker ones at high pressure.

Plot 6 corresponds to what the authors call normal operating conditions. The iso-pressure lines in the exducer are inclined under the combined effect of centrifugal and Coriolis forces with positive throughflow. A large zone of low pressure on the suction side of the inducer results from the expansion around the leading edge at high incidence.

Plots 8 and 11 are at the onset of a stalled flow. The zone of high pressure starts growing on the pressure side and moves towards the inlet.

Plot 16 shows a large area with high, almost uniform pressure corresponding to full channel stall. The iso-pressure lines in the exducer are at almost constant radius due to centrifugal pressure rise in the absence of blade loading and Coriolis forces because of zero or small throughflow velocity. The large pressure rise, observed in the inducer on plot 16, is an unsteady phenomenon resulting from the inertia of fluid when suddenly decelerated.

Plot 20 shows an opposite slope of the iso-pressure lines near the suction side of the exducer. This is the consequence of a local return flow along this side, in line with the vortex rotating in the direction of the impeller as visualized by Lenneman and Howard (Figure 8.37).

After the zone of high pressure has been blown out by the centrifugal forces the velocity suddenly increases in the whole blade channel. The high velocity gives rise to a large suction to pressure side pressure gradient and a large zone of low pressure, extending over almost the complete channel (plot 24).

The normal blade loading is gradually reestablished, first near the full blade pressure side (plot 26) resulting in a uniform inclination of the isopressure lines in the exducer. Isopressure lines at constant radius near the suction side of the full blade indicate a small velocity in that section.

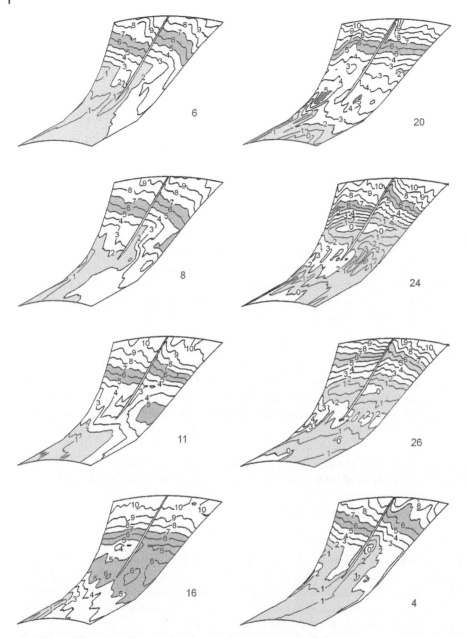

Figure 8.39 Sequence of iso-pressure lines during rotating stall in an impeller channel with splitter vane (from Chen et al. 1993).

The suction to pressure side pressure increase is gradually reestablished everywhere (plot 4). The rapid pressure rise along the suction side near the exit shown on plot 6 will provoke suction side flow separation, after which the stall pattern develops again.

Because flow separation and return flow play an important role in the onset of rotating stall, it is not surprising that a limit value of the outlet-to-inlet velocity ratio is used as a criterion for the onset of impeller rotating stall. The data in Figure 8.40 are obtained from a large number of different impellers (Rodgers 1977, 1978). They relate the maximum overall diffusion to (a)

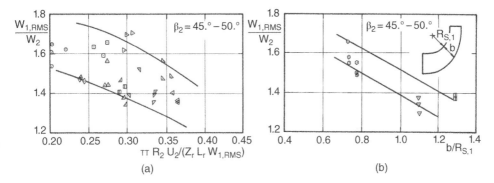

Figure 8.40 Inlet-to-outlet velocity ratio at the onset of PIRS as a function of (a) overall diffusion and (b) meridional curvature (from Rodgers 1978).

Figure 8.41 Throat over leading edge velocity ratio at onset of PIRS (from Kosuge et al. 1982).

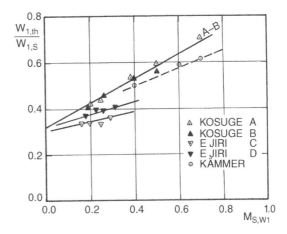

the blade loading or (b) the meridional curvature. The correlations show a trend but the large scatter does not allow a reliable prediction of the stability limit.

Impeller rotating stall is probably more dependent on separation in the inducer than on diffusion in the exducer. Separating both parts of the impeller could therefore result in a better prediction. Yoshinaka (1977) relates impeller stall to the non-dimensional static pressure rise between the impeller leading edge and the throat. This is similar to the semi-vaneless space stall criterion proposed by Kenny (1970) for vaned diffusers (Figure 8.46).

A similar approach by Kosuge et al. (1982) relates onset of impeller rotating stall to the inducer leading edge-to-throat velocity ratio as a function of the inlet Mach number at the shroud (Figure 8.41). Experimental data from two impellers show a good agreement. However, additional data of Kämmer and Rautenberg (1986) and Ejiri et al. (1983), also shown on Figure 8.41, cast some doubt on the general validity of this correlation.

A correlation proposed by Japikse (1996), relating the stability limit to the inducer shroud incidence, makes a distinction between advanced and common compressor designs. One observes again a large scatter (Figure 8.42) that cannot be explained by the reported differences in diffuser and outlet collector geometry. The largest incidence should be interpreted as an indication of the achievable limits for a good design with favorable (uniform) inlet conditions. This type of correlation does not explicitly account for the impact of the blade thickness and leading edge angle on the optimum and hence also on the maximum incidence (Figure 2.25).

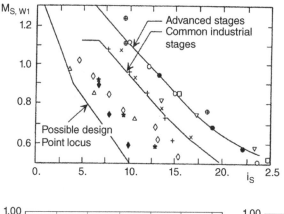

Figure 8.42 Incidence at onset of PIRS (from Japikse 1996).

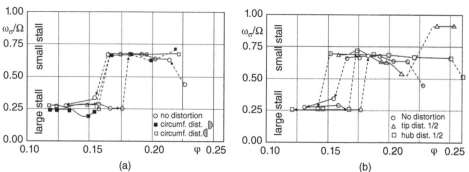

Figure 8.43 Variation of absolute propagation speed with flow coefficient and (a) circumferential and (b) spanwise inlet distortion (from Ariga et al. 1987).

The influence of inlet distortions on impeller rotating stall has been examined by Ariga et al. (1987). Figure 8.43 displays the variation in the rotational speed for (a) circumferential inlet distortion and (b) spanwise distortion. In the absence of distortion one observes a rapid increase in rotational speed from 50% at onset to $\approx 70\%$ at decreasing flow coefficient. It is reported by the authors that the amplitude of the velocity variations increases progressively with decreasing flow coefficient, which agrees very well with the observations by Frigne et al. (1984) for PIRS (Section 8.1). A sudden increase in amplitude occurs when the rotational speed drops to 30%. Ariga et al. (1987) consider this as "large stall", triggered by the diffuser. Indeed both the sudden increase in amplitude and the lower rotational speed are what one can expect from AIRS resulting from a strong interaction between the impeller and diffuser. A hysteresis is observed when recovering from it because, once started, large stall can temporarily suppress PIRS. The main effect of a circumferential distortion is the absence of a hysteresis at the limit between large and small stalls.

A spanwise inlet distortion at the hub or shroud shifts the stall onset to larger flow coefficients (Figure 8.43b). In the case of hub distortion the rotational speed (ω_σ/Ω) starts at 90% and suddenly decreases to almost 50%, after which it gradually increases again with decreasing mass flow. Shroud distortion also shifts the stall onset to larger flow coefficients but the transition to large stall is postponed to smaller flow coefficients. It is more difficult for an AIRS to take over from a stronger PIRS because AIRS is related to the unsteady reaction of the impeller to circumferential flow distortions.

The impeller is more sensitive to the spanwise distortion than to the circumferential one because the former appears permanently. The incidence is continuously higher and the impeller cannot profit from an unsteady temporary overshoot of maximum incidence occurring with circumferential distortion.

8.5   Vaned Diffuser Rotating Stall

Vaned diffuser rotating stall shows similarities to impeller rotating stall. However, the incoming flow is more complex because of the large spanwise non-uniformity, the very tangential flow direction, and the unsteadiness of the flow in both the fixed and rotating frames of reference. Typical temporary changes in the inlet angle are of the order of 10° to 15° in combination with a much larger spanwise variation of up to 25° (Krain 1981; Liu et al. 2010). The high frequency of the unsteadiness ($S_R > 2$) and rapid mixing of the flow explain the relative insensitivity of the vaned diffuser to the circumferential distortion of the blade to blade flow. The spanwise non-uniformity is considered to be the major destabilizing factor because of its permanent impact on the vane incidence (Section 7.2.2).

The pressure rise in the individual subcomponents of a vaned diffuser (vaneless and semi-vaneless space, diffuser channel, and diffuser exit) together with the overall pressure rise has been measured by Hunziker and Gyarmathy (1994). The results are shown in Figure 8.44. The pressure rise versus mass flow curve for the vaneless and semi-vaneless space (2–th) shows a stabilizing negative slope over the whole operating range, leveling off at minimum flow coefficient. The diffuser channel (th–4) and exit (4–5) show a positive slope and are often considered the critical part of the vaned diffuser. A positive slope of the pressure versus mass flow curve destabilizes the flow in rotating systems because any change in mass flow will add extra energy to the perturbation, which will increase its amplitude (Section 8.3.1). However, there is no energy added in non-rotating systems. Diffusers are less destabilizing than impellers because they only dissipate energy. The impact on the stability of a positive slope of the pressure rise curve of a non-rotating component is to change the slope of the overall performance curve.

The increasing pressure rise in the diffuser inlet section at decreasing mass flow results in an increase in the throat blockage (Figure 4.35) and hence in a decrease in the pressure rise in the

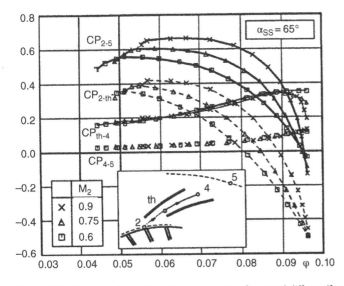

Figure 8.44 Pressure rise in the subcomponents of a vaned diffuser (from Hunziker and Gyarmathy 1994).

downstream divergent channel (Figure 4.43). A vaned diffuser is likely to remain stable as long as the increase in the pressure rise in the semi-vaneless space is larger than the decrease of the pressure rise in the diverging channel because of the increased throat blockage. Although the pressure rise between the impeller exit and the throat section has a negative slope, it is generally accepted that this is the critical part of the diffuser in terms of stability because of its impact on the downstream components.

Figure 8.18 shows how the stability of a vaneless diffuser depends on the inlet flow angle, the radial extent, and the width-to-inlet radius ratio. Although it is questionable if the correlations are still accurate when the axisymmetry of the flow is distorted by the upstream influence of the diffuser vanes, they still have some interest and can be guidelines when defining the radial extension of the vaneless space.

In analogy to axial compressors distinction is made between modal and spike type instabilities. The impact of modal perturbations on overall performance can be very small and dedicated instrumentation may be needed to detect them. Depending on the operating conditions the amplitude of an initial perturbation will grow or decrease with time in a reversible way (Day 1993). Compression system modeling techniques (Longley 1994) allow an estimate of this type of instability by considering the impeller diffuser interaction.

When throttling the compressor the waves might transform in an irreversible way into a spike-type rotating stall. The rotational speed of the latter is often much larger than that of the former. However, modal perturbations and finite stall cells are not necessarily consecutive phenomena. With early detection and depending on the growth rate of the amplitude there might be room for control of the modal perturbation.

Spike-type instabilities are likely to originate from the irreversible interaction of the incoming flow with the diffuser vane leading edge. They appear suddenly and are therefore difficult to control. Everitt and Spakovszky (2013) describe the mechanism of the spike-type stall as a self-sustained instability in the vaned diffuser without interaction with the impeller. This is illustrated by the numerical results in Figure 8.45, obtained from diffuser only calculations. The inlet conditions, defined from steady flow simulations in a single impeller channel, are axisymmetric but with a decrease in the radial velocity component near the shroud. This gives rise to a large incidence on that side of the diffuser vanes. When reducing the mass flow, a separation bubble is formed on the leading edge suction side that locally blocks the diffuser inlet, resulting in return flow in an area where the radial velocity is already small or negative. According to Everitt and Spakovszky (2013), vorticity is generated and convected backward into the vaneless space where it adds to the disturbances that are moving around and results in an increased blockage in the vaneless and semi-vaneless space. This further destabilizes the flow in the diffuser and leads to large spike-type stall cells.

This spike-type instability is more likely to occur at the highest rotational speeds, i.e. at higher incidence and a large spanwise non-uniformity of the flow. The losses have a rather large impact on the overall performance and may trigger surge where the steady overall performance curve has a small negative slope. The modal type of rotating stall is more likely to occur at lower RPM with spanwise more uniform flow.

Although the diffuser leading edge incidence is a key factor in the onset of diffuser rotating stall, it is not a common stability criterion. One reason for this may be the difficulty of defining the incidence of a fluctuating and spanwise non-uniform flow. Most stall criteria are based on the static pressure rise from the vane leading edge to the throat, which, to a large extent, depends on the incidence. At positive incidence the flow accelerates around the leading edge, requiring extra diffusion before reaching the throat section. Rodgers (1961) states that at inlet

Figure 8.45 Counter-rotating vortices near the shroud of a vaned diffuser inlet at small outflow (from Everitt and Spakovszky 2013).

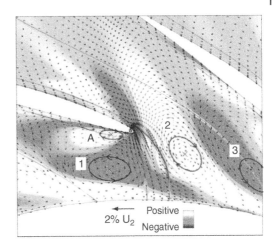

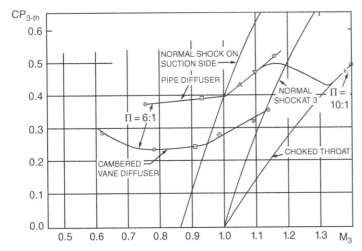

Figure 8.46 Limit of static pressure rise in the semi-vaneless space (from Kenny 1970).

Mach numbers exceeding 0.85, most diffusers for radial compressors will barely take any positive incidence and will probably stall between 0° and +2° incidence. This is confirmed by the experimental data in Figure 4.35.

The maximum static pressure rise is limited by stall, resulting in a breakdown of the performances of the whole diffuser because of excessive throat blockage. Practical prediction methods make no distinction between deep stall and surge because the former is likely to trigger the latter. The limit of the static pressure rise, shown in Figure 8.46, is from Kenny (1970). The limit static pressure rise coefficient is about 0.4 for subsonic flows and increases at supersonic flows because of the additional static pressure rise over the shock. The results in Figure 8.46 suggest that pipe diffusers perform better than vaned island ones. However, other data show a comparable maximum pressure rise with vaned island diffusers (Came and Herbert 1980) or even superior both in terms of pressure rise and operating range (Rodgers and Sapiro 1972).

Japikse and Osborne (1986) compared the experimental values of the maximum static pressure rise CP_{2-th} of different diffusers with the data in Figure 8.46 and observed important discrepancies. This reflects the influence of the inlet flow conditions, diffuser leading edge incidence, and detailed vane geometry on the flow in the vaneless and semi-vaneless space, and

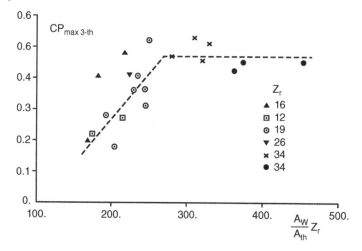

Figure 8.47 Correlation for the semi-vaneless space pressure recovery at surge (from Elder and Gill 1984).

explains the large scatter in experimental values. Conrad et al. (1980) corrected the maximum pressure rise as a function of the vane suction side angle.

Various investigators have suggested that the ratio of impeller over diffuser vane number has an influence on the flow range. The idea behind this relates to the ratio of the width of the wake emerging from the impeller over the pitch of the vane passage. Large diffuser vane pitches are less likely to be blocked by impeller blade wakes at high incidence whereas small diffuser pitches may be completely blocked. It is argued that increasing the number of impeller blades has a stabilizing effect because it results in smaller wakes and increased frequency of the diffuser inlet flow perturbations.

Elder and Gill (1984) extended the ideas of Baghdadi and McDonald (1975) and Came and Herbert (1980) to the correlation shown in Figure 8.47 relating the maximum static pressure rise at stall (and surge) to the ratio of the whetted surface of the semi-vaneless space A_W over the diffuser throat area A_{th} and the number of impeller blades including splitter blades (Z_r). The whetted surface increases with increasing diffuser leading edge radius and decreasing number of diffuser vanes. The larger this surface the more time the flow has to mix out before reaching the diffuser throat and the smaller will be the unsteadiness. The correlation also indicates that there is no reason to increase the wetted surface, i.e. reduce the number of vanes beyond some limit. A too small number of diffuser vanes increases the risk of rotating stall in a too large vaneless and semi-vaneless space. At a given diffuser length it also decreases the L/O_{th} ratio of the diverging channel.

Decreasing the number of diffuser vanes allows an increase in the vane suction side angle without changing the throat section, i.e. without changing the choking limit (Figure 4.37). Linking the diffuser stall limit to zero incidence operation means that by reducing the number of diffuser vanes, the stable operating range limit can extended to smaller mass flows. This is a possible reason for the increase in operating range experienced by Japikse (1980) when reducing the number of diffuser vanes from 34 to 17 (Figure 8.48). He also mentions the favorable impact of an increased shock pressure recovery on the maximum pressure increase upstream of the throat. It was not specified if in addition to the vane number other parameters, such as total throat area and vane setting angle, have also been changed.

The semi-vaneless space interacts with the divergent channel downstream of the throat (2D or conical) through the throat blockage. It has been experimentally demonstrated that a change in throat blockage will change the value of CP but not the locus of maximum CP (Section 4.2.4).

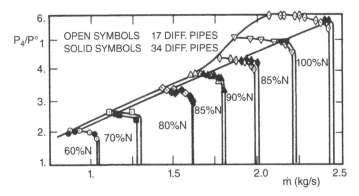

Figure 8.48 Influence of compressor vane number on stability (from Japikse 1980).

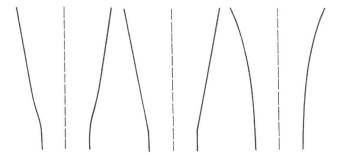

Figure 8.49 Contouring of channel diffusers.

Hence a gradual increase in throat blockage will not provoke an abrupt change in performance but a gradual decrease in the pressure rise in the downstream channel. A discontinuous increase in the throat blockage because of flow separation on the vane leading edge is considered to be the major cause of a spike-like rotating stall. The divergent channel should therefore not be considered as the major trigger of stall but rather in combination with the vaneless and semi-vaneless space.

Clements and Artt (1987a,b, 1988) mention a dependence of the stall limit on the divergence angle of the divergent channel. Increasing the opening angle allows the *CP* to be maximized (Reneau et al. 1967) but makes the diffuser more sensitive to a change in inlet flow conditions. Channel diffusers can be optimized by contouring the walls (Figure 8.49). Based on boundary layer considerations Huo (1975) concluded that the losses will be minimal when the flow is rapidly decelerated to conditions close to separation and then further decelerated at a lower rate up to the exit, i.e. controlled diffusion. This requires a bell-shaped diffuser. Carlson et al. (1967) showed a small improvement when shaping the lateral walls in this way. However, the flow is everywhere close to separation, and a small change in inlet flow conditions may cause a large decrease in the pressure rise because of flow separation near the inlet of the diverging channel. Based on stability considerations, it was recommended by Came and Herbert (1980) that a trumpet-shaped channel should be used. It was expected that this would enhance the diffuser stability at off-design operation by leaving some margin for an increase in diffuser inlet blockage. An eventual separation would then occur closer to the exit and hence have a smaller impact on stability. However, friction losses will be larger because the velocity remains higher over a longer distance. Experience has shown that there is almost no difference between the commonly used straight-walled diffuser and a bell-shaped one.

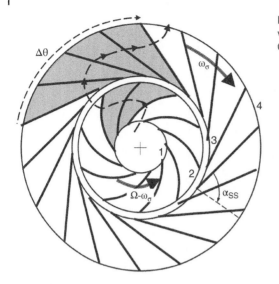

Figure 8.50 Single cell rotating stall streamlines viewed in the stall cell fixed reference frame (from Gyarmathy 1996).

The rotational speed of the stall cell(s) in centrifugal compressors with vaned diffusers can be calculated in the same way as for axial compressors (Cumpsty and Greitzer 1982; Gyarmathy 1996). Viewing the flow in a system rotating with the stall cell, the rotor will rotate with speed $\Omega - \omega_\sigma$ and the downstream diffuser in the opposite direction at speed $-\omega_\sigma$. Figure 8.50 gives a schematic view of the streamline path of the fluid trapped in a stall cell. Entering the impeller at the leading side of the stalled area the fluid undergoes a sudden deceleration because the outlet is blocked by the stalled flow in the diffuser. The consequence of this sudden deceleration is an extra pressure increase at the impeller outlet. By the time the stalled fluid has accumulated sufficient pressure rise to blow out the stalled diffuser flow, the impeller has rotated over a distance $\Delta T(\Omega - \omega_\sigma)$, the circumferential extend of the stall cell. Once the stalled fluid has entered the diffuser it starts rotating in the opposite direction into the region of the vaned diffuser stall, where it is pushed outward by the high pressure at the outlet of the impeller stall cell. At the mean time the diffuser has rotated over the extent of the stall cell now defined by $\Delta T \omega_\sigma$. The number of impeller channels required to reach the outlet pressure that is needed to blow the stalled flow out of the diffuser channels depends on the momentum of the fluid in the respective channels. This ratio defines the relative speed of the stall in the impeller and diffuser.

The unsteady incompressible and inviscid flow in a pipe is defined by

$$\frac{\partial V}{\partial t} + V\frac{\partial V}{\partial s} = -\frac{1}{\rho}\frac{dP}{dl} \tag{8.13}$$

Assuming that at the borders of the area with stalled flow the unsteady acceleration ($\frac{\partial V}{\partial t}$) is much larger than the convective one ($V\frac{\partial V}{\partial s}$), Equation (8.13) reduces to:

$$\frac{\partial V}{\partial t} = -\frac{1}{\rho_r}\frac{\Delta P}{L_r} \tag{8.14}$$

When the fluid in the impeller blade channels of length L_r passes from unstalled to stalled, the velocity changes by an amount $\Delta V = \Delta V_R / \cos\widetilde{\beta}$. Assuming further that this variation occurs in a time corresponding to the time needed by the impeller to rotate over a fraction $\Delta\theta$ of the circumference provides following approximation of the acceleration:

$$\frac{\partial V}{\partial t} \approx -\frac{\Delta V_R}{\cos\widetilde{\beta}}\frac{(\Omega - \omega_\sigma)}{\Delta\theta} \tag{8.15}$$

Eliminating $\frac{\partial V}{\partial t}$ between Equations (8.14) and (8.15) results in the following pressure pulse across the impeller when fluid is entering or leaving the stall cell:

$$\Delta P_r = \rho_r \frac{L_r}{\Delta \theta} \frac{\Delta V_R}{\cos \widetilde{\beta}} (\Omega - \omega_\sigma) \tag{8.16}$$

Note that the pressure rise by the centrifugal forces $\Delta P = \rho(U_2^2 - U_1^2)$ is not considered because it is independent of the fluid velocity and therefore equal in the stalled and unstalled zones. Measurements during rotating stall indicate a strong circumferential variation of the static pressure in the space between the impeller and diffuser.

Assuming a fixed compressor inlet and outlet pressure without circumferential distortion, any increase in the impeller static pressure must be compensated for by an opposite decrease in the vaned diffuser:

$$\Delta P_D + \Delta P_r = 0 \tag{8.17}$$

The ΔP_D of interest is the one resulting from a sudden change of velocity in the diffuser channels of length L_D by a value ΔV. This velocity along a direction close to the diffuser leading edge suction side $\alpha_{3,SS}$ is obtained from continuity defining ΔV_R:

$$\Delta P_D = \rho_D \frac{L_D}{\Delta \theta} \frac{\Delta V_R}{\cos \alpha_{3,SS}} (\omega_\sigma) \tag{8.18}$$

The impeller blade and diffuser channel length are approximated by

$$L_r = \frac{R_2 - R_1}{\cos \widetilde{\beta}} \qquad\qquad L_D = \frac{R_4 - R_2}{\cos \widetilde{\alpha}_{3,SS}}$$

Substituting Equations (8.16) and (8.18) into Equation (8.17) and because the pressure perturbation acts over the same circumferential distance $\Delta \theta$ on impeller and diffuser, one obtains the following relation between the stall cell rotational speed ω_σ and the impeller rational speed Ω:

$$\rho_D \frac{R_4 - R_2}{\cos^2 \widetilde{\alpha}_{3,SS}} \Delta V_{R,2} \omega_\sigma + \rho_r \frac{R_2 - R_1}{\cos^2 \widetilde{\beta}} \Delta V_{R,2} (\Omega - \omega_\sigma) \tag{8.19}$$

or, more explicitly

$$\frac{\omega_\sigma}{\Omega} = \frac{\rho_r \dfrac{R_2 - R_1}{\cos^2 \widetilde{\beta}} \Delta V_{R,2}}{\rho_D \dfrac{R_4 - R_2}{\cos^2 \widetilde{\alpha}_{3,SS}} \Delta V_{R,2} + \rho_r \dfrac{R_2 - R_1}{\cos^2 \widetilde{\beta}} \Delta V_{R,2}} = \frac{1}{1 + \mathcal{A}} \tag{8.20}$$

where

$$\mathcal{A} = \frac{\rho_D \Delta R_D}{\rho_r \Delta R_r} \frac{V_{R2}^2}{\cos^2 \widetilde{\alpha}_{3,SS}} \frac{\cos^2 \widetilde{\beta}}{V_{R2}^2}$$

The value of $\mathcal{A}$ is the ratio of the inertia of the fluid in the diffuser over the inertia of the fluid in the impeller. The sudden decrease in the momentum in the impeller when the outlet is blocked by the stalled diffuser creates a pressure pulse that blows the stalled flow out of the diffuser channels. The speed with which this occurs depends on the mass and velocity change at the common border.

Gyarmathy (1996) has verified this approach on different impeller diffuser combinations and found good agreement with experimental data (Figure 8.51) for a wide range of rotational speeds (between 5% and 85%).

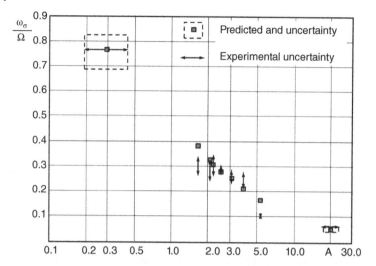

Figure 8.51 Predicted versus experimental vaned diffuser rotating stall speed (from Gyarmathy 1996).

The previous model is not directly applicable to vaneless diffusers because there is no clear coupling between the circumferential extension of the momentum exchange area of the impeller and diffuser. The fluid leaving the impeller is no longer trapped in a diffuser channel but can spread over a large circumferential distance. Hence it is not possible to estimate the amount of diffuser fluid involved in the momentum exchange. However, one may conclude that the rotational speed of impeller rotating stall decreases with increasing radial extension of the vaneless diffuser.

8.5.1   Return Channel Rotating Stall

Instabilities provoked by a vaned return channel have been presented by Bonciani et al. (1982). They observed rotating instabilities at a frequency between 9% and 15% of the impeller rotational speed and the frequency increased with decreasing flow coefficient (Figure 8.52a). In both Configurations A and B, instabilities start at the same incidence angle on the return channel blades and provoke an important decrease in return channel pressure rise (Figure 8.52b). This type of rotating stall is quite similar to vaned diffuser rotating stall because it is also the consequence of a too high incidence on the vanes at low mass flow (Figure 8.53).

8.6   Surge

Surge is an unstable operation of the complete compression system (Figure 8.54). It consists of periodic variations of the mass flow in the compressor and throttle system with periodic changes in the pressure in the plenum. The throttle system can be a valve or a turbine.

Mild surge consists of small amplitude flow oscillations over the entire annulus without net backflow at any time. Deep surge is when strong system oscillations occur, including flow reversal through the compressor. This is a damaging compressor operation and to be avoided.

Surge analysis requires the modeling of all the components, including the inlet and outlet duct, pressure vessels, and throttling system. Such a modeling, proposed by Emmons et al. (1955) and further developed by Taylor (1964) and Dussourd et al. (1977), has been extended to nonlinear systems and applied to axial compressors by Greitzer (1976, 1981).

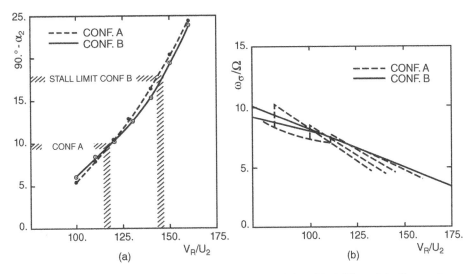

Figure 8.52 Influence of the return channel geometry on (a) the critical diffuser inlet flow angle α_{2c} and (b) the variation of rotational speed with flow coefficient for the return channel rotating stall (from Bonciani et al. 1982).

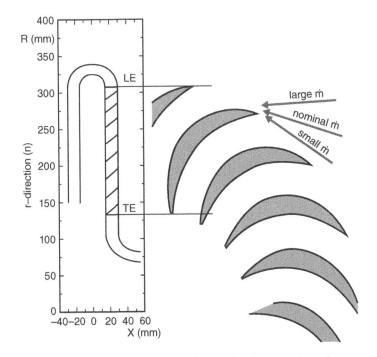

Figure 8.53 Variation of incidence with mass flow in return channels.

The compressor is replaced by a characteristic function generator and a constant area pipe. In this region the fluid has an important kinetic energy and the dimensions A_c and L_c are defined in such a way as to create the same dynamics for a given change in mass flow. The reservoir is considered to be a large capacity at variable pressure (potential energy storage) but with negligible kinetic energy. The pressure in the reservoir is controlled by a throttle valve. The analog

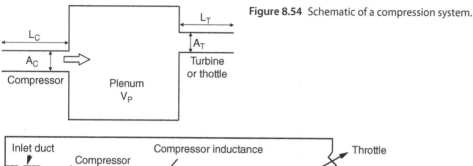

Figure 8.54 Schematic of a compression system.

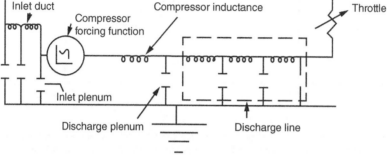

Figure 8.55 Electrical analogy of compression system (from Dean and Young 1977).

electric scheme (Figure 8.55), which also accounts for the inlet and discharge pipes, is from Dean and Young (1977).

8.6.1 Lumped Parameter Surge Model

Continuity relates the instantaneous amount of fluid in the plenum volume $\mathcal{V}$ to the incoming and outgoing mass flow:

$$\dot{m}_C - \dot{m}_T = \frac{d(\rho \mathcal{V}_P)}{dt} \tag{8.21}$$

From the isentropic compression law and state equation one knows that:

$$\frac{dP}{P} = \kappa \frac{d\rho}{\rho} = \kappa \frac{d\rho}{P} R_G T \tag{8.22}$$

hence

$$\frac{d\rho}{dt} = \frac{1}{\kappa R_G T} \frac{dP}{dt} \tag{8.23}$$

so that Equation (8.21) can be written as

$$\dot{m}_C - \dot{m}_T = \mathcal{V}_P \frac{1}{a^2} \frac{dP_P}{dt} \tag{8.24}$$

where $\dot{m}_C$ and $\dot{m}_T$ are the instantaneous mass flow in the compressor and throttle, a is the speed of sound, and P_P is the plenum pressure.

The unsteady variation of the velocity of the incompressible fluid in a compressor with cross section A_C and length L_C (Figure 8.56) can be described by a solid body acceleration or deceleration driven by the difference between the compressor outlet pressure and the plenum pressure. Referring both to the same inlet pressure results in

$$A_C \int_{in}^{out} \frac{\partial \dot{m}_C}{\partial t} \frac{dl}{A} = (\Delta P_C - \Delta P_P) A_C = \rho A_C L_C \frac{dV_C}{dt} \tag{8.25}$$

Figure 8.56 Unsteady incompressible flow in a pipe.

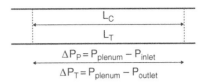

where ΔP_C is the instantaneous pressure rise in the compressor and ΔP_P is the difference between the plenum and inlet pressure. They are the same during steady operation but can temporarily be different from each other during unsteady and transient operation. L_C, defined by Equation (7.5), is the equivalent compressor length that accounts for the local accelerations in a real compressor due to local changes of the cross section area. A_C is the reference compressor (inlet or outlet) cross section area.

The change in mass flow can then be written as

$$\frac{d\dot{m}_C}{dt} = (\Delta P_C - \Delta P_P)\frac{A_C}{L_C} \tag{8.26}$$

Applying the same theory to a turbine or throttle valve of equivalent length L_T and cross section A_T, the instantaneous change of mass flow in the throttling system can be described by

$$\frac{d\dot{m}_T}{dt} = (\Delta P_P - \Delta P_T)\frac{A_T}{L_T} \tag{8.27}$$

where ΔP_T is the instantaneous pressure drop in the turbine or throttle and ΔP_P is the difference between the plenum and outlet pressure. Assuming that the throttle outlet pressure equals the compressor inlet pressure, the value of ΔP_P is the same as in Equation (8.25).

Substituting $\phi = \dot{m}/(A_C\rho U_2)$ and $\psi = \Delta P/(\rho U_2^2/2)$ into Equation (8.24) one obtains

$$\phi_C - \phi_T = \frac{V_P U_2}{2a^2 A_C}\frac{d\psi_P}{dt} \tag{8.28}$$

Splitting the flow coefficients into a time-averaged value and a fluctuating one, and because the averaged flow coefficients $\widetilde{\phi}_C = \widetilde{\phi}_T$, one obtains

$$\frac{d\psi_P}{dt} = \frac{2a^2 A_C}{V_P U_2}(d\phi_C - d\phi_T) \tag{8.29}$$

Substituting the flow and work coefficient also in Equation (8.26), the variation of the compressor flow coefficient is given by

$$\frac{d\phi_C}{dt} = \frac{U_2}{2L_C}(\psi_C - \psi_P) \tag{8.30}$$

In the same way the change in flow coefficient in the throttle valve at a fixed throttle valve cross section area is defined by

$$\frac{d\phi_T|_{(A_T=C^{te})}}{dt} = \frac{A_T}{A_C}\frac{U_2}{2L_T}(\psi_P - \psi_T) \tag{8.31}$$

Equations (8.29), (8.30), and (8.31) are the three basic equations describing the unsteady behavior of the compression system.

However, as we are interested only in defining the stability limit, the analysis can be based on the linearized quasi steady compressor performance curve. This allows a prediction of the onset of surge but not the amplitude of the unsteady flow and pressure changes in the destabilized region (Gysling et al. 1991). Calculation of the unsteady pressure and mass flow variation during surge would require an additional equation to account for the time delay between the unsteady flow variations and the quasi steady ones. Local linearization of the compressor and throttle valve characteristic around a mean value results in:

$$\psi_C = \widetilde{\psi}_C + \psi'_C d\phi_C \tag{8.32}$$

$$\psi_T = \widetilde{\psi}_T + \psi'_T d\phi_T|_{(A_T=C^{te})} \tag{8.33}$$

where ψ'_C and ψ'_T are the slope of the non-dimensionalized compressor and throttle versus mass flow curves (Figures 8.58 and 8.59):

$$\psi_P = \widetilde{\psi_P} + d\psi_P \tag{8.34}$$

Assuming that the plenum pressure and the flow in the compressor and throttle change at the same frequency,

$$d\psi_P = \mathcal{P}e^{St} \tag{8.35}$$

$$d\phi_C = \mathcal{C}e^{St} \tag{8.36}$$

$$d\phi_T = \mathcal{T}e^{St} \tag{8.37}$$

where S is a complex number $A_\sigma + i\omega_\sigma$ and e^{St} can also be written as

$$e^{St} = e^{(A_\sigma)t}\,e^{(i\omega_\sigma)t} \tag{8.38}$$

As illustrated on Figure (8.57), such a system is stable if the real part of S, i.e. A_σ, < 0 because any perturbation will decrease in amplitude with time. $A_\sigma > 0$ results in an increase in the amplitude with time and hence an unstable system. The imaginary part of S, i.e. ω_σ, defines the frequency of the periodic change. $\omega_\sigma = 0$ results in an aperiodic growth or decay of an initial perturbation.

Substituting Equations (8.35), (8.36), and (8.37) in Equations (8.29), (8.30), and (8.31), and putting the equivalent length of the throttle valve equal to zero, results in a system of equations defining $\mathcal{P}$, $\mathcal{C}$, and $\mathcal{T}$. It has a solution if the determinant is zero, i.e. when S satisfies following

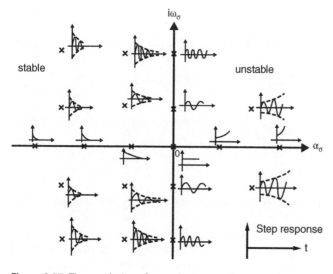

Figure 8.57 Time evolution of perturbation as a function of α_σ.

Figure 8.58 Static stability condition.

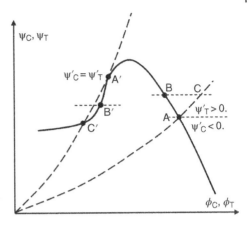

Figure 8.59 Dynamic stability condition.

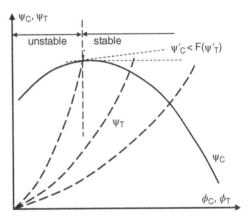

second-order equation, relating the geometry and the local compressor and throttle character-istics defined by ψ'_C and ψ'_T:

$$S^2 + \left(\frac{2A_C \, a^2}{V_P \, U_2} \frac{1}{\psi'_T} - \frac{U_2}{2l_C} \psi'_C \right) S + \frac{A_C \, a^2}{L_C \, V_P} \left(1 - \frac{\psi'_C}{\psi'_T} \right) = 0 \tag{8.39}$$

or in brief

$$S^2 + BS + D = 0 \tag{8.40}$$

The system will be stable if both roots of S

$$S_{1,2} = -\frac{-B + \sqrt{B^2 - 4D}}{2} \tag{8.41}$$

have a negative real value, hence when $D > 0$ and $B > 0$.

The first condition requires that the slope of the throttle line must be larger than the slope of the compressor performance curve ($\psi'_T > \psi'_C$). This is the static stability condition and is illustrated in Figure 8.58.

An accidental increase in plenum pressure A will result in a decrease in the mass flow in the compressor to point B and an increase in the mass flow in the throttle to point C. As more flow will be leaving than entering the plenum, the plenum pressure will decrease to its equilibrium position A. Hence any accidental change in the plenum pressure is annihilated by a change in the mass flow in the compressor and throttle.

This is no longer the case when the compressor operates at a mass flow between points A′ and C′. At plenum pressure B′, the compressor will provide a mass flow that is larger than that going out at the throttle. Hence, the plenum pressure will increase until the equilibrium is reestablished again at point A′. Stable operation is possible only at mass flows below C′ and larger than A′.

The second stability condition is more restrictive and requires that $\mathcal{B} > 0$, or

$$\psi'_C \leq \frac{4\,\mathcal{V}_C}{M^2_{u,2}}\,\frac{1}{\mathcal{V}_P}\,\frac{1}{\psi'_T} = \frac{1}{B^2}\,\frac{1}{\psi'_T} \tag{8.42}$$

where

$$B^2 = \frac{M^2_{u,2}\mathcal{V}_P}{4\mathcal{V}_C} \tag{8.43}$$

$\mathcal{V}_C = L_C A_C$, the compressor volume, and $M_{u,2}$ is the Mach number based on the peripheral velocity U_2.

This dynamic stability criterion predicts the onset of surge when the slope of the compressor pressure rise curve exceeds a positive value function of the slope of the throttle characteristics and of what is commonly called the Greitzer B^2 factor. The latter is proportional to the peripheral Mach number and the ratio of the plenum over the compressor volume.

Compact compressors operating at a high peripheral speed and with a large plenum will be less stable than large compressors operating at low peripheral speed and a small plenum. It is common practice in industry to limit the operating range of a compressor to the point of maximum pressure rise because the range with positive slope is potentially unstable. Fixing the stability limit at the point of maximum pressure ratio is on the safe side if it is not a priori known if the compressor will operate with a large or small plenum.

The accuracy of this lumped parameter model to predict the onset of surge depends on the accuracy by which the slope of the compressor pressure rise curve can be defined. The latter is a function of the work input and losses, hence is subject to uncertainties. The slope of the pressure rise curve will be smooth only if there is no sudden change in losses or energy input, such as that at the onset of AIRS. Progressive stall in the diffuser or impeller may trigger surge by making the slope of the compressor curve more positive. One should not conclude from this that rotating stall is a necessary condition for surge.

The lumped parameter model allows the experimental pressure rise curve shown in Figure 8.60 to be explained. The compressor cross section has a throttling system at the exit of the vaneless diffuser to avoid the upstream propagation of the volute pressure distortions. The plenum volume is very small because it is limited to the vaneless diffuser. The performance map illustrates how the use of a close-coupled throttling valve allows stable operation also to the left of the point of maximum pressure rise.

The impact of the Mach number on stability is illustrated in Figure 8.61. Surge occurs on the left side of the maximum pressure rise ($\psi'_C > 0$) for tests at low RPM because of the small B^2 value at low M_2 value. This does not exclude the possibility that rotating stall occurs, as suggested by the drop in efficiency at 14 000 RPM. The system becomes less stable and the surge and stall line coincide at higher rotational speed.

The destabilizing effect of a positive slope of the pressure rise curve is similar to that explained in Section 8.3.1 for AIRS. It is analogue to what occurs in a mass spring system (Figure 8.62). The oscillation amplitude increases when the excitation force $F(t)$ is in phase with the movement $V(t)$, i.e. when energy is added to the system. Otherwise the vibration will quickly damp out.

The previous stability model also explains why pumps do not experience surge ($M_2 = 0$) if the whole system is filled with incompressible fluid. An important condition for a compression

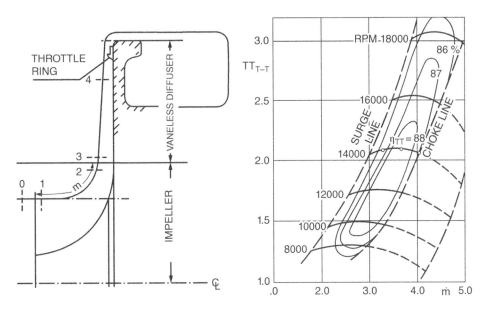

Figure 8.60 Diffuser throttling valve and corresponding performance curve (from Eckardt 1976).

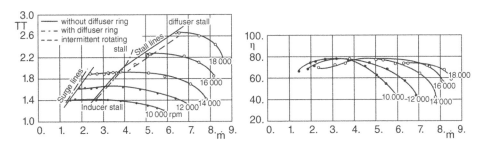

Figure 8.61 Impact of Mach number on surge limit (from Kämmer and Rautenberg 1986).

Figure 8.62 Mechanical oscillating system.

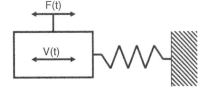

system to surge is the capacity to temporarily store energy and re-inject it in the system. Compressors (pumps) operating with non-compressible fluids will not go into surge unless there is a cavity somewhere with a compressible fluid (air) that acts like a spring, where energy can be stored by compression and returned later by expansion, as illustrated in Figure 8.63, where the plenum contains a large quantity of air.

8.6.2 Mild Versus Deep Surge

The lumped parameter model not only predicts the onset of surge, it also allows a rather accurate prediction of the surge frequency, as illustrated by the experimental results of Rothe and

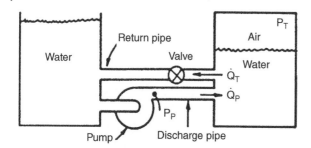

Figure 8.63 Surge in pumps (from Rothe and Runstadler 1978).

Runstadler (1978). The surge frequency is defined by the imaginary part of S (Equation (8.39)):

$$f_\sigma = \frac{\omega_\sigma}{2\pi} = \frac{a}{2\pi L_C}\sqrt{\frac{V_C}{V_P}} = \frac{U_2}{2\pi L_C}\frac{1}{2B} \tag{8.44}$$

This frequency, also called the Helmhoz freqency, is proportional to the ratio of the compressor over plenum volume and inversely proportional to the time needed by a pressure perturbation to travel over the compressor length L_C at the speed of sound.

A nonlinear model that also considers damping is required to predict the amplitude of the flow oscillations during surge. However, the amplitude of the flow oscillations is also related to the frequency at which the instabilities occur. A short compressor with a small plenum over compressor volume results in small amplitude high frequency mass flow and pressure oscillations called mild surge. It is characterized by a rough operation of the compressor with production of a chugging noise. The main difference with rotating stall is that the pressure variations in different circumferential locations are in phase. The oscillations start at small amplitude near the maximum of the pressure rise curve because of dynamic instability. The amplitude is small because only small variations of the mass flow are needed to created large variations of the pressure in a small plenum and to invert the driving forces. The average mass flow follows a quasi steady operating curve on which oscillations are superposed (small B value on Figure 8.64). The mass flow variations are sinusoidal and the amplitude can go up to 20% of the mean flow.

Dean and Young (1977) describe mild surge as an oscillation around the point of average mass flow whereby vibrational energy is added on the unstable side and dissipated on the stable side (Figure 8.65). The lower the average mass flow, the more time is spend in the unstable side, the more unsteady energy is added, and the larger will be the amplitude of the oscillation. The amplitude of the surge cycle increases with decreasing mass flow until complete blow-down.

A small compressor with a large plenum operating at high speed results in large amplitude oscillations at low frequency, as shown for large B values in Figure 8.64. Every point on the

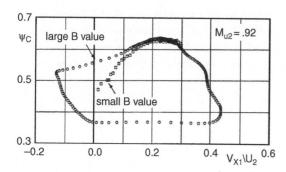

Figure 8.64 Large and mild surge in radial compressors (from Fink et al. 1992).

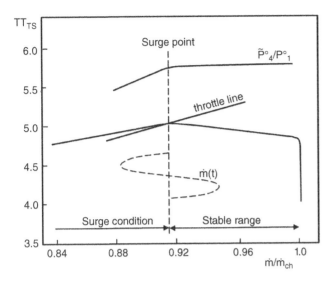

Figure 8.65 Flow oscillations during quasi steady operations (from Toyama et al. 1977).

large amplitude cycle is at the same time interval, which allows an estimation of the speed by which the instantaneous mass flow changes. Fink et al. (1992) distinguish four regimes in a deep surge cycle in a compressor with vaneless diffuser. The first one, when decreasing the mass flow towards the maximum pressure rise, is at increasing impeller tip incidence and Mach number. In a second phase, when the time average mass flow is below that at maximum pressure rise, this gives rise to mild surge with the amplitude growing in time. It is followed by a complete flow reversal (blow-down) in the third regime, resulting in a large pressure drop in the plenum. The last regime is the reestablishment of a positive throughflow followed by a gradual increase in the pressure rise along the quasi steady performance curve. The plenum pressure starts increasing with decreasing mass flow until it reaches the maximum, after which the cycle restarts. Although the vaneless diffuser is a destabilizing element (positive slope of the pressure rise curve) it was in this case not the triggering mechanism for surge. The onset of the blow-down was attributed to a sudden drop in impeller pressure rise because of the impeller leading edge stall during mild surge. The time interval between the occurrence of a blow-down decreases with decreasing average mass flow and increasing B^2 values.

Mild surge as a precursor for deep surge has also been experienced by Toyama et al. (1977) on a high pressure ratio centrifugal compressor with vaned diffuser. Compressors operating at high Mach numbers have a more violent surge behavior than those operating at low Mach number because of the higher pressures involved. This is illustrated by the variation in the pressure at different locations between the inlet and outlet of the compressor (Figure 8.66). One observes a low frequency oscillation with small amplitude in the precursor period before the onset of the deep surge at time 0. Mass flow oscillations of ±8% of the average mass flow occurred near the maximum pressure rise point. All signals are in phase at low frequency (≈ 10 Hz), which is close to the Helmholtz resonance frequency of the test loop. Shortly before deep surge occurs the diffuser throat pressure suddenly increases to almost the collector pressure, indicating that the flow is stalled in the divergent channels downstream of the throat.

Deep surge starts at $t = 0$ with return flow. This is confirmed by the total pressure measured in the throat section, by a probe pointing towards the exit, and indicating the same pressure

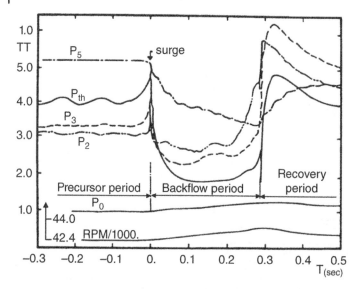

Figure 8.66 Variation of the pressure during surge at different locations between compressor inlet and outlet (from Toyama et al. 1977).

(not shown on the figure) as in the collector. The drop of static pressure at the vaned diffuser throat and in the semi-vaneless space is much more abrupt than in the collector because of the large backward velocity.

Back flow through the diffuser and impeller emptys the collector and fills the inlet plenum. The latter has a larger volume than the outlet collector and the pressure increase is much smaller. The pressure difference between the collector and inlet plenum, which is the driving force for the return flow, decreases until the impeller can invert the flow direction. The impeller rotates faster, the inlet pressure is higher, and the diffuser inlet pressure is low, hence the recovery from surge is very abrupt. This creates a large overshoot of the pressure in the vaneless and semi-vaneless space because the flow is limited by choking in the diffuser throat. Equilibrium is restored by a decrease in the impeller RPM and an increase in the back pressure in the collector. The RPM returns to normal while the collector pressure increases further until the initial conditions for mild surge are reestablished.

The recovery period between two events of deep surge depends on the speed at which the critical operating conditions are restored, i.e. the distance of the operating point from maximum pressure rise, the geometry of the compressor, and its environment (the volume and inertia of the compressor are expressed by the B^2 factor).

Based on more detailed measurements and calculations Toyama et al. (1977) postulated that surge is triggered by the flow in the vaned diffuser inlet region. Although gross separation was not observed the deterioration of the flow in this section was at the origin of the breakdown of the channel diffuser performance because of an increase in the throat blockage. Such a deterioration of the diffuser inlet flow with decreasing $\dot{m}$ is described by Everitt and Spakovszky (2013) in relation to diffuser rotating stall (Section 8.5).

Ribi and Gyarmathy (1993) show data where the mild surge oscillations have been at the origin of impeller rotating stall. They claim that deep surge is triggered only if the rotating stall exists for at least one and a half rotations, i.e. the time needed to destabilize the vaned diffuser.

Large amplitude transients between two unstable operating points, described by Greitzer (1976), on an axial compressor are the consequence of the particular shape of the

Figure 8.67 Small surge at high frequency (from Greitzer 1976).

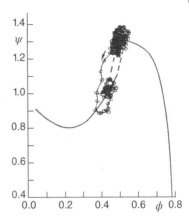

performance curve (Figure 8.67). As explained in Figure 8.58 the static stability criterion does not allow operation between the points A' and C'. Radial compressor performance curves rarely show a sufficiently large pressure drop for this phenomenon to occur because a very large part of the pressure rise is due to the centrifugal force and is less affected by impeller stall.

8.6.3 An Alternative Surge Prediction Model

Although it is not based on a rigorous theoretical model, as is the previous one, the surge prediction model of Yoshinaka (1977) is quite interesting. It deserves a more detailed discussion because it gives remarkable results and provides some insight into how the operating map can be modified.

Based on the observation that the diffuser tends to control compressor surge at lower than design speed and that the inducer controls surge at overspeed operation, Yoshinaka developed a graphical representation that allows the critical component to be defined. For this purpose the compressor characteristic lines are plotted as a function of the inlet flow coefficient (ϕ_1) and the diffuser leading edge flow angle (α_3).

Figure 8.68 shows a typical graph in which the impeller and diffuser performance are presented by constant speed lines in the ϕ, α_3 plane. They all converge at $\alpha_3 = 90°$ for $\phi_1 = 0$ and higher RPM speed lines have higher α_3 values than the ones at low RPM. Also indicated are:

- the diffuser stall limit, defined by the critical static pressure rise criterion in Figure 8.46
- the diffuser choke limit, based on the assumption of sonic velocity at the throat section
- the inducer stall limit, calculated from the critical static pressure rise coefficient between the impeller leading edge and the throat
- the inducer choke limit, calculated from the classical inducer choke prediction assuming sonic velocity in the throat.

The inducer choke and stall lines show an operating range that decreases with inlet Mach number, hence with RPM, in a way similar to the one presented in Figure 2.23.

According to the Yoshinaka model, stable operation is possible in zones 1 (because only the impeller stalls), 2 (no component stalls), and 3 (because only the diffuser stalls). Surge occurs in zone 4, where both the impeller and the diffuser stall. No operation is possible in zone 5 because the impeller and/or the diffuser choke.

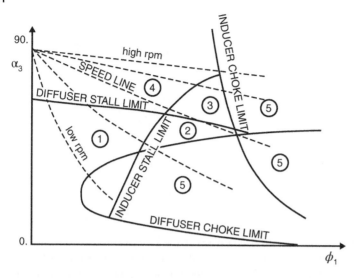

Figure 8.68 Compressor performance map in an α_3, ϕ_1 graph (from Yoshinaka 1977).

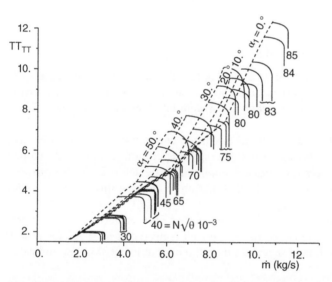

Figure 8.69 Compressor performance maps for different settings of the IGV (from Yoshinaka 1977).

This approach is illustrated by the performance map in Figure 8.69 for different settings of the IGVs. Superposing the experimental data on the α_3 versus ϕ graph shows good agreement with the model. Figure 8.70 is for $\alpha_1 = 0$, an IGV setting which is well suited for high RPM (Figure 8.69). The experimentally defined surge and choke limits agree well with the predictions. Only the experimental data at $N/\sqrt{\theta_1} = 65\ 000$ to $75\ 000$ show a smaller range than predicted. However, it was pointed out by the author that at these values the impeller static

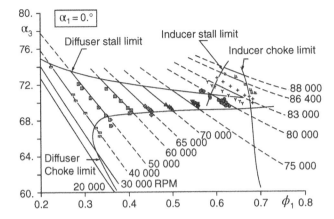

Figure 8.70 Compressor performance map in an α_3, ϕ_1 graph for large mass flow IGV setting, comparison between theory and experimental data (from Yoshinaka 1977).

Figure 8.71 Compressor performance map in an α_3, ϕ_1 graph for small mass flow IGV setting, comparison between theory and experimental data (from Yoshinaka 1977).

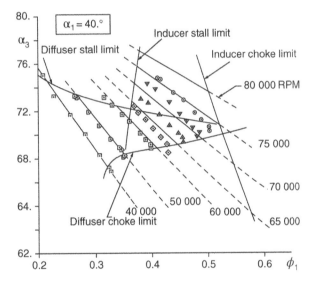

pressure rise decreased with decreasing mass flow and that this occurred only at this RPM. Hence the model is valid as long as the inducer stall is not too strong in order not to perturb the whole impeller flow field.

Figure 8.71 shows the comparison for $\alpha_1 = 40$, an IGV setting that is better suited for reduced mass flow. The agreement between the predicted and experimental operating limit is rather good. All experimental points fall in the predicted stable flow zone.

This approach also allows prediction of the impact of the IGV setting angle on the shape of the performance map by shifting the impeller choke and surge line accordingly.

9

Operating Range

The fastest way to obtain a large operating range is with a good design, with a correct blade loading distribution providing a negative slope of the pressure rise curve. The outcome of different optimizations is illustrated in Figure 9.1 and shows that maximizing the operating range may be at the expense of a lower efficiency.

This chapter discusses some specific actions and geometrical modifications that aim to extend the operating range of centrifugal compressors. A first group of range extension actions intends to shift the surge and/or stall limit to smaller mass flows or to limit the amplitude of the instability such that the impact on performance or vibrations is negligible. Other actions aim to prevent the compressor from crossing the surge or stall limit and start operating in the unstable region. The following list of actions is not exhaustive and the order of presentation is not a ranking in terms of importance. One should also keep in mind that some spectacular improvements that are presented in the literature may be corrections only for shortcomings of the initial geometry.

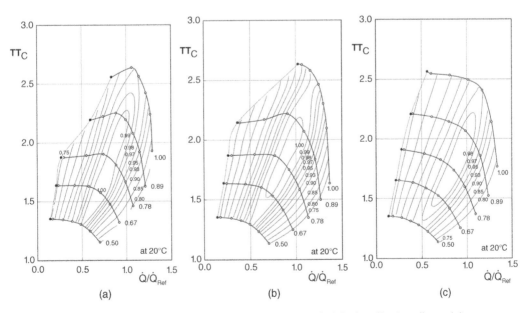

Figure 9.1 Comparison between experimental performance map of (a) the baseline impeller and the ones optimized for (b) maximum efficiency and (c) maximum range (from Ibaraki et al. 2014).

Design and Analysis of Centrifugal Compressors, First Edition. René Van den Braembussche.
© 2019, The American Society of Mechanical Engineers (ASME), 2 Park Avenue, New York, NY, 10016, USA (www.asme.org).
Published 2019 by John Wiley & Sons Ltd.

9.1 Active Surge Control

Active surge control prevents an initial flow perturbation being amplified to large amplitude instabilities. It consists of a sensor, to measure unsteady flow or pressure variations, an actuator, to influence the flow, and a controller connecting the two. Possible sensors measure the variation of the mass flow, the plenum pressure, the compressor inlet total, or static pressure. Actuators drive a close-coupled valve at the diffuser exit, a fast response throttle valve downstream of the plenum, a movable wall to change the plenum volume, or a tip clearance control system.

The increase in the operating range by active control is achieved by modifying the system dynamics, rather than by a modification of the compressor flow. The compressor characteristics are almost unchanged. The amplitude of the unsteady phenomena is reduced but there will always remain a small instability because without input signal the control system stops working and returns to the unstable uncontrolled situation.

The main reason for using active surge control is to allow safe operation of the compressor at the point of maximum efficiency, when this point is near to the surge limit, i.e. to guarantee a level of robustness such that small perturbations would not drive the compressor into surge (Yoon et al. 2012). It is not the purpose to extend the normal operating range for continuous operation beyond the natural stability limit because efficiencies will anyway be lower on the left side of the point of maximum pressure rise. Simpler systems of variable geometry may be much more appropriate for such a permanent extension of the operating range.

Different strategies for active control have been examined by Simon et al. (1993) combining one of the four sensors with one of the four actuators shown in Figure 9.2 by a simple proportional control G (without phase shift). The conclusions are summarized in Table 9.1. It is shown that a close-coupled valve, driven by the variations of the compressor mass flow, allows an unlimited range increase. However, very large values of the gain G may be required as it increases proportionally with ψ_C'. Monitoring the close-coupled valve by the inlet static pressure variations has a similar effect. Driving such a valve as a function of the plenum pressure oscillations does not enhance stability because the plenum is downstream of the valve and Table 9.1 shows the same stability limit as for an uncontrolled system.

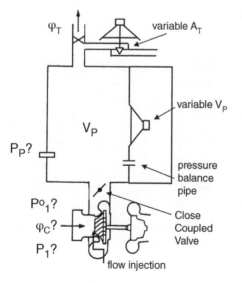

Figure 9.2 Actuators and sensors.

Table 9.1 Limitations on compressor range extension with active control (from Simon et al. 1993).

Actuator sensor	Close-coupled valve	Plenum bleed valve	Movable plenum wall
$\dot{m}_C$	Unlimited range increase $$G \sim \frac{\psi'_C}{\mathcal{T}} \text{ as } \psi'_C \to \infty$$ $$G \sim \frac{\psi'_C}{\mathcal{T}} \text{ as } B \to \infty$$	Limited range increase $$\psi'_C < \frac{1}{B^2 \Psi'_T}$$	Limited range increase $$\Psi'_C < \frac{1}{B^2 \psi'_T}$$
P_P	No range increase $$\psi'_C < \frac{1}{B^2 \psi'_T}$$	Limited range increase $$\psi'_C < \frac{1}{B}$$	Limited range increase $$\psi'_C < \frac{1}{B}$$
P_1^o	Limited range increase $$\psi'_C < \psi'_T$$	Limited range increase $$\psi'_C < \psi'_T$$	Limited range increase $$\psi'_C < \psi'_T$$
P_1	Unlimited range increase $$G \sim \frac{\psi'_C}{2\mathcal{T}\phi} \text{ as } \psi'_C \to \infty$$ $$G \sim \frac{\psi'_C}{2\mathcal{T}\phi} \text{ as } B \to \infty$$	Unlimited range increase $$G \sim \frac{\psi'_C B^2}{\mathcal{U}} \text{ when } \frac{1}{2\phi B^2 \psi'_T} < 1$$ $$G \sim \frac{\psi'_C}{2\mathcal{U}\phi \psi'_T} \text{ else as } \psi'_C \to \infty$$ $$G \sim \frac{B^2 \psi'_C}{\mathcal{U}} \text{ as } B \to \infty$$	Unlimited range increase $$G \sim \frac{\psi'_C B}{\mathcal{W}} \text{ when } \frac{1}{2\phi B^2 \psi'_T} < 1$$ $$G \sim \frac{\psi'_C}{2\mathcal{W}\phi \psi'_T B} \text{ else as } \psi'_C \to \infty$$ $$G \sim \frac{B\psi'_C}{\mathcal{W}} \text{ as } B \to \infty$$

$\mathcal{T} = -d\psi_{CC}/dA_{CC}$, change of closed coupled valve pressure drop with close-coupled valve area.
$\mathcal{U} = d\phi_T/dA_T$, change of valve flow with throttle valve area.
$\mathcal{W} = 2P_P/(P_1 M_2^2)$, non-dimensionalized change of plenum pressure with plenum volume.

The plenum movable wall stability criteria are similar to those of the throttle valve except when driven by the inlet static pressure. However, experimental results (Simon et al. 1993) show a larger range extension with a movable plenum wall than with a variable throttle (Figure 9.5).

Most stability limits define a maximum value of ψ'_C that decreases with increasing B values. Centrifugal compressors are more compact than axial ones (smaller $\mathcal{V}_C$) and B values can be as high as 4, which makes stabilization more difficult.

A major condition for active control to be effective is a smooth continuation of the pressure rise curve into the region of surge. As all criteria define a limit for ψ'_C, one cannot expect that a control system will be able to compensate for a discontinuous change in pressure characteristic.

9.1.1 Throttle Valve Control

This approach stabilizes the flow by modulating the mass flow in the throttle valve as a function of the pressure fluctuations in the plenum. Plenum pressure variations, due to incipient surge, are measured by a fast response pressure transducer and fed to the controller, who makes a suitable transformation of the input signal. It can be a simple proportional relation (defined by a real number) or a more complex one also including a phase shift (characterized by a complex number). The output signal is applied to a fast response regulation valve. The purpose is to reestablish the dynamic stability by modifying the throttle characteristics at the operating point by a fast response valve, the operating point itself being set by a slowly reacting throttling system.

The variation of the flow is due to the difference between the plenum pressure ψ_P and the instantaneous throttle valve pressure drop ψ_T (at constant throttle valve cross section area), as specified in Equation (8.33), plus an extra flow accounting for the change in throttle valve area A_T defined by

$$d\phi_T|_{(\psi_P=\psi_T)} = \widetilde{\phi_T} \frac{dA_T}{A_T} \tag{9.1}$$

The latter is related to the change in plenum pressure by means of a transfer function Z:

$$\frac{dA_T}{A_T} = Z \frac{d\psi_P}{\psi_P} = G\, e^{-i\beta_\sigma}\, \frac{d\psi_P}{\psi_P} \tag{9.2}$$

where G is the gain and β_σ is the phase shift. Combining this with the variation of the throttle valve mass flow at constant cross section A_T defined in Equation (8.31) results in the following expression for the total change of throttle valve flow coefficient:

$$\frac{d\phi_T}{dt} = \frac{A_T}{A_C} \frac{U_2}{2L_T}(\psi_P - \psi_T) + Z \frac{\widetilde{\phi_T}}{\widetilde{\psi_P}} \frac{\widetilde{d\psi_P}}{dt} \tag{9.3}$$

Substituting this into the surge analysis model (Equations (8.24) to (8.27)) results in the following expression for the growth rate S of the perturbation:

$$S^2 + \left(\frac{2A_C a^2}{V_P U_2} \frac{\widetilde{\phi_T}}{\widetilde{\psi_P}} Z - \frac{U_2}{2L_C}\psi_C' + \frac{2A_C\, a^2}{V_P\, U_2} \frac{1}{\psi_T'} \right) S$$

$$+ \frac{A_C a^2}{L_C V_P}\left(1 - \psi_C'\left(\frac{1}{\psi_T'} + \frac{\widetilde{\phi_T}}{\widetilde{\psi_P}} Z \right) \right) = 0 \tag{9.4}$$

One can conclude from this that any positive value of Z will make the first-order term more positive and enhance the dynamic stability but makes the third term more negative and decreases the static stability. As the last one is less restrictive than the former, the stability can be enhanced. Equation (9.4) reduces to an uncontrolled system for $Z = 0$.

The effect of a throttle valve control has been evaluated by introducing the parameters of a typical compressor and throttle characteristic into Equation (9.4) and calculating the variation of A_σ and ω_σ as a function of the amplification G and phase shift β_σ. The results shown in Figure 9.3 indicate that stability is enhanced ($A_\sigma < 0$) for a phase shift between 260° and 440° with maximum damping at $\beta_\sigma = 330°$. One can conclude from this that a simple

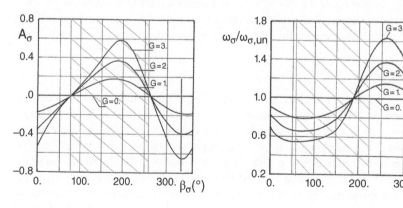

Figure 9.3 Variation of the compressor stability limit with amplification G and phase shift β_σ (from Van Bael et al. 1994).

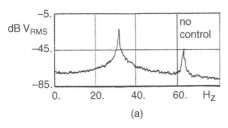
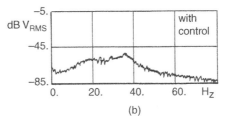

(a)   (b)

Figure 9.4 Plenum pressure power spectrum (a) without and (b) with controller (no inlet pipe) (from Van Bael et al. 1994).

proportional controller ($\beta_\sigma = 0$) will also stabilize the system, i.e. when the throttle opens at increasing plenum pressure and closes at decreasing pressure. However, real sensors and actuators create a non-negligible phase shift that is dependent on the frequency. The controller should correct for this.

Dynamic stability is assured when the first-order term is positive, i.e. when

$$\psi'_C < \frac{1}{B^2}\left(\frac{1}{\psi'_T} + \frac{\widetilde{\phi_T}}{\widetilde{\psi_P}}Z\right) \tag{9.5}$$

The effectiveness of the control system depends on the operating point defined by $\widetilde{\psi_P}$ and $\widetilde{\phi_T}$. Increasing Z is equivalent to a decrease in the throttle curve slope and allows a proportional increase in the admissible positive slope of the compressor pressure rise curve. However, there is a maximum limit for Z imposed by the static stability criterion

$$1 - \psi'_C\left(\frac{1}{\psi'_T} + \frac{\widetilde{\phi_T}}{\widetilde{\psi_P}}Z\right) > 0 \tag{9.6}$$

i.e. when the modified slope of the throttle line becomes smaller than that of the compressor pressure rise curve.

The frequency of surge varies with the controller phase shift and gain (Figure 9.3b). It equals the frequency of an uncontrolled system ($\omega_\sigma/\omega_{\sigma,un} = 1$) only at ($\beta_\sigma = 180°$ and $330°$).

A simple analog controller providing the correct phase shift at frequencies up to 40 Hz can suppress surge at 31 Hz in a turbocharger compressor, operated at 20 000 RPM, connected to a 6 liter plenum followed by a throttle valve (Van Bael et al. 1994) (Figure 9.4).

The same control system and test configuration but with a 3.6 m long inlet pipe was less successful. Surge was suppressed but pressure fluctuations at 45 Hz, corresponding to resonance in the inlet pipe, were amplified. A more performant digital controller, compensating the actuator phase shift also at higher frequencies, is needed to suppress all instabilities (Di Liberti et al. 1996).

9.1.2 Variable Plenum Control

Figure 9.2 is a schematic view of a variable plenum control system. In this laboratory set-up the wall movement is by means of a large loudspeaker separating the plenum from the auxiliary plenum. They are connected by a small tube to reduce the energy requirements by minimizing the time-averaged pressure difference over the loudspeaker. After a suitable transformation of the plenum pressure variations, the controller displaces the loudspeaker with surface A_P over a distance

$$dy = Zd\psi_P \tag{9.7}$$

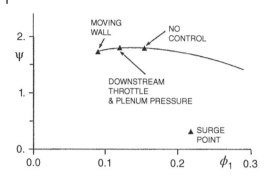

Figure 9.5 Radial compressor range extension with different control systems (from Simon et al. 1993).

An extra term has to be added to the continuity Equation (8.21) to account for the corresponding change in plenum volume:

$$\dot{m}_C - \dot{m}_T = \frac{d(\rho V_P)}{dt} + \rho A_P Z \frac{d\psi_P}{dt} \tag{9.8}$$

Experimental data collected by Simon et al. (1993) indicate that a periodic change of the plenum volume (V_P) driven by the plenum pressure oscillations has more potential for an increased range than a throttle valve control (Figure 9.5). The main drawback of this approach is the need for major modifications of the plenum chamber and the larger energy that is required to move the wall.

Gysling et al. (1991) and Arnulfi et al. (1999, 2001) propose a passive damping device to reduce the amplitude of the unsteady flow variations at deep surge by dissipating the unsteady energy added by the compressor. The system consists of a movable wall of mass m_w and surface A_P, separating the primary plenum from the auxiliary one V_{aux}, a damper with coefficient ζ, a spring with stiffness K_S, and a pressure equalizing tube of sufficiently small diameter in order not to influence the dynamic characteristics of the damper (Figure 9.6).

When using passive control, the following equations must be combined with Equations (8.26) and (8.27):

$$\dot{m}_C - \dot{m}_T = V_P \frac{d\rho}{dt} + \rho A_P \frac{dy}{dt} \tag{9.9}$$

$$m_w \frac{d^2 y}{dt^2} + \zeta \frac{dy}{dt} + K_S y = (P_P - P_{aux}) A_P \tag{9.10}$$

A parametric study is required to define the optimal values of m_w, V_{aux}, ζ, and K_S as a function of the factor B, i.e the values that maximize the slope of the compressor characteristic that allows stable operation. Arnulfi et al. (1999) show that these parameters are constant except the optimal damping ζ, which decreases for $B < 0.6$. The same study shows that the maximum stable slope of the compressor characteristic is only slightly smaller than that of an active controlled system (Figure 9.7) and that both allow a similar extension of the operating range.

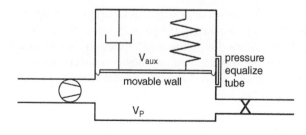

Figure 9.6 Schematic view of a passive control system with movable wall (Arnulfi et al. 2001).

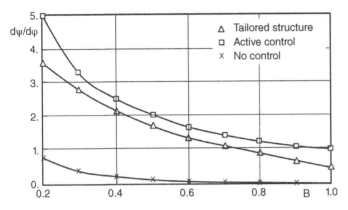

Figure 9.7 Maximum compressor characteristic slope when using active and passive movable wall control (from Arnulfi et al. 1999).

The main drawback of a passive control system is the size and practical construction of the damper. The passive control system of Gysling (1991) for a small turbocharger with an impeller diameter of 5.5 cm requires an auxiliary volume that is more than three times larger than the primary one, a 6.2 kg movable wall, and a damping coefficient between 1000 and 2000 N/(m/s). Arnulfi et al. (2001) mention a movable wall of 1200 kg with a 1 m^2 surface and a $\mathcal{V}_{aux}$ of 17.5 m^3 for a four-stage compressor with a 46.5 cm impeller diameter and 8 mm blade height. They also present a hydraulic oscillator, which should be technically more feasible.

9.1.3 Active Magnetic Bearings

Yoon et al. (2012) and Ahn et al. (2009) use active magnetic bearings (AMB) as an actuator for surge control. The control is obtained by changing the impeller tip clearance. Centrifugal compressors are well suited for this because tip clearance is easily modified by an axial displacement of the shaft by means of the trust bearing (Figure 9.8). An attempt to use clearance control to avoid stall in axial compressors turned out to be much more difficult (Spakovsky and Paduano 2000).

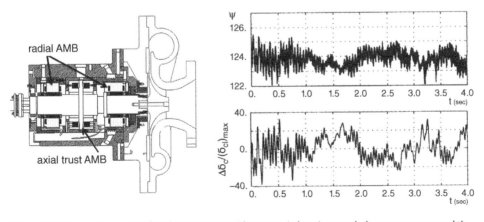

Figure 9.8 Single-stage centrifugal compressor with magnetic bearings and plenum pressure and tip clearance variation during active control (from Yoon et al. 2012).

An increase in the tip clearance decreases the pressure rise by a decrease in the slip factor and an increase in the losses (Equations (3.85) and (3.55)). The impact on stability is accounted for by adding an extra term in Equation (8.30) expressing the influence of tip clearance on the compressor pressure rise:

$$\frac{d\phi_C}{dt} = \frac{U_2}{2L_C} \left(\psi_C + 2.43\frac{\Delta\delta_{cl}}{b_2} \left(1 - \frac{R_{1S}^2}{R_2^2}\right) - \psi_P \right)$$

(9.11)

where $\Delta\delta_{cl} = \tilde{\delta}_{cl} - \delta_{cl}(t)$, the change in tip clearance ($\Delta\delta_{cl} > 0$ for decreasing clearance).

The system has been applied to a compressor of 0.25 m diameter in combination with a plenum of 0.7 m² and a 5.2 m long inlet duct. When closing the throttle valve, instabilities occur first at 21 Hz, corresponding to inlet pipe acoustic resonance. At smaller mass flows the frequency drops to 7 Hz, corresponding to surge.

A first series of tests, sensing the plenum pressure, resulted in an overcompensating control system. This was due to an incorrect estimation of the compressor mass flow variation from the measured pressure oscillations and the linearized pressure rise curve. This turned out to be very inaccurate when operating in the critical region near the point of maximum pressure rise. Adding a direct measurement of the instantaneous compressor mass flow oscillations as an additional control input resulted in a suppression of the low frequency oscillations. However, the high frequency oscillations, corresponding to inlet pipe resonance, were only postponed to a slightly lower mass flow (throttle valve opening from 17.3% to 15.8%).

Adding the dynamics of the inlet pipe to the model and improving the capturing of high frequency oscillations also allowed an extension of the stable operating range by 21.3%. The signal in Figure 9.8 shows a stabilization of the compressor by an increase in the tip clearance in phase with the increase in the plenum pressure.

9.1.4 Close-coupled Resistance

The analysis by Simon et al. (1993) (Table 9.1) indicates an unlimited range increase when the throttle valve is driven by the instantaneous change of mass flow or inlet static pressure. Prediction of the possible extension of the operating range requires the knowledge of the pressure rise curve to the left of the uncontrolled surge limit. Theoretical predictions are possible but may be insufficiently accurate because the performance correlations are not well known at these off-conditions.

The pressure rise curve can be measured after stabilizing the flow by means of an uncontrolled close-coupled resistance (CCR). This rather simple device to suppress surge consists of a throttle between the impeller and outlet plenum (Figure 9.9a for a vaned diffuser and Figure 8.60 for a vaneless diffuser).

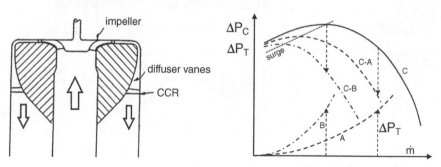

Figure 9.9 Close-coupled resistance and the influence of on performance map (from Dussourd et al. 1977).

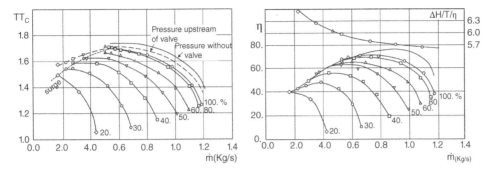

Figure 9.10 Compressor performance with close-coupled resistance (from Dussourd et al. 1977).

The influence of a CCR is twofold. The throttle dynamically separates the compressor from the outlet plenum. This results in a very small B^2 factor, which, according to Equation (8.42), allows stable operation at low mass flows where the original compressor is unstable because of the positive slope of the performance curve. The quasi steady impeller pressure rise curve, needed to investigate active control, can be measured upstream of the throttle valve up to very small mass flow (Fink et al. 1992).

The CCR creates a steady pressure drop ΔP_T that increases quadratically with volume flow (Figure 9.9b). As a consequence the pressure rise curve, measured downstream of the throttle, is lower and the maximum shifts to smaller mass flows. The amount of pressure and efficiency drop depends on the valve position.

The main complication of a CCR is the need for an annular valve to avoid circumferential pressure distortions that could trigger other types of instabilities. The dashed line on Figure 9.10 shows the experimental results measured upstream of the CCR located directly downstream of the diffuser vanes (Figure 9.9a). This "true compressor" pressure rise curve shows a smooth variation with stable flow up to 20% of the valve opening. The pressure is slightly lower than the one measured without a valve because of an interaction between the diffuser vanes and the throttle valve. The pressure downstream of the fully open valve (full lines) at 100% open is slightly lower than upstream because the valve (two perforated disks next to each other) allows only a maximum opening of 50%. These losses could be avoided by using other valve designs, such as the one shown in Figure 8.60. The other curves show the downstream pressure and efficiency for different settings of the valve. A significant increase in the stable operating range is obtained, but at the cost of a decreasing pressure ratio and efficiency. However maximum pressures are not much lower than the ones upstream of the valve.

9.2 Bypass Valves

The compressor surge limit is characterized at each RPM by the pressure ratio and mass flow. Surge can be avoided by ensuring that the mass flow is not reduced below this limit. However, the surge limit may be crossed when changing the operating condition of a gas turbine or turbocharger too rapidly. Injecting more fuel to increase the power output or RPM of a gas turbine easily provokes choking of the turbine. The corresponding increase in the turbine inlet temperature results in a decrease in the density $\rho = P/(R_G T)$ that cannot be fully compensated for by the increase in the sonic velocity $a = \sqrt{\kappa R_G T}$. At constant pressure the choking mass flow $\rho a A_{th} = P \sqrt{\kappa/(R_G T)} A_{th}$ decreases with increasing turbine inlet temperature. This can be compensated for by opening the turbine nozzle (variable geometry) or by an increase in the

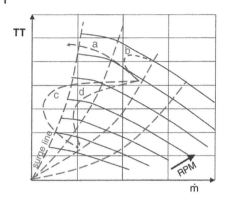

Figure 9.11 Variation of the compressor operating point.

compressor outlet pressure (higher RPM). Depending on the surge margin the combination of these may be insufficient to prevent the compressor from going into surge, as shown by path a in Figure 9.11. If the temperature rise occurs sufficiently slowly the gas turbine may have the time to increase the RPM and the compressor can stay out of surge by providing the pressure increase needed to operate on the same throttle line (path b).

Similar phenomena can occur when changing the operating conditions of a turbocharger or pipeline compressor. In order to assure a rapid reacceleration of the piston engine, one may want to keep the turbocharger RPM high even when only a low mass flow is required by the piston engine running idle. Rapidly reducing the mass flow in the compressor may leave insufficient time to adjust the RPM and the compressor may run into surge (path c in Figure 9.11).

The simplest surge control is by a bleed valve, blowing off the excess mass flow to the atmosphere (Figure 9.12). This results in large energy losses if the energy of the bleed is not recuperated by a turbine driving the compressor. However, the large investment of a turbine is justified only if the bleed-off is not accidental but has a more permanent character.

Severe surge problems can also occur during an emergency shut down (ESD) of a pipeline compressor because of power failure. The rapid decrease in RPM requires a fast action to keep the compressor out of the critical conditions. A typical anti-surge control system consists of hot (a) and cold (b) recycle valves between the compressor outlet and inlet (Figure 9.13). The cold valve, connecting the cooler exit to the compressor inlet, is modulated for surge control during start-up and turn-down, and may have a slower response than the hot valve. The latter is a quick-opening on–off valve, connecting the compressor exit (before the cooler) to the compressor inlet, and is used exclusively for ESD. It can operate for only a short period because recirculating hot outlet gases results in a rapid increase in the inlet temperature. The cold valve

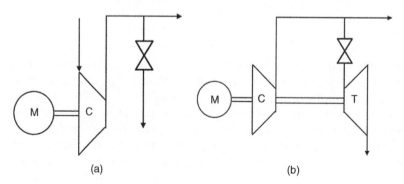

Figure 9.12 Compressor anti-surging valves.

Figure 9.13 Anti-surge arrangement with (a) hot and (b) cold recycle valves.

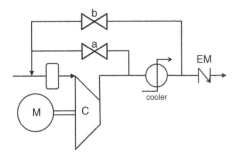

can be used for a longer period without risk of overheating the compressor. In the case of ESD it can also be opened in parallel with the hot one so that the latter can be smaller. The valve EM is a shut-off valve that will rapidly close in the case of a power failure to avoid reflux of gas from the downstream pipeline.

Important issues are the minimum operating margin that is required to guarantee a safe operation, in combination with the response time of the recirculation system. This depends on the characteristics of the control system as well as on the layout of the pipes and volumes. The valves should be sufficiently large to pass the complete mass flow at all speed lines and open sufficiently fast to ensure that the compressor mass flow is at any RPM larger than the one at surge. Such systems have been analyzed by Kurz and White (2004), White and Kurz (2006), Schmitz and Fitzky (2004), and Kurz et al. (2006).

After a power failure the compressor is driven only by inertia. The torque generated by a decelerating rotor is:

$$torque = J\frac{d\Omega}{dt} \tag{9.12}$$

where J is the moment of inertia of the gas turbine, gear box or electrical motor driving the compressor. The power that keeps the compressor running is:

$$Pw = torque\ \Omega = J\Omega\frac{d\Omega}{dt} \tag{9.13}$$

Observations have shown that compressors driven by gas turbines lose about 25% speed in the first second. Those driven by an electrical motor may loose around 30%. The latter results in a 50% reduction in the pressure rise. The short duration of the emergency stop allows the change in heat transfer in the cooler to be neglected.

Assuming quasi steady flow in the compressor one can derive the mass flow corresponding to a given back pressure and RPM from the steady performance map. The power required to compress the fluid in that operating point scales with Ω^3:

$$\dot{m}\Delta H \approx \psi\phi U^3 \approx C^{te}\Omega^3 \tag{9.14}$$

and equals the one recuperated by inertia

$$C^{te}\Omega^3 = J\Omega\frac{d\Omega}{dt} \tag{9.15}$$

After integration in time one obtains:

$$\Omega(t) = 1/\left(\frac{C^{te}t}{J} + \frac{1}{\Omega_{t=0}}\right) \tag{9.16}$$

This allows the variation of the rotational speed to be calculated and hence the time-wise variation of the surge limit from the performance map.

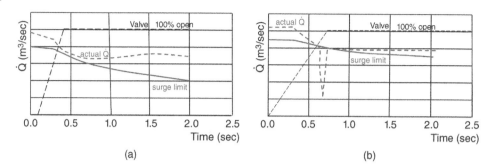

Figure 9.14 Variation of volume flow and surge limit during an ESD (from White and Kurz 2006).

The compressor outlet pressure should also be known to find out how the actual mass flow on the operating line of the corresponding RPM compares to the mass flow at surge.

The time-dependent change in the compressor back pressure is defined by applying continuity to the volumes and pipes (V_p). It is a function of the instantaneous difference between the incoming volume flow of the compressor and the outgoing one at the hot valve (a):

$$\frac{dP_6}{dt} = \frac{\kappa P_6}{V_p}(\dot{Q}_{comp} - \dot{Q}_{valve})$$

(9.17)

Assuming the same pressure at the compressor exit and valve inlet, the mass flow in the valve can be calculated as a function of its position.

Stability is assured if at all instants of time, i.e. at all RPM, the instantaneous mass flow in the compressor is larger than the one corresponding to surge (Figure 9.14a). Such a calculation allows verification if the size of the valves and opening speed are sufficiently large to ensure that at each instant of time sufficient mass flow is passing through the valves. The larger the volume (cooler and pipes) the slower will be the pressure decrease and the larger the risk that the compressor will surge (Figure 9.14b). Alternative arrangements of the recycle valves are discussed by White and Kurz (2006).

9.3 Increased Impeller Stability

The overall impeller diffusion, the leading edge to throat velocity ratio, and the incidence have been mentioned as critical parameters for the impeller stability. Variable inlet guide vanes allow an extension of the stable operating range to much smaller mass flows by an adjustment of the inlet flow angle to the blade leading edge angle (Rodgers 1977; Shouman and Anderson 1964). Besides the mechanical complication the main problem with prerotation is a decrease in pressure rise when increasing the prerotation. However, this can be compensated for by a sufficient backsweep of the impeller vanes, as explained in Section 2.1.1, and illustrated by the experimental results of Figure 9.15.

Variable guide vanes are an expensive mechanical device increasing the compressor volume and weight. Kyrtatos and Watson (1980) propose a tangential injection of recirculating flow at the impeller inlet. Prerotation then depends on the position and direction of the injection and is proportional to the amount of air blown. Disadvantages are the energy loss by recirculating compressed fluid and an eventual tendency to become unstable by a back-coupling between the pressure rise and prerotation.

The influence of the blade loading on the surge margin has been studied by Flynn and Weber (1979). They claim that leading edge separation results in impeller rotating stall and surge when

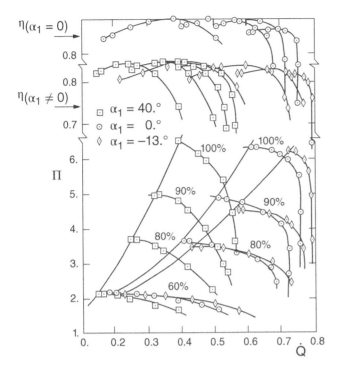

Figure 9.15 Influence of a variable inlet guide vane on a performance map (from Rodgers 1977).

the separated zone extends sufficiently far downstream to merge with the exducer separation. By increasing the blade thickness, they prevented the two separated flow zones from communicating with each other. However, the considerable increase in the stable operation range was at the cost of large wake losses downstream of the thick trailing edges. Extending the blade length to smooth the trailing edges resulted in a small improvement in the efficiency but also in a much larger and heavier impeller (Figure 9.16).

Thick wedge-shaped impeller blades with larger blade height have been used by Rusak (1982) and Sayed (1986) to reduce the friction losses by increasing the hydraulic diameter of very low specific impellers. The sudden expansion at the thick trailing edge is not considered as an extra source of losses but as a replacement of the jet–wake pattern in the exducer. The trailing edge is rounded off to reduce the sudden expansion dump losses at the impeller exit (Figure 9.17a). Shop performance tests indicated an overall stage efficiency increase of up to 5% (Sayed 1986). The extra losses that may result from a sudden spanwise contraction that may be needed at the impeller exit to maintain stable flow in the diffuser are not discussed.

An extremely low NS impeller is shown in Figure 9.17b, i.e. a disk with holes intersecting each other at the impeller inlet. Flow channels have an optimum shape (maximum hydraulic diameter, DH) but the extra losses that result from a sudden expansion and spanwise contraction that may be needed at the diffuser inlet to maintain stable flow in the diffuser are not discussed. The latter could eventually be avoided by placing the volute much closer to the impeller exit and compensating for the absence of a vaneless diffuser by a larger exit diffuser downstream of the volute, where the aspect ratio can be close to one.

In a discussion on a paper by Reddy and Kar (1971), Wiesner cites the work of Anisimov et al. (1962) claiming that impellers with 14–18 blades have the best performance while 10–12 blades

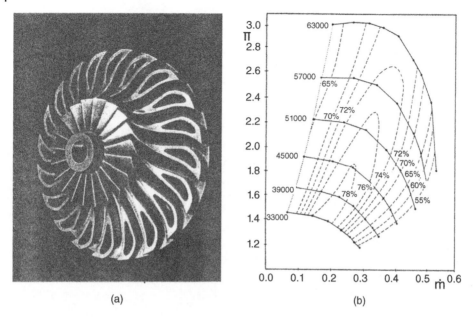

(a) (b)

Figure 9.16 (a) A thick-bladed impeller and (b) the increased range for a thick-bladed impeller (from Flynn and Weber 1979).

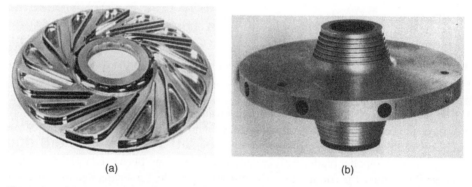

(a) (b)

Figure 9.17 (a) Low *NS* (without shroud) (from Rusak 1982) and (b) extremely low *NS* impeller (from Schmalfuss 1972).

result in a more stable range. Reducing the blade number results in a smaller slip factor, which has the same effect as bending the blades more backward.

9.3.1 Dual Entry Compressors

In order to handle very large volume flows it is not uncommon to place two identical impellers back to back (Figure 9.18). The advantage is the elimination of the disk friction losses and leakage flow on the back side of the impellers. There is, however, a risk that the inlet geometry may not be exactly symmetric, by the presence of a driving shaft on one side, resulting in slightly different performance curves on both sides (Figure 9.19). Another reason for a difference in pressure rise can be the unequal tip clearance because of manufacturing or mounting inaccuracies, or thermal dilatation of the shaft.

Figure 9.18 Dual entry compressor.

Figure 9.19 Flow unbalance and instability in an asymmetric dual entry compressor.

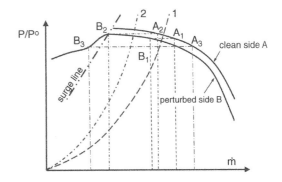

The parallel operation of two impellers with slightly different characteristics is illustrated in Figure 9.19. As they are subject to the same outlet pressure, the mass flows will be different. Both impellers operate in a stable way at throttle position 1. In spite of the small differences in the pressure rise curve, the mass flow is not very different and in the region of stable operation.

As the throttle valve is closed to position 2, the outlet pressure is increased and the B compressor will be at the limit of stability. Only a small perturbation is required for the compressor to end up in the unstable zone. The flow may stabilize in new operating points with B_3 to the left of the surge line. As the mass flow is drastically decreased, only a lower pressure is needed to pass the amount of fluid in the throttle. The outlet pressure of both impellers is the same and impeller A starts operating at a lower pressure ratio but larger mass flow. Depending on the shape of the operating curves the B impeller might also go into deep surge, resulting in violent oscillations in both impellers.

Other consequences are that the axial forces are no longer balanced and the full operating curve cannot be used, as illustrated by the experimental data in Figure 9.20. This figure also illustrates that the impact of an inlet distortion strongly depends on the slope of the performance curve. Compressors with a steeper characteristic curve are less sensitive to asymmetries.

Remediation is obtained by eliminating the difference in performance. This can be done by reducing the differences in inlet shape or clearance. If the geometrical differences cannot be eliminated one may weaken the difference by putting perturbations (screens) at the inlet to make the two performance curves match. Lei (2011) installed a ported shroud to extend the stable operating range of the weakest impeller and reestablish overall stability. Other measures may be to adapt the impeller outlet radius or to modify the compressor geometry to obtain a more negative slope of the performance curve.

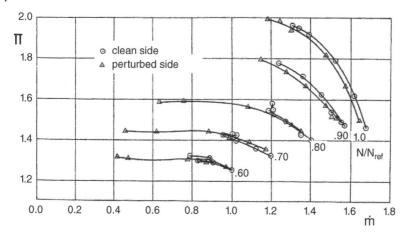

Figure 9.20 Impact of inlet asymmetry on the pressure rise curves of a dual entry compressor (from Dussourd and Putman 1960).

9.3.2 Casing Treatment

Shroud casing treatments consist of small cavities (honeycomb) or groves in the shroud (Figure 9.21). This can be efficient in delaying rotating stall, if the stall is associated with the boundary layer on the shroud casing. It is ineffective if the rotating stall is due to excessive blade loading, i.e. in the case of a low number of highly loaded impeller blades, because modification of the shroud will have little or no impact on the suction side separation (Greitzer et al. 1979).

Some successful applications of casing treatment at the centrifugal impeller inlet, outlet, and vaned diffuser have been reported by Jansen et al. (1980) and Fisher (1989). The flow enters the inclined axial grooves at the impeller inlet on the pressure side and is reinjected on the suction

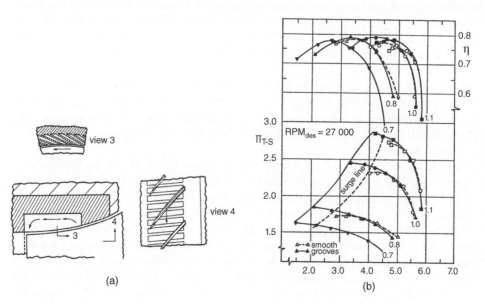

Figure 9.21 (a) Inclined axial grooves and (b) comparison between performances with smooth and treated inducers (from Jansen et al. 1980).

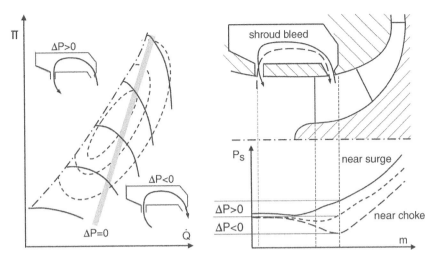

Figure 9.22 Flow in the shroud recirculation system (from Dickmann et al. 2005).

side (Figure 9.21a). This results in a remarkable improvement in the range and efficiency of a medium pressure ratio compressor (Figure 9.21b). The same casing treatment applied to a five to one pressure ratio compressor provided a similar extension of the operating range, but with a reduction in the pressure ratio and several points of efficiency loss.

Elder (1994) mentions that circumferential groves over a small range of the shroud casing demonstrated a modest range increase but that efficiency penalties generally ensued unless the original efficiency was poor. The minimal impact of circumferential grooves on range is confirmed by a numerical study of Du and Seume (2017), who predict an increase in the pressure ratio with axial grooves but with a penalty on efficiency.

A ported shroud is a more elaborated version of the casing treatment, extending upstream of the leading edge The way this device influences the flow at choke and near surge operation is illustrated in Figure 9.22.

Near choke, the flow first accelerates in the inducer because of negative incidence. The shroud pressure decreases below the upstream level ($\Delta P < 0$) and fluid flows from the inlet at pressure P_1 to the downstream slot on the shroud wall. It bypasses the throat section and increases the mass flow in the impeller beyond the original choking limit. At low mass flow, the pressure increases rapidly in the inducer to a value above the inlet pressure ($\Delta P > 0$) and fluid flows backwards from the shroud to the inlet. Recirculating the flow results in an increase in the axial velocity at the shroud leading edge without an increase in the overall mass flow. The amount of recirculation depends on the difference between the inducer pressure rise and the friction and mixing losses in the ported shroud.

In absence of vanes in the return channel, the swirl at the downstream slot is transmitted to the impeller inlet and creates a prerotation in the direction of the impeller rotation. Both phenomena, increased axial velocity and positive swirl, contribute to a decrease in the incidence and an extension of the operating range to lower mass flows. However, this is at the cost of a lower pressure ratio and lower efficiency because of the extra friction losses in the ported shroud (Figure 9.23a). The latter may be compensated for by the improved performance of the exducer because low energy fluid on the shroud has been blown off and the flow starts with a new boundary layer.

Chen and Lei (2013) give an overview of the problems involved in the design and application of ported shrouds to turbochargers. The struts fixing the shroud wall on the compressor casing

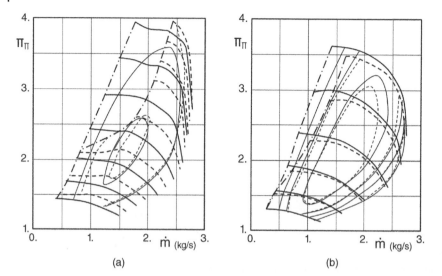

(a)   (b)

Figure 9.23 Effect of (a) a vaneless and (b) a vaned ported shroud on compressor performance (from Chen and Lei 2013).

create wakes at the impeller inlet, resulting in extra noise by the interaction with the impeller leading edge. Together with the lower efficiency and increased manufacturing complexity, these are the major disadvantages of this type of ported shrouds.

Installing vanes that create a counterswirl at the impeller inlet (Figure 9.24) results in a higher efficiency and a shift of the surge limit to smaller mass flows (Tamaki 2011; Sivagnanasundaram et al. 2010, 2012). A first consequence is an increase in the amount of recirculating flow and hence of the axial velocity at the leading edge. Tamaki (2011) demonstrates that this is not the reason for the increased operating range because the more flow is recirculating, the larger will be the negative prerotation and hence the incidence. The increase in the relative flow angle β_1 results in an extra pressure rise in the inducer upstream of the downstream slot. The relative Mach number is higher and the shock moves backward compared to the non-recirculating shroud. This weakens the tip leakage vortex, known to have an important impact on impeller stability. The thicker or even separated boundary layer resulting from the higher incidence and extra pressure rise does not influence the flow in the exducer because it is extracted in the downstream slot and recirculated. Extracting fluid also decreases the meridional velocity and at unchanged relative flow angle (fixed by the blade angle) this results in an increase in the absolute tangential velocity and hence an extra energy input immediately downstream of the slot. The exducer starts at a higher pressure with a

Figure 9.24 Ported shroud with swirl vanes.

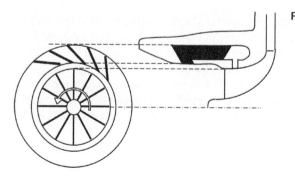

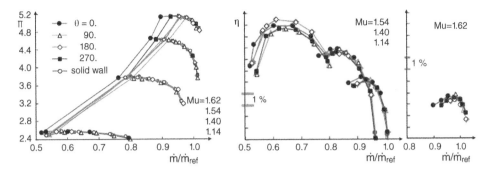

Figure 9.25 Impact of the circumferential location of an asymmetric ported shroud on compressor characteristic (from Tamaki et al. 2012).

new boundary layer, allowing a further increase in pressure rise with improved efficiency and enhanced stability.

All this contributes to an extra increase in the pressure rise in the impeller at decreasing mass flow, i.e. in a more negative slope of the pressure rise curve and a shift of the stability limit to smaller mass flows (Figure 9.23b). The improved performance of the exducer largely compensates for the friction and mixing losses in the ported shroud and is at the origin of the increased efficiency. Chen and Lei (2013) recommend installing a proper vaned diffuser in the ported shroud to maximize the recirculation effect and minimize the losses. Tamaki et al. (2012) also provide a 1D model to estimate the amount of recirculating flow and the increase in the energy input upstream of the shroud slot.

The swirl vane setting angles, and the location and dimension of the ports have been subject of intensive research (Tun et al. 2016; Ishida 2004, 2005; Sivagnanasundaram et al. 2013). Recirculating devices are designed to have almost zero recirculation at the design point to avoid deterioration of the performance. The axial location of the downstream port is important.

Zheng et al. (2010) proposed using a non-axisymmetric ported shroud, extending over only 90° of the circumference, to compensate for the asymmetric flow distortion imposed by the volute. Experimental optimization by Tamaki et al. (2012) shows a considerable extension of the operating range, especially at high inlet Mach numbers when the impeller is sensitive to incidence changes due to the volute pressure distortion (Figure 9.25).

9.4 Enhanced Vaned Diffuser Stability

Casing treatment at the impeller exit and vaned diffuser wall treatment with radial slots in the extended hub section by Jansen et al. (1980) gave very unsatisfactory results (Figure 9.26). The considerable increase in the choking mass flow and operating range results from the increase in the leakage flow in the cavities bypassing the diffuser throat. The extra energy added by these rotating cavities does not contribute to the pressure rise but is dissipated. It is estimated that roughly 70% of the loss increase is caused by the vaned diffuser.

A more successful application of diffuser casing treatment by Amann et al. (1975) is by a circumferential slit at the diffuser inlet, connected to an annular chamber. This reduces the circumferential pressure distortion over the diffuser vane pitch and has a favorable effect on the surge limit (Figure 9.27).

Ono (2013) realized a 6% reduction of the surge mass flow in a five to one pressure ratio compressor by placing the circumferential slit and cavity on the shroud wall near the diffuser

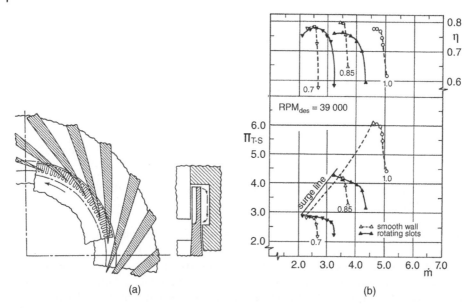

Figure 9.26 (a) Impeller exit casing treatment and (b) effect on performance (from Jansen et al. 1980).

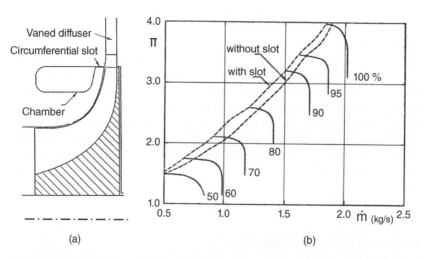

Figure 9.27 (a) Circumferential slit and (b) effect on operating range (from Amann et al. 1975).

throat. It is believed that in this case the range extension is the consequence of an attenuation of the impact of the volute pressure distortion on the diffuser inlet flow. Any local increase in the diffuser exit pressure propagates upstream to the diffuser throat. The diffuser channels with a high outlet pressure have a higher throat static pressure and hence a smaller mass flow. A direct consequence is a local increase in the incidence and the pressure rise between the impeller exit and the throat section. As shown in Figures 4.33 and 4.35 this results in a larger throat blockage and according to Figure 4.42 in a decrease in the static pressure rise in the divergent channel downstream of the throat, which in term leads to a further increase in the throat static pressure. Any pressure distortion at the diffuser exit is amplified and the critical pressure rise for stall (Figure 8.47) may locally be reached.

The main consequence of the slit is a circumferential uniformization of the diffuser inlet flow. Experiments with individual cavities at each throat section did not show any range extension. The flow enters the slit at locations with high throat static pressure and exits at locations with low throat pressure. The throat blockage decreases in locations with high back pressure and the pressure rise increases in the downstream channel. The mass flow upstream of the throat increases at locations with high incidence and decreases at locations with low incidence. Gallaway et al. (2017) calculated a reduction of the circumferential variation of the incidence angle from 10° on a solid shroud to only 2° on a slotted one. An additional way the slit might stabilize the flow is by aspirating the low energy flow in the shroud wall boundary layer in the "critical" throat sections. It is likely that the slit and cavity will be most effective when placed on the shroud side because, as shown by steady and unsteady calculations, the risk of flow separation is highest at that side.

It could be concluded from this that the favorable impact of the slit and cavity might disappear when there is no circumferential distortion imposed by the volute. However, Eisenlohr and Benfer (1988) showed a similar shift of the surge limit in a five to one pressure ratio compressor by perforating the shroud wall between the diffuser throat area and a circumferential cavity. In the absence of a volute they attributed the range extension to the stabilizing effect of the perforated wall on the shocks in the semi-vaneless space. The improvement in the surge margin was not affected by the introduction of a circumferential inlet distortion.

An optimization of the diffuser vanes by adapting the leading edge angle to the spanwise variation of the inlet flow angle has also been attempted. However, optimum incidence cannot be achieved at all mass flows. Goto et al. (2009) indicate an improvement in the performance and a shift of the surge line to smaller mass flows by tapering the diffuser vanes on the hub side (Figure 9.28). Dussourd et al. (1977) claim that such a vane tapering promotes a gradual stall of the diffuser by unloading the diffuser leading edge. Tapering the vanes on the shroud side resulted in a lower pressure rise.

An important characteristic of the pipe diffuser is a spanwise non-uniform leading edge with sweep resulting from the intersection of two adjacent conical inlet sections (Figure 4.22). The interaction of the impeller wakes with the curved leading edge is not clearly defined and may have a favorable impact on vane vibrations and noise generation. Near the hub and shroud, the diffuser leading edge extends to close to the impeller exit and it could be expected that stall could be postponed by preventing return flow of the wall boundary layers. However, Rodgers (1961) observed a larger range with a vaned island diffuser than with a pipe diffuser.

Yoshida et al. (1991) have studied the influence of the radial extend of the vaneless space on the type of instabilities in vaned diffusers. All measurements were made with a lowly loaded impeller $\beta_{2,bl} = 70°$ and a wide diffuser ($b_2/R_2 = 0.2$). The results are shown in Figure 9.29. The difference between D30G20, D20G12, and D20G05 geometry is the location of the diffuser leading edge at $R_3/R_2 = 1.40, 1.24$, and 1.01, respectively. The onset of VDRS, with a rotational frequency $\omega_\sigma/\Omega\lambda < 0.1$, is postponed from $\phi = 0.1$ to $\phi = 0.08$ when reducing the radial extend of the vaneless space and completely disappears at the minimal impeller diffuser gap. This confirms that the theory of Senoo and Kinoshita (1978) is also valid for the vaneless space of vaned diffusers.

Figure 9.28 Tapered vaned diffuser vane (from Goto 2009).

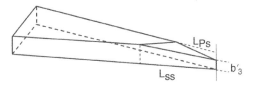

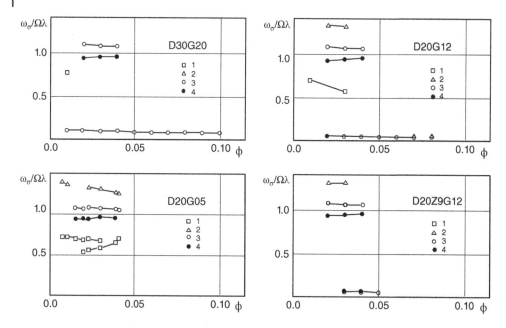

Figure 9.29 Rotating stall propagation speed in a diffuser with different extents of the vaneless space (from Yoshida et al. 1991).

The only difference between D20G12 and D20Z9G12 is a reduction in the number of diffuser vanes from 18 to 9. The shift of the VDRS limit to smaller mass flows is the consequence of the larger amplitude of the non-rotating distortion by the lower number of diffuser vanes. This suggests that rotating instabilities can be suppressed by a non-rotating distortion of sufficiently large amplitude. A similar effect was obtained by turning the diffuser vanes 10° more radial.

The single stall cell rotating at $0.5 < \omega_\sigma/\Omega\lambda < 0.75$ was also measured in the geometry without a vaned diffuser and is considered as PIRS. The other rotating distortions with three to four cells at $0.9 < \omega_\sigma/\Omega\lambda < 1.1$ always appear as a couple. The sum of the λ values equals the number of impeller blades and their average frequency is the impeller rotational frequency. They are considered to be the result of a nonlinear interaction of the other instabilities with the blade passing frequency. The measurements did not explain the origin of the rotating stall pattern with low amplitude at $\omega_\sigma/\Omega\lambda \approx 1.3$.

The surge limit of a vaned diffuser can be shifted to lower mass flow by stabilizing the flow in the diffuser channel. Marsan et al. (2012, 2015) proposed bleeding-off the separated flow in the diffuser channel. As discussed in Section 7.2.2, the numerical predictions of the diffuser flow are not always very accurate and bleeding is efficient only if made at the right location. The results in Figure 9.30 are the outcome of a numerical analysis and show a considerable extension of the stable operating range because of improved diffuser performance, but at a cost of a relatively small efficiency penalty. The favorable impact of a diffuser vane suction side bleed on the surge limit is experimentally confirmed by Rodgers (1982).

Skoch (2003, 2005) tried to stabilize the flow in a vaned island diffuser by air jets on the hub or shroud side near the leading edge. Nozzle orientations were set to inject flow at −8°, 0°, and +8° relative to the vane mean camber line. Several injection flow rates were tested using both an external air supply and recirculation of flow from the diffuser exit. The compressor operating range did not improve at any injection flow rate and even decreased with increasing injection rate.

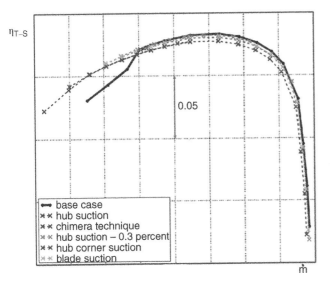

Figure 9.30 Range extension by an aspirated diffuser (from Marsan et al. 2012).

9.5 Impeller–diffuser Matching

Varying the diffuser vane setting does not significantly influence the work input by the impeller. In combination with variable IGV and backward leaning blades (Section 2.1) it allows a constant output pressure over a wide range of mass flows. A major problem of variable vaned diffusers is the complex geometrical relation between the vane setting or suction side angle α_{SS}, the radial extension of the vaneless space, and the throat section. As already explained in Section 4.2.3 any modification of one parameter has an impact on the other.

Different ways to change the diffuser geometry can be found in the literature. Wedge-type vanes may rotate around a point near the leading edge (Figure 9.31a). The orientation is defined by a gliding ring near the trailing edge. The movable part may also be limited to adapt only the leading edge area to the incoming flow. Curved vanes often rotate around a point nearer to the blade mid-chord to limit the torque (Figure 9.31b). It is recommended that the torque is always in the same direction to avoid oscillations of the vanes due to loose fittings.

Rotating the vanes shifts the surge point, a function of incidence, whereas the change of throat section has a direct influence on the choking mass flow. This results in a change in the performance curve, as illustrated in Figure 9.32. The model of Yoshinaka described in Section 8.6.3

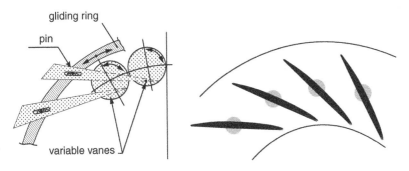

Figure 9.31 Mechanisms for variable vaned diffuser geometry (from Vignau et al. 1987).

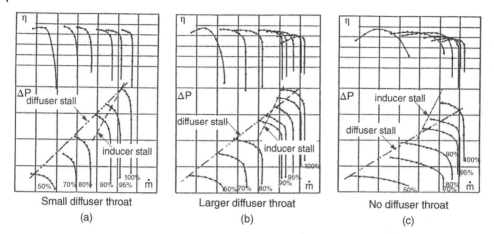

Figure 9.32 Compressor performance maps: (a) vaned diffuser with small throat, (b) vaned diffuser with large throat, and (c) vaneless diffuser.

is well suited to evaluate the effect of a geometrical change of the diffuser on the operating range and to indicate what measures are appropriate to achieve a required change in the performance map.

Turning the vanes more tangentially results in a reduction in the choking mass flow because of a reduction in the throat section and a shift of the diffuser stall limit to smaller mass flow. As illustrated in Figure 9.32a, the impeller stalls at every speed line and the surge limit is defined by the diffuser stall. The operating range is very small because the diffuser incidence at choking is already close to the one at stall and only a small reduction in the mass flow is possible before the diffuser also stalls and surge occurs.

Turning the vanes more radially opens the diffuser throat section and has a favorable effect on choking mass flow (Figure 9.32b). However, it goes with a decrease in the vane suction side angle α_{ss} and a shift of the diffuser stall limit to larger mass flows. The shift of the diffuser stall limit to higher mass flows is compensated for by a larger shift of the choking limit, resulting in an increase in the stable operating range. At higher than 95% nominal speed the diffuser stalls before the impeller stalls and surge is controlled by the latter. This is reflected by the discontinuity in the surge limit. Changing the vaned diffuser setting angle allows monitoring of the intersection point of the impeller and diffuser stall limits. In terms of maximum range, the optimal diffuser throat section area is the one where, at design RPM, diffuser and impeller choking coincide.

Taking out the diffuser vanes (Figure 9.32c) shifts the diffuser stall limit to the critical inlet flow angle for VDRS. According to the Yoshinaka model the latter defines the surge limit at low RPM. The impeller stalls at low RPM (below 90%) well before surge occurs. The surge line is almost unchanged at high RPM because it is controlled by inducer stall. The choking mass flow in the compressor with a vaneless diffuser is large because it is only limited by the impeller choking. The vaneless diffuser stalls before the compressor surges at the impeller stall limit. The consequence is a large decrease in efficiency when approaching the surge limit. The choking limit at high speed is identical for the large vaned and vaneless diffuser because in both cases it is defined by impeller choking.

A major problem with variable vaned diffusers is the clearance between the vanes and the diffuser wall required to allow the rotation of the vanes (Rodgers 1968). These leakage losses may result in a non-negligible penalty in efficiency, as shown in Figure 9.33.

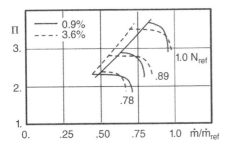
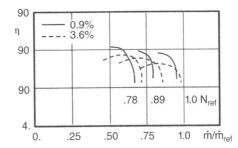

Figure 9.33 Typical compressor performance deterioration as a function of percentage of variable diffuser vane clearance (from Rodgers 1968).

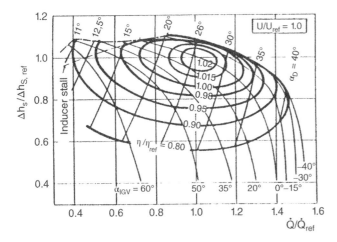

Figure 9.34 Variation of head and efficiency for an optimum combination of IGV and diffuser vanes (from Simon et al. 1986).

A combination of variable IGV and diffuser vanes for industrial compressors is presented by Simon et al. (1986) and Harada (1996). The optimal impeller and diffuser inlet flow angle can be established for a wide range of mass flows. As explained in Section 2.1.1, a variation of the mass flow by prerotation has an impact on the velocity triangle at the diffuser inlet. The vaned diffuser may no longer be optimal and the increase in the impeller stability range is likely to be limited by the unchanged diffuser surge limit (Figure 2.6). An adjustment of the diffuser vane setting angle to the incoming flow allows profiting from an extension of the impeller operating range with limited penalty on efficiency (Figure 9.34).

Salvage (1996, 1998) proposes a split ring pipe diffuser to adapt the throat section without changing the leading edge suction side angle α_{ss}. The diffuser is divided into two concentric rings (Figure 9.35a) with the dividing radius close to the throat. The effective throat section can be decreased when rotating the front or rear part. This device acts as a variable close-coupled resistance to stabilize the flow in the vaneless and semi-vaneless space. Although the fully open diffuser is geometrically similar to the baseline geometry, experiments indicate a loss in operating range by a shift in the surge margin to larger mass flow. The latter was attributed to leakage flow between the diffuser channels at the interface between the rotating and fixed part of the diffuser. Reducing the diffuser throat area in combination with variable IGV resulted in a significant improvement of the surge limit at an efficiency level in line with off-design operation.

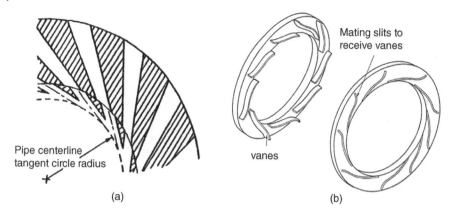

Figure 9.35 (a) Variable pipe diffuser throat section (from Salvage 1998) and (b) variable vaned diffuser width (from Baljé 1981).

Recirculating the flow from the diffuser exit to the diffuser inlet in a controlled way allows the same flow to be maintained in the diffuser passages while decreasing the impeller mass flow at no extra energy cost but increased losses. The extension of the operating range is larger than with the split diffuser, but at a lower efficiency. The re-injection angle turns out to be an important parameter confirming the impact of the vane incidence on the diffuser stability (Salvage 1996).

Another way to adapt the varying flow angle to the diffuser blade angle is by changing the diffuser width (Figure 9.35b). Decreasing the diffuser width will decrease the diffuser inlet flow angle α_2 and allows it to operate at lower mass flow. In addition to a sudden change in passage width at the diffuser inlet, the main problem is again the leakage flows in addition to the non-negligible increase in manufacturing cost.

9.6 Enhanced Vaneless Diffuser Stability

A decrease in the diffuser width at a constant mass flow results in

- a reduction in α_2 because of an increased radial velocity component with unchanged V_u
- an increase in α_{2c} because of the smaller b/R_2 ratio, according to the theory of Senoo and Kinoshita (1978).

Both contribute to a more stable flow by shifting the diffuser inlet condition from the unstable to the stable area (Figure 9.36).

For narrow diffusers, the critical zone for return flow is close to the diffuser inlet (Figure 8.18) and the reduction of the diffuser width should start close to the inlet. Discontinuous variations of the flow area, however, must be avoided because they reduce the diffuser stability. The curvature at the diffuser inlet should be carefully selected, as shown by the detailed study presented in Section 8.2.3.

For wide diffusers, the critical zone for flow reversal is at a larger radius ratio (defined from Figure 8.18). It is therefore recommended that a gradual decrease in diffuser width is used up to the critical zone followed by a parallel diffuser downstream. A decrease in the diffuser width immediately at the inlet would have the same effect in terms of flow stability but, as shown by Lüdtke (1983), would generate larger friction losses because of the smaller hydraulic diameter. Figure 9.37 illustrates the increased stability obtained with a tapered diffuser (b) compared to

Figure 9.36 Diffuser flow stabilization by contraction.

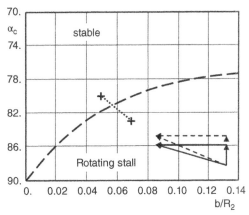

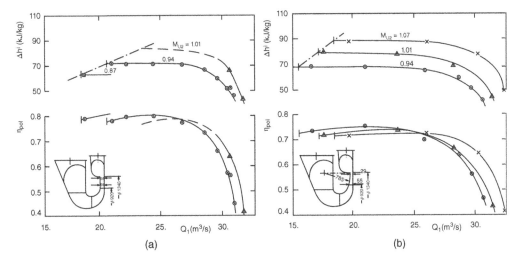

Figure 9.37 Flow stability in (a) a parallel wall diffuser and (b) a tapered diffuser (from Lüdtke 1983).

a parallel wall diffuser (a). However, a narrower diffuser exit results in a slightly lower static pressure rise and a larger radial velocity at the volute inlet, hence a lower efficiency because of higher volute losses.

The diffuser width can also be adjusted as shown in Figure 8.26a. A large round-off at the inlet ensures a smooth transition from the impeller to the diffuser at operating points (small mass flow) where the inlet geometry is critical. The large cavity at the wide open position of the diffuser is acceptable because this position is of interest only at maximum mass flow when stability is not an issue.

Ishida et al. (2001) have stabilized a vaneless diffuser by increasing the wall roughness on the hub wall over different radial extensions to an equivalent sand grain size of 0.55 mm. A considerable extension of the stable operating range was observed, but at the cost of a lower pressure rise in the diffuser because of increased friction losses (Figure 9.38).

Axisymmetric diffuser exit throttling, by retractable diffuser exit rings, has a favorable effect on vaneless diffuser stability, as shown by Abdelhamid (1982) and Watanabe et al. (1994). As long as the impeller flow is stable (low loaded impellers) VDRS can be delayed to almost zero mass flow, but at the cost of a considerable drop in efficiency (Figure 9.39). Part of it is due to the losses in the throttle valve and, as was derived from the variation of the enthalpy input

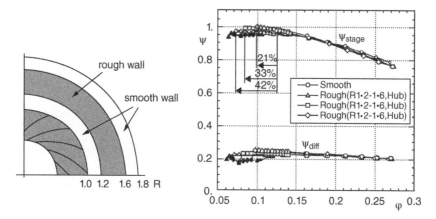

Figure 9.38 Influence of diffuser wall roughness on the stability of vaneless diffusers (from Ishida et al. 2001).

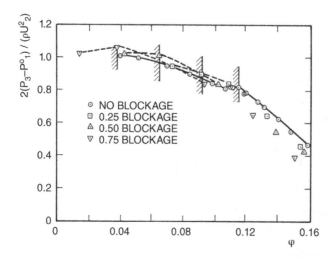

Figure 9.39 Effect of diffuser outlet throttling on vaneless diffuser stability (from Abdelhamid 1982).

$\Delta H^i/(T\eta)$, part of it is the consequence of the considerable increase in the work input probably because of increasing internally recirculating flows. Similar results were obtained by Imaichi and Tsurusaki (1979) by blocking off the diffuser exit area with porous belts of different area ratios.

A similar increase in diffuser stability, but with higher efficiency, can be obtained by means of small vanes installed at the diffuser exit (Abdelhamid 1987). An appropriate setting of the exit vanes results in an increase in diffuser static pressure rise over an enlarged range. For each value of the diffuser inlet flow angle α_2, a minimum value of α_{ss} exists below which stall occurs again (Figure 9.40). A maximum value of α_{ss} can be defined on the basis of diffuser performance deterioration because of reacceleration of the flow.

9.6.1 Low Solidity Vaned Diffusers

The advantages of a low solidity vaned diffuser (LSD) are improved performance and a more stable flow without limitation of the operating range by vaned diffuser choking (Senoo et al. 1983; Senoo 1984). Several different low solidity airfoil diffusers with various channel widths

Figure 9.40 Variation of the stable operating range as a function of diffuser outlet vane setting (from Abdelhamid 1987).

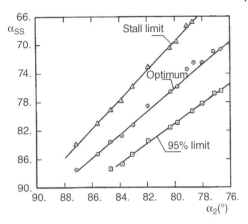

(2.0 < b_3 < 6.0 mm), inlet radius ratios (1.03 < R_3/R_2 < 1.20), and solidity (0.69 < σ < 0.87) have been tested by Nishida et al. (1991) on the compressors described in Section 8.2.3. The most important design parameter appeared to be the radial position of the diffuser vane leading edge. Stall is suppressed when the R_3/R_2 ratio is smaller than the one at which backflow starts. A sample of the results is shown in Figure 9.41, where VD stands for a low solidity vaned diffuser and VL stands for vaneless diffuser. The geometry of the different diffusers is specified in Table 9.2.

Narrow diffusers require the vanes to be close to the impeller exit to be effective. However, the minimum value of R_3/R_2 is limited by the increase in the noise level due to the unsteady interaction between impeller and diffuser vanes (Figure 4.39). The results shown in Figure 9.42 compare the performance of a vaneless diffuser with that of two low solidity diffusers, designed by conformal transformation of a linear cascade with a chord over pitch ratio of 0.69. The maximum mass flow is unchanged because it is limited by choking of the transonic impeller. The largest shift of the stall limit is at high RPM where the diffuser inlet flow is more tangential and the LSD is most useful.

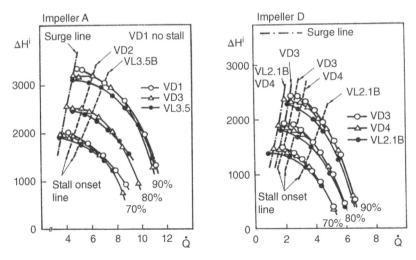

Figure 9.41 Effect of vaned diffuser leading edge diameter ratio on a wider (A) and a narrow (D) impeller (from Nishida et al. 1991).

Table 9.2 Vaned diffuser data (from Nishida et al. 1991).

	b_3 (mm)	R_3/R_2	σ	$\alpha_{3,D}$	$\Re$ (mm)
VD1	3.65	1.067	0.85	79.4	3.0
VD2	3.65	1.200	0.85	79.4	3.0
VD3	2.00	1.030	0.69	73.0	2.5
VD4	2.00	1.150	0.69	73.0	2.5
VD5	3.50	1.030	0.87	80.0	2.5
VD6	6.00	1.067	0.78	77.0	3.0

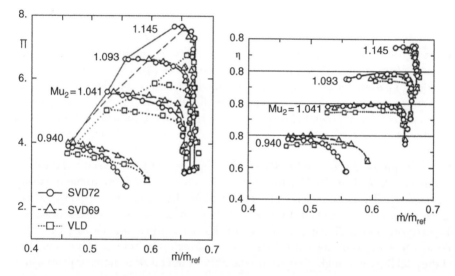

Figure 9.42 Increased operating range obtained with low solidity vaned diffusers (from Hayami et al. 1990).

A parametric study of LSDs by Mukkavilli et al. (2002) demonstrate the direct impact of the vane setting angle and solidity on the performance curve (Figure 9.43). A correct setting angle and proper solidity allow the performance to improve over the whole operating range. Considerable improvements can be obtained at low mass flow without penalty at maximum mass flow (Figure 9.44).

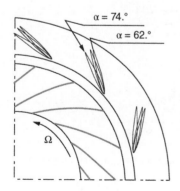

Figure 9.43 Various of LSD vane settings (from Mukkavilli et al. 2002).

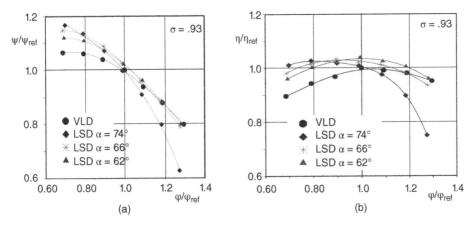

Figure 9.44 Influence of LSD vane setting angle on (a) pressure rise and (b) efficiency (from Mukkavilli et al. 2002).

9.6.2 Half-height Vanes

Half-height vanes or "ribs" have been used to stabilize the flow in vaneless diffusers. Extensive work by Yoshinaga et al. (1985) has led to some considerable performance improvements, as shown in Figure 9.45. It is important to note that in addition to an improved range the pressure rise and efficiency have also increased. The function of these ribs is to stabilize the flow on the shroud, where the velocity is more tangential. Reorienting the flow in a more radial direction allows flow separation on the most critical side to be avoided without creating a throat section.

9.6.3 Rotating Vaneless Diffusers

Free rotating diffusers (Figure 9.46a) have been studied by Rodgers (1970) and Rodgers and Mnew (1975). These diffusers are driven by the friction on the shroud and hub walls, and by the struts connecting those walls. The diffuser wall velocity increases with the radius whereas the fluid velocity decreases with increasing radius. Minimum friction losses occur when the diffuser rotates at the average tangential velocity of the flow, i.e. when energy is added to the diffuser near the inlet, where the fluid tangential velocity is larger than the diffuser velocity, and partially

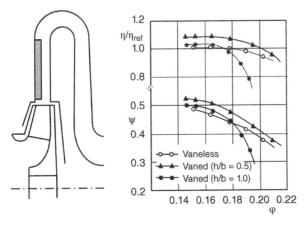

Figure 9.45 Difference in performance between a vaneless diffuser and a diffuser with half-height blades (from Yoshinaga et al. 1985).

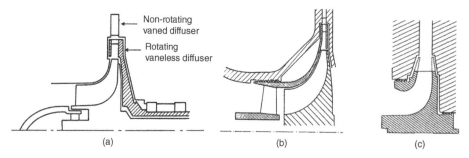

Figure 9.46 Rotating shroud and hub contours (from Rogers 1970; Sapiro 1983).

recuperated near the outlet, where the fluid peripheral velocity has decreased below the diffuser speed. The higher tangential velocity on the walls creates extra centrifugal forces leading to an improved stability of the vaneless diffuser by avoiding return flow. This allows an extension of the stable operating range and also improves the diffuser pressure recovery. The latter is particularly true when the diffuser can be extended radially to allow a shock-free deceleration of the supersonic flow before reaching the diffuser vane leading edge of high pressure ratio compressors (Rodgers and Sapiro 1972).

The estimated gain in diffuser performance is high (three to four points in high pressure ratio compressors with radial ending blades) but must be balanced against a much more complex geometry and the extra energy dissipation resulting from the struts that keep the diffuser walls together, and from the windage losses on the diffuser outer walls and the bearing losses of the rotating support.

Figure 9.46b shows a combination of a preinducer with a rotating diffuser. The preinducer has a lower inlet Mach number and incidence which is favorable both at choking and surge. However, vanes are required in the rotating diffuser to provide sufficient energy to drive the pre-inducer.

Cutting back the blade trailing edge radius is a technique used to adjust the impeller head after some preliminary performance testing. In a multistage compressor this is often used to improve the stage matching and to better realize the required overall performance curve. The question is: should the trimming be limited to the blades or also include the shrouds, resulting in some cavity at the exit of the trimmed impeller (Figure 9.46c)?

Lindner (1983) analyzed geometries with different cut-back of the blades while keeping the hub and shroud walls unchanged. Experiments did not show a change in stability limit but a 4.5% cut-back showed a 1% increase in efficiency whereas a 3–5% drop in efficiency was observed at a 9% cut-back because of the increasing windage losses on the disk extensions.

Sapiro (1983) analyzed the impact of a 20% reduction in the blade outlet radius. Experimental results for the lower specific speed impeller ($N_{SC} = 90$) showed a 3.5% reduction of the peak efficiency when keeping the hub and shroud wall at the original radius with a 7% larger head than when the disks were also cut to the smaller radius. This is the consequence of the additional pumping by friction on the extended disks, which was estimated to be responsible for a 1% decrease in the impeller efficiency. It is postulated that in low N_{SC} impellers the extended disks result in higher windage losses in the side cavities and make the flow more tangential at the vaneless diffuser inlet. The consequences are larger vaneless diffuser losses because of a longer flow path length (Figure 9.47a), and a shift of the rotating stall limit to a 6% larger mass flow.

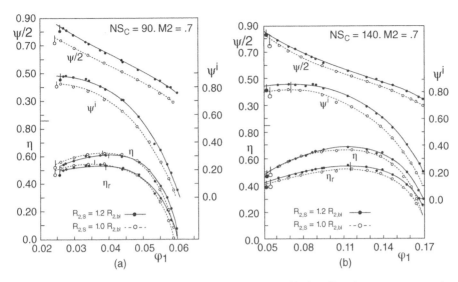

Figure 9.47 Impact of the radial extension of the shroud and hub wall on the compressor performance at (a) $NS_C = 90$ and (b) $NS_C = 140$ (from Sapiro 1983).

Maintaining the 20% extension of the shrouds for a high specific compressor ($N_C = 140$) resulted in a 3% increase in efficiency, a 9.5% larger head at peak efficiency and a 5% decrease in the mass flow at stability limit (Figure 9.47b). The shroud extension increased the head and the windage and vaneless space losses. However, the latter have only a limited effect because of the higher mass flow in high NS_C compressors.

Bibliography

Abdelhamid AN 1980 Analysis of rotating stall in vaneless diffusers of centrifugal compressors. ASME 80-GT-184.

Abdelhamid AN 1981 Effects of vaneless diffuser geometry on flow instability in centrifugal compression systems. ASME 81-GT-10; *Canadian Aeronautics and Space Journal*, 29, 259–288.

Abdelhamid AN 1982 Control of self-excited flow oscillations in vaneless diffuser of centrifugal compression systems. ASME 82-GT-188, *Canadian Aeronautics and Space Journal*, 29, 336–345.

Abdelhamid AN 1987 A new technique for stabilizing the flow and improving the performance of vaneless radial diffusers. *Trans. ASME, Journal of Turbomachinery*, 109, 36–40.

Abdelhamid AN and Bertrand J 1979 Distinction between two types of self excited gas oscillations in vaneless radial diffusers. ASME 79-GT-58; *Canadian Aeronautics and Space Journal*, 26, 105–117.

Abdelhamid AN, Colwill WH, and Barrows JF 1979 Experimental investigation of unsteady phenomena in vaneless radial diffusers. *Trans. ASME, Journal of Engineering for Power*, 101, 52–60.

Abdelwahab A 2010 Numerical investigation of the unsteady flow fields in centrifugal compressor diffusers. ASME GT-2010-22489.

Agostinelli A, Nobles D and Mockridge CR 1960 An experimental investigation of radial thrust in centrifugal pumps. *Trans. ASME, Journal of Engineering for Power*, 82, 120–126.

Ahn HJ, Park MS, Sanadgol D, Park I-H, Han D-C and Maslen EH 2009 A pressure output feedback control of turbo compressor surge with a thrust magnetic bearing actuator. *Journal of Mechanical Science and Technology* 23, 1406–1414. DOI: 10.1007/s12206-008-1125-y.

Aksoy HG 2002 Unsteady incompressible flow in pump impeller with outlet pressure distortion. von Karman Institute PR 2002-27.

Alsalihi Z and Van den Braembussche RA 2002 Evaluation of a design method for radial impellers based on artificial neural network and genetic algorithm. *Proceedings of ESDA 2002, 6th Biennial Conference on Engineering Systems Design and Analysis*, Istanbul. ASME, ESDA2002/ATF-069.

Amann CA, Nordenson GE and Skellenger GD 1975 Casing modification for increasing the surge margin of a centrifugal compressor on an automotive turbine engine. *Trans. ASME, Journal of Engineering for Power*, 97, 329–336.

Anisimov et al. 1962 The influence of the number of vanes on the efficiency of the centrifugal wheel with a single stage cascade. Energomaschinostroyeniye No 221, translated. FTD-TT62-1816 Power machine construction, Foreign technology division, Air Force Systems Command Wright-Patterson Air Force Base, Ohio.

Aoki M 1984 Instantaneous interblade pressure distributions and fluctuating radial thrust in a single-blade centrifugal pump. *Bulletin of the Japan Society of Mechanical Engineering*, 233, 2413–2420.

Design and Analysis of Centrifugal Compressors, First Edition. René Van den Braembussche.
© 2019, The American Society of Mechanical Engineers (ASME), 2 Park Avenue, New York, NY, 10016, USA (www.asme.org).
Published 2019 by John Wiley & Sons Ltd.

Arbabi A and Ghaly W 2013 Inverse design of turbine and compressor stages using a commercial CFD program. ASME GT-2013-96017.

Ariga I, Masuda S and Ookita A 1987 Inducer stall in a centrifugal compressor with inlet distortion. *Trans. ASME, Journal of Turbomachinery*, 109, 27–35.

Arnulfi GL, Giannattasio P, Giusto C, Massardo AF, Micheli D and Pinamonti P 1999 Multistage centrifugal compressor surge analysis: Part II – Numerical simulation and dynamic control parameters evaluation. *Trans. ASME, Journal of Turbomachinery*, 121, 312–320.

Arnulfi GL, Micheli D and Pinamonte P 2001 An innovative device for passive control of surge in industrial compression systems. *Trans. ASME, Journal of Turbomachinery*, 123, 473–482.

Aungier R 2000 *Centrifugal Compressors: A Strategy for Aerodynamic Design and Analysis*. ASME Press, ISBN 0-7918-0093-8.

Ayder E 1993 Experimental and Numerical Analysis of the Flow in Centrifugal Compressor and Pump Volutes. PhD Thesis, von Karman Insititute – Universiteit Gent.

Ayder E and Van den Braembussche RA 1991 Experimental study of the swirling flow in the internal volute of a centrifugal compressor. ASME 91-GT-7.

Ayder E and Van den Braembussche RA 1994 Numerical analysis of the three-dimensional swirling flow in centrifugal compressor volutes. *Trans. ASME, Journal of Turbomachinery*, 116, 462–468.

Ayder E, Van den Braembussche RA and Brasz JJ 1993 Experimental and theoretical analysis of the flow in a centrifugal compressor volute. *Trans. ASME, Journal of Turbomachinery*, 115, 582–589.

Bäck TH 1996 *Evolutionary Algorithms in Theory and Practice*. Oxford University Press, New York.

Baghdadi S 1977 The effect of rotor blade wakes on centrifugal compressor diffuser performance – A comparative experiment. *Trans. ASME, Journal Fluids Engineering*, 99, 45–52.

Baghdadi S and McDonald AT 1975 Performance of three vaned radial diffusers with swirling transonic flow. *Trans. ASME, Journal of Fluid Engineering*, 97, 155–173.

Baljé O 1961 A study of design criteria and matching of turbomachines. Part B: Compressor and pump performance and matching of turbocomponents. *Trans. ASME, Journal of Engineering for Power*, 83, 103–114.

Baljé O 1970 Loss and flow path studies in centrifugal compressors – Part II. ASME 70-GT-12-b. *Trans. ASME, Journal Engineering for Power*, 92(3), 275.

Baljé O 1981 *Turbomachines – A Guide to Design, Selection and Theory*. John Wiley and Sons, New York.

Bammert K, Jansen M, Knapp P and Wittekindt W 1976 Strömungsuntersuchungen an beschaufelten Diffusoren für Radialverdichter. *Konstruktion*, 28, 313–319.

Bammert K, Jansen M and Rautenberg M 1983 The influence of the diffuser inlet shape on the performance of a centrifugal compressor stage. ASME 83-GT-09.

Baun DO and Flack RD 1999 A Plexiglas research pump with calibrated magnetic bearings/load cells for radial and axial hydraulic force measurements. *Trans. ASME, Journal of Fluids Engineering*, 121, 126–132.

Benichou E and Trébinjac I 2015 Steady/unsteady diffuser flow in a transonic centrifugal compressor. 11th European Conference on Turbomachinery, Fluid Dynamics and Thermodynamics (ETC11), Madrid, 23–27 March, 2015.

Benichou E and Trébinjac I 2016 Comparison of steady and unsteady flows in a transonic radial vaned diffuser. *Trans. ASME, Journal of Turbomachinery*, 138, 121002-1 to 10.

Bennet I, Tourlidakis A and Elder R 1998 Detailed measurements within a selection of pipe diffusers for centrifugal compressors. ASME 98-GT-92.

Benvenuti E 1977 Development testing of stages for centrifugal process compressors. In: *Industrial Centrifugal Compressors*, VKI Lecture Series 95, von Karman Institute, LS 95.

Benvenuti E, Bonciani L and Corradini U 1980 Inlet flow distortions on industrial compressor stages – Experimental investigations and evaluations of effect on performance. In: *Centrifugal Compressors, Flow Phenomena and Performance*, AGARD CP-282, paper 4.

Bertini L, Neri P, Santus C, Guglielmo A and Mariott G 2014 Analytical investigation of the SAFE diagram for bladed wheels, numerical and experimental validation. *Journal of Sound and Vibration*, 333, 4771–4788.

Bhinder F and Ingham D 1974 The effect of inducer shape in the performance of high pressure ratio centrifugal compressors. ASME 74-GT-122.

Biheller HJ 1965 Radial force on the impeller of centrifugal pumps with volute, semivolute and fully concentric casings. *Trans. ASME, Journal Engineering for Power*, 87, 319–323.

Binder RC and Knapp RT 1936 Experimental determination of the flow characteristics in the volutes of centrifugal pumps. *Trans. ASME*, 58, 649–663.

Block JA, Rundstadler PW and Dean R 1972 Modeling of a high pressure ratio centrifugal compressor using a low speed of sound gas. Japan Society of Mechanical Engineering, published by CREARE Inc., Hannover, New Hampshire.

Bonaiuti D, Arnone A, Hah C and Hayami H 2002 Development of secondary flow field in a low solidity diffuser in a transonic centrifugal compressor. ASME GT-2002-30371.

Bonciani L, Ferrara PL and Timori A 1980 Aero-induced vibrations in centrifugal compressors. In: *Proceedings of Rotordynamic Instability Problems in High Performance Turbomachinery*, Texas A&M University, NASA CP 2133, 85–94.

Bonciani L, Terrinoni L and Tesei A 1982 Unsteady flow phenomena in industrial centrifugal compressor stage. In: *Proceedings of Rotordynamic Instability Problems in High Performance Turbomachinery*, Texas A&M University, NASA CP 2250, 344–364.

Borges JE 1990 A Three-Dimensional Inverse Method for Turbomachinery: Part I – Theory. *Trans. ASME, Journal of Turbomachinery*, 112, 346–354.

Bowermann R and Acosta A 1957 Effect of the volute on performance of a centrifugal pump impeller. *Trans. ASME*, 79, 1057–1069.

Breugelmans F 1972 The Supersonic Compressor Stage. In: *Industrial Turbo-compressors*, von Karman Institute Lecture Series 47.

Brown WB and Bradshaw GR 1947 Design and performance family of diffusing scrolls with mixed flow impeller and vaneless diffuser. NACA TR 936.

Brownell RB and Flack RD 1984 Flow Characteristics in the Volute and Tongue Region of a Centrifugal Pump. ASME 84-GT-82.

Busemann A 1928 Das Förderhöhenverhaltnis radialer Kreiselpumpen mit logarithmisch-spiraligen Schaufeln. *Zeitschrift für Angewandte Mathematik und Mechanik*, 8, 372–384.

Came P and Herbert M 1980 Design and experimental performance of some high pressure ratio centrifugal compressors. In: *Centrifugal Compressors, Flow Phenomena and Performance*, AGARD-CP-282, paper 15.

Carlson J, Johnston J and Sagi C 1967 Effects of wall shape in flow regimes and performance in straight two-dimensional diffusers. *Trans. ASME, Journal of Basic Engineering*, 89, 151–160.

Carnevale EA, Ferrara G, Ferrari L and Baldassarre L 2006 Experimental characterization of vaneless diffuser rotating stall, Part VI: Reduction of three impeller results. ASME GT-2006-90694.

Carrard A 1923 On calculations for centrifugal wheels. La technique Moderne, T. XV. No. 3, translated by John Moore (1975), Cambridge University Engineering Department, CUED/A-TURBO/TR 73.

Casey M 1985 The effect of Reynolds number on the efficiency of centrifugal compressor stages. *ASME, Journal of Engineering for Power*, 107, 541–548.

Casey M and Robinson CJ 2011 A unified correction method for Reynolds number, size and roughness effects on the performance of compressors and pumps. *Proc. IMechE, Part A. Journal of Power and Energy*, 225, 864–876.

Cellai A, Ferrara G, Ferrar L, Mengoni CP and Baldassarre L 2003a Experimental investigation and characterization of the rotating stall in a high pressure centrifugal compressor: Part III: Influence of diffuser geometry on stall inception and performance (2nd impeller tested). ASME GT-2003-38390.

Cellai A, Ferrara G, Ferrar L, Mengoni CP and Baldassarre L 2003b Experimental investigation and characterization of the rotating stall in a high pressure centrifugal compressor: Part IV: Impeller influence on diffuser stability. ASME GT-2003-38394.

Chamieh DS, Acosta AJ, Brennen CE and Caughe T 1985 Experimental measurements of hydrodynamic radial forces and stiffness matrices for centrifugal pump impellers. *Trans. ASME, Journal of Fluids Engineering*, 107, 307–315.

Chapman DR 1954 Some possibilities of using gas mixtures other than air in aerodynamic research. NACA TN 3226.

Chen H and Lei VM 2013 *Casing treatment and inlet swirl of centrifugal compressors Trans. ASME, Journal of Turbomachinery*, 135, 0401011–8.

Chen J, Hasemann H, Seidel U, Jin D, Huang X and Rautenberg M 1993 The interpretation of internal pressure patterns of rotating stall in centrifugal compressor impellers. ASME 93-GT-192, 503–529.

Chen J, Hasemann H, Shi L and Rautenberg M 1994 Stall inception behavior in a centrifugal compressor. ASME 94-GT-159.

Chew H and Vaughan C 1988 Numerical predictions for the flow induced by an enclosed rotating disc. ASME 88-GT-127.

Childs D 1986 Force and moment rotordynamic coefficients for pump impeller shroud surfaces. In: *Rotordynamic Instability Problems in High Performance Turbomachinery*, NASA CP 2443, 503–529.

Childs D 1992 Pressure oscillations in the leakage annulus between a shrouded impeller and its housing due to pressure discharge – Pressure disturbances. *Trans. ASME, Journal of Fluids Engineering*, 114, 61–67.

Childs PRN and Noronha MB 1999 The impact of machining techniques on centrifugal compressor impeller performance. *Trans. ASME, Journal of Turbomachinery*, 121, 637–643.

Chochua G, Koch JM and Sorokes JM 2005 Analytical and computational study of radial loads in volutes and collectors. ASME GT-2005-68822.

Clements WW and Artt DW 1987a Performance prediction and impeller–diffuser matching for vaned diffuser centrifugal compressors. IMechE C256/87, 183–198.

Clements WW and Artt DW 1987b The influence of diffuser channel geometry on the flow range and efficiency of a centrifugal compressor. *Proc. IMechE*, 201 A2, 145–152.

Clements WW and Artt DW 1988 The influence of diffuser channel length–width ratio on the efficiency of a centrifugal compressor. *Proc. IMechE*, 202 A3, 163–169.

Clements WW and Artt DW 1989 The influence of diffuser vane leading edge geometry on the performance of a centrifugal compressor. ASME 89-GT-163.

Colebrook C 1939 Turbulent flow in pipes with particular reference to the transition region between the smooth and rough pipe laws. *Journal of the Institute of Civil Engineers*, 11 (4), 133–156. doi:10.1680/ijoti.1939.13150. See also: V.L. Streeter, Chapter VI: Steady flow in pipes and conduits. In *Engineering Hydraulics*, H. Rouse (ed.), New York, 1950.

Conrad O, Raif K and Wessels M 1980 The calculation of performance maps for centrifugal compressors. In: *Performance Prediction of Centrifugal Pumps and Compressors*, Special Publication of the ASME Gasturbine Conference, March 10–13, 1980, New Orleans.

Cumpsty NA and Greitzer EM 1982 A simple model for compressor stall cell propagation. *ASME, Journal of Engineering for Power*, 104, 170–176.

Daily JW and Nece RE 1960 Chamber dimension effects on reduced flow and frictional resistance of enclosed rotating discs. *Journal of Basic Engineering*, 82, 217–232.

Dallenbach F 1961 The aerodynamic design and performance of centrifugal and mixed flow compressors. *SAE Technical Progress Series*, 3, 2–30.

Daneshkhah K and Ghaly W 2007 Aerodynamic inverse design for viscous flow in turbomachinery blading. *AIAA, Journal of Propulsion and Power*, 23, 814–820.

Dang TG 1992 A fully three-dimensional inverse method for turbomachinery blading in transonic flows. ASME 92-GT-209.

Davis R and Dussourd J 1970 A unified procedure for the calculation of off-design performance of radial turbomachinery. ASME 70-GT-64.

Dawes W 1995 A simulation of the unsteady interaction of a centrifugal impeller with its vaned diffuser: Flow analysis. *ASME, Journal of Turbomachinery*, 117, 213–222.

Day IJ 1993 Stall inception in axial flow compressors. *Trans. ASME, Journal of Turbomachinery*, 115, 1–9.

Dean R 1972 Advanced radial compressors. In: *Advanced Radial Compressors*, von Karman Institute Lecture Series 50.

Dean R and Senoo Y 1960 Rotating wakes in vaneless diffusers. *Trans. ASME, Journal of Basic Engineering*, 82, 563–570.

Dean RC Jr. and Young LR 1977 The time domain of centrifugal compressor and pump stability and surge. *Trans. ASME, Journal of Fluids Engineering*, 99, 53–63.

Demeulenaere A 1997 Conception et Développement d'une Méthode Inverse pour la Génération d'Aubes de Turbomachines. PhD thesis, von Karman Institute – Unversité de Liège.

Demeulenaere A and Van den Braembussche RA 1996 Three-dimensional inverse method for turbomachinery blading design. In: *3rd Eccomass Computational Fluid Dynamic Conference, Paris, France, September 9–13, 1996*.

Demeulenaere A, Léonard O and Van den Braembussche RA 1997 A two-dimensional Navier–Stokes inverse solver for compressor and turbine blade design. *Proceedings of the 2nd European Conference on Turbomachinery-Fluid Dynamics and Thermodynamics, March 5–7, 1997, Antwerp, Belgium*. ISBN 90-5204-32-X.

Demeulenaere A and Van den Braembussche RA 1998a Three-dimensional inverse method for turbomachinery blading design. *Trans. ASME, Journal of Turbomachinery*, 120, 247–255.

Demeulenaere A, Léonard O and Van den Braembussche RA 1998b Application of a three-dimensional inverse method to the design of a centrifugal compressor impeller. ASME 98-GT-127.

Demeulenaere A and Van den Braembussche RA 1999 A new compressor and turbine blade design method based on three-dimensional Euler computations with moving boundaries. *Inverse Problems in Engineering*, 7, 235–266.

Deniz E, Greitzer E and Cumpsty NA 2000 Effect of the inlet flow field conditions on the performance of centrifugal compressor diffusers: Part II – Straight-channel diffuser. *Trans. ASME, Journal of Turbomachinery*, 122, 11–21.

Denton JD 1986 The use of distributed body force to simulate viscous effects in 3D flow calculations. ASME 86-GT-144.

de Vito L, Van den Braembussche RA and Deconinck H 2003 A novel two-dimensional viscous inverse design method for turbomachinery blading. *ASME, Journal of Turbomachinery*, 125, 310–316.

Dick E, Vierendeels J, Serbruyns S and Vande Voorde J 2001 Performance prediction of centrifugal pumps with CFD tools. *TASK Quarterly*, 5, 579–594.

Dickmann HP, Secall Wimmel T, Szwedowicz J, Filsinger D and Roduner C 2005 Unsteady flow in a turbocharger centrifugal compressor – 3D-CFD simulation and numerical and experimental analysis of impeller blade vibration. ASME GT-200568235.

Di Liberti J-L, Van den Braembussche RA, Rasmussen S and Konya P 1996 Active control of surge in centrifugal impellers with inlet pipe resonance. Presented at the ASME International Mechanical Engineering Congress and Exhibition, Atlanta, Georgia, ASME 96-WA/PID-1.

Di Sante A and Van den Braembussche RA 2010 Experimental study of the effects of spanwise rotation on the flow in a low aspect ratio diffuser for turbomachinery applications. Experiments in Fluids, DOI 10.1007/s00348-010-0829-9.

Doulgeris and Van den Braembussche RA 2003 Optimization of a asymmetric inlet guide vane for radial compressors. von Karman Institute SR 2003-28.

Du J and Seume JR 2017 Design of casing treatment on a mixed-flow compressor. ASME GT-2017-65226.

Dussourd JL and Putman WG 1960 Instability and surge in dual entry centrifugal compressors. In: *Compressor Stall, Surge and System Response*, ASME.

Dussourd JL, Plannebecker GW and Singhania SK 1977 An experimental investigation of the control of surge in radial compressors using close coupled resistance. *Trans. ASME, Journal of Fluids Engineering*, 99, 64–76.

Eckardt D 1976 Detailed flow investigations within a high speed centrifugal compressor. *Trans. ASME, Journal of Fluids Engineering*, 98, 390–402.

Eckert B and Schnell E 1961 *Axial- und Radialkompressoren*. Springer Verlag, Berlin.

Eisenlohr G and Benfer FW 1988 Massnahmen zur Verringerung der Auswirkungen von Einlaufstörungen auf das Betriebsverhalten eines Radialverdichters [Measures for reducing the effect of inlet distortion on the characteristic of a centrifugal compressor]. *MTZ Motortechnische Zeitschrift*, 49–6, 265–270.

Ejiri E, Kosuge H and Ito T 1983 A consideration concerning stall and surge limitations within centrifugal compressors. *Proceedings of the International Gas Turbine Congress*, Tokyo, Part II, 478–485.

Elder RL 1994 Enhancement of Turbocompressor Stability. IMechE C477/029/94, pp. 145–158

Elder RL and Gill ME 1984 A discussion of the factors affecting surge in centrifugal compressors Trans. ASME, Journal of Engineering Gas Turbine and Power, 107, 499–506.

Elholm T, Ayder Z and Van den Braembussche RA 1992 Experimental study of the swirling flow in the volute of a centrifugal pump. *Trans. ASME, Journal of Turbomachinery*, 114, 366–372.

Ellis GO 1964 A study of induced vorticity in centrifugal compressors. *Trans. ASME, Journal of Engineering for Power*, 86, 63–76.

Ellis GO and Stanitz JD 1952 Comparison of two- and three-dimensional potential-flow solutions in a rotating impeller passage. NACA TN 2806.

Emmons HW, Pearson CE and Grant HP 1955 Compressor surge and stall propagation. *Trans. ASME*, 77, 455–469.

Erdos JI and Alzner E 1977 Computation of unsteady transonic flows through rotating and stationary cascades. NASA CR 2900.

Everitt JN and Spakovszky ZS 2013 An investigation of stall inception in centrifugal compressor vaned diffuser. *Trans. ASME, Journal of Turbomachinery*, 135, 011025-1-10.

Eynon A and Whitfield A 2000 Pressure recovery in a turbocharger compressor volute. *Proc. IMechE, Part A*, 214, 599–610.

Fatsis A 1995 Numerical Study of the 3D Unsteady Flow and Forces in Centrifugal Impellers with Outlet Pressure Distortion. PhD Thesis, von Karman Insititute – Universiteit Gent.

Fatsis A, Pierret S and Van den Braembussche RA 1995 3D Unsteady flow and forces in centrifugal impellers with circumferential distortion of the outlet static pressure. ASME 95-GT-33.

Ferrara G, Ferrari L, Mengoni CP, de Lucia M and Baldassarre L 2002a Experimental investigation and characterization of the rotating stall in a high pressure centrifugal compressor: Part I: Influence of diffuser geometry on stall inception. ASME GT-2002-30389.

Ferrara G, Ferrari L, Mengoni CP, de Lucia M and Baldassarre L 2002b Experimental investigation and characterization of the rotating stall in a high pressure centrifugal compressor: Part II: Influence of diffuser geometry on stage performance. ASME GT-2002-30390.

Ferrara G, Ferrari L and Baldassarre L 2006 Experimental characterization of vaneless diffuser rotating stall: Part V: Influence of diffuser geometry on stall inception and performance (3rd impeller tested). ASME GT-2006-90693.

Filipenco VG, Deniz S, Johnston JM, Greitzer EM and Cumpsty NA 2000 Effect of inlet flow field conditions on the performance of centrifugal compressor diffusers: Part I – Discrete passage diffuser, Trans. *ASME, Journal of Turbomachinery*, 122, 1–10.

Fink DA, Cumpsty NA and Greitzer EM 1992 Surge dynamics in a free-spool centrifugal compressor system. *Trans. ASME, Journal of Turbomachinery*, 114, 321–332.

Fisher FB 1989 Application of map width enhancement devices to turbocharger compressor stages. In: *Power Boost: Light, Medium and Heavy Duty Engines*, SP-780, SAE Paper No. 880794.

Flathers M and Bache G 1999 Aerodynamically induced radial forces in a centrifugal gas compressor: Part 2 – Computational investigation. *Trans. ASME, Journal for Engineering for Gas Turbine and Power*, 121, 725–734.

Flathers M, Bache G and Rainsberger R 1996 An experimental and computational investigation of flow in a radial inlet of an industrial pipeline centrifugal compressor, Trans. ASME, *Journal of Turbomachinery*, 118, 371–387.

Flynn PF and Weber HG 1979 Design and test of an extremely wide flow range compressor. ASME 79-GT-80.

Fowler H 1966 *An investigation of the flow processes in a centrifugal compressor impeller*. National Research Council, Canada, DME 230.

Fowler H 1968 The distribution and stability of flow in a rotating channel. *Trans. ASME, Journal of Engineering for Power*, 90, 229–235.

Frigne P 1982 Theoretische en Experimentele Studie van de Subsynchrone Zelfgeëxiteerde Stromingsoscillaties in Radiale Turbocompressoren [in Dutch]. PhD Thesis, von Karman Insititute – Universiteit Gent.

Frigne P and Van den Braembussche RA 1979 One-dimensional design of centrifugal compressors taking into account flow separation in the impeller. von Karman Institute TN 129.

Frigne P and Van den Braembussche RA 1984 Distinction between types of impeller and diffuser rotating stall in centrifugal compressors with vaneless diffusers. *Trans. ASME, Journal of Engineering for Gas Turbines and Power*, 106, 468–474.

Frigne P and Van den Braembussche RA 1985 A theoretical model for rotating stall in the vaneless diffuser of centrifugal compressors. *Trans. ASME, Journal of Engineering for Gas Turbines and Power*, 107, 468–474.

Galloway L, Spence S, Kim SI, Rusch D, Vogel K and Hunziker R 2017 An investigation of the stability enhancement of a centrifugal compressor stage using a porous throat diffuser. ASME GT-2017-63071.

Galvas MR 1973 FORTRAN program for predicting off-design performance of centrifugal compressors. NASA TN D-7487.

Gauger NR 2008 Efficient deterministic approaches for aerodynamic shape optimization. In: *Optimization and Computational Fluid Dynamics*, D Thevenin and GG Janiga (eds), DOI: 10.1007/978-3-540-72153-6.

Giachi M, Ramalingam V, Belardini E, De Bellis F, and Reddy C 2014 Parametric perfomance of a class of standard discharge scrolls for industrial centrifugal compressors. ASME GT-2014-26831.

Giannakoglou KC, Papadimitriou DI and Papoutsis-Kiachagias EM 2012 The continuous adjoint method: theory and industrial applications. In: *Introduction to Optimization and Multidisciplinary Design in Aeronautics and Turbomachinery*, J Périaux and T Verstraete (eds), VKI LS 2012-03, ISBN-13 978-2-87516-032-4.

Giles MB 1988 Calculation of unsteady wake rotor interaction. *Journal of Propulsion Power*, 4, 356–362.

Giles MB 1991 *UNSFLO: A numerical method for unsteady flow in turbomachinery*. Gas Turbine Laboratory Report GTL 205, MIT Department of Aeronautics and Astronautics.

Goldberg D 1989 *Genetic Algorithms in Search, Optimization and Machine Learning*, Addison-Weslet, Stuttgart.

Gong Y, Sirakov BT, Epstein AH and Tan CS 2004 Aerothermodynamics of micro-turbomachinery. ASME GT-2004-53877.

Goto T, Ohmoto E, Ohta Y and Outa E 2009 Noise reduction and surge margin improvement using a tapered diffuser vane in a centrifugal compressor. Proceedings of the 9th International Symposium on Experimental and Computational Aerothermodynamics of Internal Flow, ISAIF9-3A-2, Gyeongju, Korea.

Graham RW and Guentert EC 1965 Compressor stall and blade vibration. In *Aerodynamic Design of Axial-flow Compressors*, IA Johnsen and RO Bullock (eds), NASA SP-36, NTIS N65-23345.

Greitzer EM 1976 Surge and rotating stall in axial flow compressors. *Trans. ASME, Journal of Engineering for Power*, 98, 190–217.

Greitzer EM 1981 The stability of pumping systems. The 1980 Freeman Scholar lecture. *Trans. ASME, Journal of Fluids Engineering*, 103, 193–242.

Greitzer EM, Nikkanen JP, Haddad DE, Mazzawy RS and Joslyn HD 1979 A fundamental criterion for the application of rotor casing treatment. *Trans. ASME, Journal of Fluids Engineering*, 101, 237–243.

Gyarmathy G 1996 Impeller diffuser momentum exchange during rotating stall. ASME International Mechanical Engineering Congress and Exhibition, paper 96-WA/PID-6.

Gysling DL, Dugundji J, Greitzer EM and Epstein AH 1991 Dynamic control of centrifugal compressor surge using tailered structures. *Trans. ASME, Journal of Turbomachinery*, 113, 710–722.

Hagelstein D, Hillewaert K, Van den Braembussche RA, Engeda A, Keiper R and Rautenberg M 2000 Experimental and numerical investigation of the flow in a centrifugal compressor volute. *Trans. ASME, Journal of Turbomachinery*, 122, 22–31.

Han F, Qi D, Tan J, Wang L and Mao Y 2012 Experimental and numerical investigation of the flow in the radial inlet of a centrifugal compressor. ASME GT2012-69353.

Harada H 1996 Study of a surge-free centrifugal compressor with automatically variable inlet and diffuser vanes. ASME 96-GT-153.

Harinck J, Alsalihi Z, Van Buijtenen JP and Van den Braembussche RA 2005 Optimization of a 3D radial turbine by means of an improved genetic algorithm. Proceedings of the 6th European Conference on Turbomachinery, Fluid Dynamics and Thermodynamics, 7–11 March 2005, Lille, France, Paper 162.

Harley P, Spence S, Filsinger D, Dietrich M and Early J 2012 An evaluation of 1D design methods for the off-design performance prediction of automotive turbocharger compressors. ASME GT2012-69743.

Hartmann MJ and Wilcox WW 1957 Problems encountered in the translation of compressor performance from one gas to another. *Trans. ASME*, 79, 887–897.

Hasegawa Y, Kikuyama K. and Maeda T 1990 Effects of blade number on hydraulic force perturbation on impeller of volute-type centrifugal pump. *JSME International Journal, Series II*, 33, 736–742.

Haseman H, Chen J, Jin D and Rautenberg M 1993 *Internal transient pressure patterns during rotating stall in centrifugal impellers*. Proceedings of the 2nd International Symposium on Experimental and Computational Aerothermodynamics of Internal Flow, ISAIF2, Prague, pp. 291–302.

Haupt U, Bammert K and Rautenberg M 1985a Blade vibration on centrifugal compressors: Fundamental considerations and initial measurements. ASME 85-GT-92.

Haupt U, Bammert K and Rautenberg M 1985b Blade vibration on centrifugal compressors: Blade response to different excitation conditions. ASME 85-GT-93.

Hawthorn WR 1974 *Secondary vorticity in stratified compressible fluids in rotating systems.* Cambridge University Engineering Department, CUED/A-Turbo/TR 63.

Hawthorn WR, Tan CS, Wang C and McCune JE 1984 Theory for blade design for large deflections: Part 1 – Two-dimensional cascades. *Trans. ASME, Journal of Engineering Gas Turbines and Power*, 106, 346–353.

Hayami H, Senoo Y and Utsunomiya K 1990 Application of low-solidity cascade diffuser to transonic centrifugal compressors. *Trans. ASME, Journal of Turbomachinery*, 112, 25–29.

Hazby HR and Xu L 2009 Role of tip leakage in stall of a transonic centrifugal impeller. ASME GT-2009-59372.

Hazby H, Robinson C, Casey M, Rusch D and Hunziker R 2017 Free-form versus ruled inducer design in a transonic centrifugal impeller. ASME GT-2017-63538.

He L and Ning W 1998 Efficient approach for analysis of unsteady viscous flows in turbomachines. *AIAA Journal*, 36, 2005–2012.

Hehn A, Mosdzien M, Grates D and Jeschke P 2017 Aerodynamic optimization of a transonic centrifugal compressor by using arbitrary blade surfaces. ASME GT-2017-63470.

Heinrich M and Schwarze R 2017 Genetic optimization of the volute of a centrifugal compressor. Proceedings of the 12th European Conference on Turbomachinery Fluid Dynamics and Thermodynamics, Paper ETC2017-072.

Herbert M 1980 A method of centrifugal compressor performance prediction. In: *Performance Prediction of Centrifugal Pumps and Compressors*, ASME International Gas Turbine Conference.

Herrig L, Emery J and Erwin J 1957 Systematic 2D cascade tests of NACA 65-series compressor blades at low speeds. NACA TN 3916.

Hillewaert K and Van den Braembussche RA 1999 Numerical simulation of impeller–volute interaction in centrifugal compressors. *Trans. ASME, Journal of Turbomachinery*, 121, 603–608.

Hillewaert K and Van den Braembussche RA 2000 Comparison of frozen rotor to unsteady computations of rotor-stator interaction in centrifugal pumps, Proceedings of the 5th Nationaal Congres over Theoretische en Toegepaste Mechanica, Louvain-la-Neuve, Belgium.

Hirsch Ch, Kang S and Pointel G 1996 A numerically supported investigation of the 3D flow in centrifugal impellers – Part II: Secondary flow structure. ASME 96-GT-152.

Holmes DG 2008 Mixing plane revisited: A steady mixing plane approach designed to combine high levels of conservation and robustness. ASME GT-2008-51296.

Hübl HP 1975 Beitrag zur Berechnung des Spiralgehauses von Radialverdichtern und Vorherbestimmung seines Betriebsverhaltens. Mitteilungen des Institutes für Dampf- und Gasturbiben, Nr. 7, Technische Universität Wien.

Hunziker R and Gyarmathy G 1994 The operational stability of a centrifugal compressor and its dependence on the characteristics of the subcomponents. *Trans. ASME, Journal of Turbomachinery*, 116, 250–259.

Huo S 1975 Optimization based on boundary layer concept for compressible flow. *Trans. ASME, Journal of Engineering for Power*, 97, 195–206.

Iancu F, Trevino J and Sommer S 2008 Low solidity cascade diffuser and scroll optimization for centrifugal compressors. ASME GT-2008-50132.

Ibaraki S, Matsuo T and Yokoyama T 2007 Investigation of unsteady flow in a vaned diffuser of a transonic centrifugal compressor. *Trans. ASME, Journal of Turbomachinery*, 129, 686–693.

Ibaraki S, Van den Braembussche RA, Verstraete T, Alsalihi Z, Sugimoto K and Tomita I 2014 Aerodynamic design optimization of a centrifugal compressor impeller based on an artificial neural network and genetic algorithm. IMechE 11th International Conference on Turbochargers and Turbocharging, Paper 0058, pp. 65–77, DOI: 10.1533/978081000342.65.

Imaichi K and Tsurusaki H 1979 Rotating stall in a vaneless diffuser of a centrifugal fan. In: *Flow in Primary, Non-Rotating Passages in Turbomachines*, ASME Winter Annual Meeting 1979.

Imbach H 1964 Die Berechnung der kompressiblen, reibungsfreien Unterschallströmung durch räumliche Gitter aus Schaufeln auch grosser Dicke und starker Wölbung. PhD Thesis, ETH Zürich. Prom 3401.

Inoue M 1978 Radial vaneless diffusers: a re-examination of the theories of Dean and Senoo and of Johnston and Dean. ASME 78-GT-186.

Ishida M, Sakaguchi D and Ueki H 2001 Suppression of rotating stall by wall roughness control in vaneless diffusers of centrifugal blowers. *Trans. ASME, Journal of Turbomachinery*, 123, 64–72.

Ishida M, Surana T, Ueki H and Sakaguchi D 2004 Suppression of unstable flow at small flow rates in a centrifugal blower by controlling tip leakage flow and reverse flow. ASME GT-2004-53400.

Ishida M, Sakaguchi D and Ueki H 2005 Optimization of inlet ring groove arrangement for suppression of unstable flow in a centrifugal impeller. ASME GT-2005-68675.

Isomura K, Murayama M and Kawakubo T 2001 Feasibility study of a gas turbine at micro scale. ASME 2001-GT-101.

Iverson H, Rolling R and Carlson J 1960 Volute pressure distribution, radial force on the impeller and volute mixing losses of a radial flow centrifugal pump. *Trans. ASME, Journal of Engineering for Power*, 82, 136–144.

Jaatinen A, Backman J and Turunen-Saaresti T 2009 Radial forces in a centrifugal compressor equipped with vaned diffusers. ASME GT-2009-60130.

Jansen W 1964a Steady flow in a radial vaneless diffuser. *Trans. ASME, Journal of Basic Engineering*, 86, 607–619.

Jansen W 1964b Rotating stall in radial vaneless diffusers. *Trans. ASME, Journal of Basic Engineering*, 86, 750–758.

Jansen W 1970 A method for calculating the flow in a centrifugal impeller when entropy gradients are present. IMechE, *Internal Aerodynamics-Turbomachinery*, Paper 12.

Jansen W, Carter AF and Swarden C 1980 Improvements in surge margin for centrifugal compressors. In: *Centrifugal Compressors, Flow Phenomena and Performance*, AGARD CP-282, Paper 19.

Japikse D 1980 The influence of diffuser inlet pressure fields on the range and durability of centrifugal compressor stages. In: *Centrifugal Compressors, Flow Phenomena and Performance*, AGARD CP-282, Paper 13.

Japikse D 1982 Advanced diffusion levels in turbocharger compressors and component matching. Proceedings of the 1st International Conference on Turbocharging and Turbochargers, London IMechE, pp. 143–155.

Japikse D 1985 Assessment of single and two zone modeling of centrifugal compressors – Studies in component performance, Part 3. ASME 85-GT-73.

Japikse D 1996 *Centrifugal Compressor Design and Performance*. Concepts ETI, ISBN 0-933283-03-2.

Japikse D and Baines N 1994 *Introduction to Turbomachinery*. Concepts ETI, Inc and Oxford University Press.

Japikse D and Osborne C 1986 Optimization of industrial centrifugal compressors. ASME 86-GT-222.

Javed A and Kamphues E 2014 Evaluation of the influence of volute roughness on turbocharger compressor performance from a manufacturing perspective. ASME GT-2014-26949.

Johnsen JA and Ginsburg A 1953 Some NACA research on centrifugal compressors. *Trans. ASME*, 75, 805–817.

Johnston JP 1960 On the 3D turbulent boundary layer generated by secondary flow. *Trans. ASME, Journal of Basic Engineering*, 82, 233–250.

Johnston JP 1974 The effects of rotation on boundary layers in turbomachine rotors. In: *Fluid Mechanics, Acoustics and Design of Turbomachinery*, NASA SP 304, Part I, 207–242.

Johnston JP and Dean R 1966 Losses in vaneless diffusers of centrifugal compressors and pumps. *Trans. ASME, Journal of Engineering for Power*, 88, 49–62.

Kämmer N and Rautenberg M 1982 An experimental investigation of rotating stall flow in centrifugal compressors. ASME 82-GT-82.

Kämmer N and Rautenberg M 1986 A distinction between different types of stall in a centrifugal compressor stage. *Trans. ASME, Journal of Engineering for Gas Turbines and Power*, 108, 83–92.

Kang Jeong-Seek, Cho Sung-Kook and Kang Shin-Hyoung 2000 Unsteady flow phenomena in a centrifugal compressor channel diffuser. ASME 2000-GT-451.

Kannemans H 1977 Calculation of 3D incompressible potential flow by means of singularities in mixed flow machines. VKI PR 1977-01.

Katsanis Th 1964 Use of arbitrary quasi-orthogonals for calculating the flow distribution in the meridional plane of a turbomachine. NASA TN-D-2546.

Kawata Y, Kanki H and Kawakami T 1983 Experimental research of the radial force on centrifugal pumps – 1st report influence of impeller vane number casing type. *Trans. Japanese Society of Mechanical Engineering, Series C*, 49(437), 31–38.

Kennedy J and Eberhart R 1995 Particle swarm optimization. In: *Proceedings of IEEE International Conference on Neural Networks*, IV, pp. 1942–1948.

Kenny D 1970 Supersonic radial diffusers. In: *Advanced Compressors*, AGARD-LS-39-70.

Kenny D 1972 A comparison of the predicted and measured performance of high pressure ratio centrifugal compressor diffusers. In: *Advanced Radial Compressors*, VKI LS 50.

Kim Y, Engeda A, Aungier R and Direnzi G 2001 The influence of inlet flow distortion on the performance of a centrifugal compressor and the development of an improved inlet using numerical simulations. Proceedings of the Institution of Mechanical Engineers, **215**, Part A, A04700, 323–338.

Kinoshita Y and Senoo Y 1985 Rotating stall induced in vaneless diffusers of very low specific speed centrifugal blowers. *Trans. ASME, Journal of Engineering for Gas Turbine and Power*, 107, 514–521.

Kobayashi H, Nishida H, Takagi T and Fukushima Y 1990 A study on the rotating stall of centrifugal compressors. 2nd Report, Effect of vaneless diffuser inlet shape on rotating stall [Enshinasshukuki no senkaishissoku nikansuru kenkyu (Dai2 ho, hanenashideifuza iriguchikeijonoeikyo)], Nihon Kikaigakkai Ronbunshu (B hen). *Trans. Japan Society of Engineers* (B edition), 56 (529), 98–103.

Koch JM, Chow PN, Hutchinson BR and Elias SR 1995 Experimental and computational study of a radial compressor inlet. ASME 95-GT-82.

Kostrewa K, Alsalihi Z and Van den Braembussche RA 2003 Optimization of radial turbines by means of design of experiment. VKI PR-2003-17.

Kosuge H, Ito T and Nakanishi K 1982 A consideration concerning stall and surge limitations within centrifugal compressors. *Trans. ASME, Journal of Engineering for Power*, 104, 782–787.

Kovats A 1979 Effect of non rotating passages on performance of centrifugal pumps and subsonic compressors. In: *Flow in Primary Non-rotating Passages in Turbomachines*, ASME.

Koyama H, Masuda S, Ariga J and Watanabe J 1978 Stabilizing and destabilizing effects of Coriolis force on 2D laminar and turbulent boundary layers. ASME 78-GT-01.

Krain H 1981 A study on centrifugal impeller and diffuser flow. *Trans. ASME, Journal of Engineering for Power*, 103, 688–697.

Krain H, Hoffmann B and Beversdorff M 2007 Flow pattern at the inlet of a transonic centrifugal rotor. Proceedings of the International Gas Turbine Congress, Tokyo, Gas Turbine Society of Japan, IGTC2007-TS-034.

Kramer J, Osborn W and Hamrick J 1960 Design and tests of mixed flow and centrifugal impellers. *Trans. ASME, Journal of Engineering for Power*, 82, 127–135.

Krige DG 1951 A statistical approach to some mine valuations and allied problems at the Witwatersrand. PhD thesis, University of Witwatersrand.

Kurz R and White RC 2004 Surge avoidance in gas compression systems. *Trans. ASME, Journal of Turbomachinery*, 126, 501–506.

Kurz R, McKee R and Brun K 2006 Pulsations in centrifugal compressor installations. ASME GT-2006-90070.

Kyrtatos M and Watson N 1980 Application of aerodynamically induced preswirl to a small turbocharger compressor. *Trans. ASME, Journal of Engineering for Power*, 102, 943–950.

Lei V-M 2011 Aerodynamics of a centrifugal compressor with a double sided impeller. ASME GT-2011-45215.

Lennemann E and Howard JHG 1970 Unsteady flow phenomena in rotating centrifugal impeller passages. *Trans. ASME, Journal of Engineering for Power*, 92, 65–72.

Léonard O 1992 Conception et Développement d'une Méthode Inverse de Type Euler et Application á la Génération de Grilles d'Aubes Transoniques. PhD thesis, von Karman Institute – Universite de Liege.

Léonard O and Demeulenaere A 1997 A Navier–Stokes inverse method based on moving wall strategy. ASME 97-GT-416.

Léonard O and Van den Braembussche RA 1992 Design Method for Subsonic and Transonic Cascade with Prescribed Mach number Distribution. *Trans. ASME, Journal of Turbomachinery*, 114, 553–560.

Lieblein S 1960 Incidence and deviation angle correlations for compressor cascades. *Trans. ASME, Journal of Basic Engineering, Series D*, 82, 575–587.

Lieblein S, Schwenk F and Broderick R 1953 Diffusion factor for estimating losses and limiting blade loadings in axial compressor blade elements. NACA RM E53D01.

Lighthill JM 1945 A new method of two-dimensional aerodynamic design. ARC R&M 2112.

Lindner P 1983 Aerodynamic tests on centrifugal process compressors – Influence of diffuser diameter ratio, axial stage pitch and impeller cut-back. ASME 83-GT-172.

Lipski W 1979 The influence of shape and location of the tongue of spiral casing on the performance of single stage radial pumps. Proceedings of the 6th Conference on Fluid Machinery, Budapest, pp. 673–682.

Liu Y, Liu B and Lu L 2010 Investigation of unsteady impeller–diffuser interaction in a transonic centrifugal compressor stage. ASME GT-2010-22737.

Longley JP 1994 A review of non-steady flow models for compressor stability. *Trans. ASME, Journal of Turbomachinery*, 116, 202–215.

Lorett JA and Gopalakrishnan S 1986 Interaction between impeller and volute of pumps at off design conditions. *Trans. ASME, Journal of Fluids Engineering*, 108, 12–18.

Lüdtke K 1983 Aerodynamic tests on centrifugal process compressors – the influence of the vaneless diffuser shape. *Trans. ASME, Journal Engineering for Power*, 105, 902–909.

Lüdtke K 1985 Centrifugal process compressors – radial vs. tangential suction nozzles. ASME 85-GT-80.

Lüdtke K 2004 *Process centrifugal compressors – Basics, function, operation, design, application.* Springer Verlag, ISBN 3-540-40427-9.

Lyman FA 1993 On the conservation of rothalpy in turbomachines. *Trans. ASME, Journal of Turbomachinery*, 115, 520–526.

Marsan A, Trébinjac I, Coste S and Leroy G. 2012 Numerical investigation of the flow in a radial vaned diffuser without and with aspiration. ASME GT-2012-68610.

Marsan A, Leroy G, Trébinjac I and Moreau S 2015 A multi-slots suction strategy for controlling the unsteady hub–corner separation in a radial vaned diffuser. Proceedings of the 12th International Symposium on Experimental and Computational Aerothermodynamics of Internal Flow, ISAIF12-75, Lerici, Italy.

Matthias HB 1966 The design of pump suction bends. IAHR Symposium, Braunschweig, VDI-Verlag, pp. 21–30.

Meier-Grotrian J 1972 Untersuchungen der Radialkraft auf das Laufrad einer Kreiselpumpe bei verschiedenen Spiralgehäuseformen. PhD thesis, TU Braunschweig.

Michelassi V, Pazzi S, Giangiacomo P, Martelli F and Giachi M 2001 Performances of centrifugal compressor impellers in steady and unsteady flow regimes under inlet flow distortions. ASME 2001-GT-0325.

Mishina H and Gyobu I 1978 Performance investigations of large capacity centrifugal compressors. ASME 78-GT-3.

Mischo B, Ribi B, Seebass-Linggi C and Mauri S 2009 Influence of labyrinth seal leakage on centrifugal compressor performance. ASME GT-2009-59524.

Mizuki S, Kawashima Y and Ariga I 1978 Investigation concerning rotating stall and surge phenomena within centrifugal compressor channels. ASME 78-GT-9.

Mohseni A, Goldhahn E, Van den Braembussche RA and Seume JR 2012 Novel IGV design for centrifugal compressors and their interaction with the impeller. *Trans. ASME, Journal of Turbomachinery*, 134, 021006-(1-8). DOI: 10.1115/1.4003235.

Mojaddam M, Benisi A and Movahhedy M 2012 Investigation on effect of centrifugal compressor volute cross-section shape on performance and flow field. ASME GT-2012-69454.

Moore FK 1989 Weak rotating flow disturbances in a centrifugal compressor with a vaneless diffuser. *Trans. ASME, Journal of Turbomachinery*, 111, 442–449.

Moore FK 1991 Theory of finite disturbances in a centrifugal compressor system with a vaneless radial diffuser. ASME 91-GT-82.

Moore JJ and Flathers MB 1998 Aerodynamically induced radial forces in a centrifugal gas compressor Part 1: Experimental measurement. *Trans. ASME, Journal of Engineering for Gas Turbines and Power*, 120, 383–390.

Moore J, Moore JG and Timmis PH 1983 Performance evaluation of centrifugal compressor impellers using 3D viscous flow calculations. ASME 83-GT-63.

Morris RE and Kenny DP 1972 High pressure ratio centrifugal compressors for small gas turbine engines. ASME, 118–146.

Mukkavilli R, Raju GR, Dasgupta A, Murty GV and Shary KV 2002 Flow studies on a centrifugal compressor stage with low solidity diffuser vanes. ASME GT-2002-30386.

Nece RE and Daily JW 1960 Roughness effects on frictional resistance of enclosed rotating discs. *Trans. ASME, Journal of Basic Engineering*, 82, 553–562.

Neumann B 1991 *The Interaction between Geometry and Performance of a Centrifugal Pump.* Mechanical Engineering Publications Ltd, London.

Nishida H, Kobayashi H, Takagi T and Fukushima Y 1988 A study on the rotating stall of centrifugal compressors. 1st Report, Effect of vaneless diffuser width on rotating stall) [Enshinasshukuki no senkaishissoku nikansuru kenkyu (Dai1 ho, hanenashideifuza ryurohaba no eikyo)], Nihon Kikaigakkai Ronbunshu (B hen), *Trans. Japan Society of Mechanical Engineers*, 54(499), 589–594.

Nishida H, Kobayashi H, Takagi T and Fukushima Y 1989 A study on the rotating stall of centrifugal compressors. Proceedings of the International Symposium on Unsteady Aerodynamics and Aeroelasticity of Turbomachines and Propellers, Beijing, Pergamon Press.

Nishida H, Kobayashi H and Fukushima Y 1991 A study on the rotating stall of centrifugal compressors. 3rd Report, Rotating stall suppression method [Enshinasshukuki no senkai shissoku ni kansuru kenkyu (Dai-3-po, senkai shissoku no yokuseiho)], Nihon Kikaigakkai Ronbunshu (B hen), Trans. Japan Society of Mechanical Engineers (B edition), 543, 3794–3800.

Obayashi S, Jeong S and Chiba K 2005 Multi-objective design exploration for aerodynamic configurations. AIAA2005-4666.

Oh HW, Yoon ES and Chung MK 1997 An optimum set of loss models for performance prediction of centrifugal compressors. Proceedings of the IMechE, Part A, *Journal of Power and Energy*, **211**.

Oh J, Buckley CW and Agrawal G 2011a Numerical investigation of low solidity vaned diffuser performance in a high-pressure centrifugal compressor. Part III: Tandem vanes. ASME GT-2011-43582.

Oh J, Buckley CW and Agrawal G 2011b Numerical study on the effects of blade lean on high-pressure centrifugal impeller performance. ASME GT-2011-43583.

Okamura T 1980 Radial thrust in centrifugal pumps with a single-vane impeller. *Bulletin of The Japan Society of Mechanical Engineering*, 23 (180), 895–901.

Ono Y 2013 Solutions for better engine performance at low load by Mitsubishi turbochargers. Proceedings of the CIMAC Congres 2013, Paper 15, Shanghai.

Pampreen R 1972 The use of cascade technology in centrifugal compressor vaned diffuser design. *Trans. ASME, Journal of Engineering for Power*, 94, 187–192.

Pampreen R 1981 A blockage model for centrifugal compressor impellers. ASME 81-GT-11.

Pampreen R 1989 Automotive research compressor experience. ASME 89-GT-61.

Papailiou KD 1971 Boundary layer optimization for the design of high turning axial flow compressor blades. *Trans ASME, Journal of Engineering for Power*, 93, 147–155.

Passrucker H and Van den Braembussche RA 2000 Inverse design of centrifugal impellers by simultaneous modification of blade shape and meridional contour. ASME 2000-GT-457.

Pazzi S and Michelassi V 2000 Analysis and design of centrifugal compressor inlet volutes. ASME 2000-GT-0464.

Peck JF 1951 Investigations concerning flow conditions in a centrifugal pump, and the effect of blade loading on head slip. *Proceedings of the Institute of Mechanical Engineers*, 164, 1–30.

Perrone A, Ratto L, Ricci G, Satta F and Zunino P 2016 Multi-disciplinary optimization of a centrifugal compressor for micro-turbine applications. ASME GT-2016-57278.

Pfau H 1967 Temperaturmessungen zur Strömungsuntersuchung, insbesondere an Radialverdichter. *Konstruktion*, 12, 478–484.

Pierret S 1999 Designing Turbomachinery Blades by means of the Function Approximation Concept based on Artificial Neural Network, Genetic Algorithm and the Navier–Stokes Equations. PhD thesis, von Karman Institute – Faculté Polytechnique de Mons.

Pierret S and Van den Braembussche RA 1999 Turbomachinery blade design using a Navier–Stokes solver and artificial neural network. *Trans. ASME, Journal of Turbomachinery*, 121, No.2.

Pinckney S 1965a Optimized turning-vane design for an intake elbow of an axial-flow compressor. Report NASA TN D-3083.

Pinckney S 1965b Use of a single turning vane to eliminate flow separation in space limited 90° intake elbow of an axial flow compressor. NASA TM-X-1110.

Poloni C 1999 Multi-objective aerodynamic optimization by means of robust and efficient genetic algorithm. *Notes on Numerical Fluid Mechanics*, 68, 124.

Poulain J and Janssens G 1980 Roue de compresseur centrifuge sans flexions dans les ailes. In: *Centrifugal Compressors, Flow Phenomena and Performance*, AGARD CP-282, Paper 17.

Price K and Storn N 1997 Differential evolution. Dr. Dobbs's Journal, pp. 18–24.

Rautenberg M, Mobarak A and Malobabic M 1983 Influence of heat transfer between turbine and compressor on the performance of small turbochargers. GTSJ International Gasturbine Congress, Tokyo.

Rebernik B 1972 Investigation of induced vorticity in vaneless diffusers of radial pumps. Proceedings of the 4th Conference on Fluid Machinery, Budapest, pp. 1129–1139.

Rechenberg I 1973 *Evolutionsstrategie: Optimierung technischer Systeme nach Prinzipien der biologischen Evolution*. Fromman-Holzboog, Stuttgart.

Reddy YR and Kar S 1971 Optimum vane number and angle of centrifugal pumps with logarithmic vanes. *Trans. ASME, Journal of Basic Engineering*, 93, 411–425.

Reichl A, Dickmann HP and Kühnel J 2009 Calculation methods for the determination of blade excitation due to suction elbows in centrifugal compressors. ASME GT-2009-59178.

Reneau L Johnston JP and Kline S 1967 Performance and design of straight, 2D diffusers. *Trans. ASME, Journal of Basic Engineering*, 89, 1412–1150.

Reunanen A 2001 Experimental and Numerical Analysis of different Volutes in a Centrifugal Compressor. PhD thesis, Acta Universitatis Laperantaensis.

Ribi B and Dalbert P 2000 One-dimensional performance prediction of subsonic vaned difffusers. *Trans. ASME, Journal of Turbomachinery*, 122, 494–504.

Ribi B and Gyarmathy G 1993 Impeller rotating stall as a trigger for the transition from mild to deep surge in a subsonic centrifugal compressor. ASME 93-GT-234.

Roberts SK and Sjolander SA 2005 Effect of specific heat ratio on the aerodynamic performance of turbomachinery. *Trans. ASME, Journal of Engineering for Gas Turbines and Power*, 127, 773–780.

Robinson C, Casey M Hutchinson B and Steed R 2012 Impeller–diffuser interaction in centrifugal compressors. ASME GT-2012-69151.

Rodgers C 1961 Influence of impeller and diffuser characteristics and matching on radial compressor performance. SAE Technical Progress Series, Vol. 3, Centrifugal compressors, pp. 31–42.

Rodgers C 1962 Typical performance characteristics of gas turbine radial compressors. *Trans. ASME, Journal of Engineering for Power*, 86, 161–175.

Rodgers C 1968 Variable geometry gas turbine radial compressor. ASME 68-GT-63.

Rodgers C 1970 *Rotating Vaneless Diffuser Study*. US Army Equipment Research and Development Center.

Rodgers C 1977 Impeller stalling as influenced by diffuser limitations. *Trans. ASME, Journal of Fluids Engineering*, 99, 84–97.

Rodgers C 1978 A diffusion factor correlation for centrifugal impeller stalling. *Trans. ASME, Journal of Engineering for Power*, 100, 592–602.

Rodgers C 1980 Specific speed and efficiency of centrifugal impellers. In: *Performance Prediction of Centrifugal Pumps and Compressors*, ASME International Gas Turbine Conference.

Rodgers C 1982 The performance of centrifugal compressor channel diffusers. ASME 82-GT-10.

Rodgers C 1991 The efficiency of single stage centrifugal compressors for aircraft applications. ASME 91-GT-77.

Rodgers C 1998 The centrifugal compressor inducer. ASME 98-GT-32.

Rodgers C and Mnew H 1975 Experiments with model free rotating vaneless diffuser. *Trans. ASME, Journal of Engineering for Power*, 97, 231–242.

Rodgers C and Sapiro L 1972 Design considerations for high pressure ratio centrifugal compressors. ASME 72-GT-91.

Rothe PH and Runstadler PW 1978 First order pump surge behaviour. *Trans. ASME, Journal of Fluids Engineering*, 100, 459–466.

Runstadler PW and Dean R 1969 Straight channel diffuser performance at high inlet Mach numbers. *Trans. ASME, Journal of Basic Engineering*, 91, 397–422.

Rusak V 1982 Development and performance of the wedge-type low specific speed compressor wheel. ASME 82-GT-214.

Sagi C and Johnton JP 1967 The design and performance of 2D, curved diffusers. *Trans. ASME, Journal of Basic Engineering*, 89, 715–731.

Salvage JW 1996 Variable geometry pipe diffusers. ASME 96-GT-202.

Salvage JW 1998 Development of a centrifugal compressor with a variable geometry split-ring pipe diffuser. ASME 98-GT-7.

Sanz JM, McFarland ER, Sanger NL, Gelder T and Cavicchi RH 1985 Design and performance of a fixed, non-accelerating guide vane cascade that operates over an inlet flow angle range of 60 deg. *Trans. ASME, Journal of Engineering for Gas Turbines and Power*, 107, 477–484.

Saphar S 2004 Design of experiment, screening and response surface modeling to minimize the design cycle time. In: *Optimization Methods and Tools for Multi-Criteria/Multi-Disciplinary Design*, VKI LS 2004-07.

Sapiro L 1983 Effect of impeller-extended shrouds on centrifugal compressor performance as a function of specific speed. *Trans. ASME, Journal of Engineering for Power*, 105, 457–465.

Sayed S 1986 Selection of high performance low flow compressors. ASME 86-GT-220.

Schmalfuss H 1972 Design and constructional aspects of contemporary centrifugal compressors. In: *Industrial Turbo-compressors*, VKI Lecture Series 47.

Schmitz MB and Fitzky G 2004 Surge cycle of turbochargers: Simulation and comparison to experiments. ASME GT-2004-53036.

Schnell E 1965 Flow through radial wheels. In: *Theoretical and Experimental Research on Limit Loading Radial Compressors*, Part A, VKI CN 53.

Schuster P and Schmidt-Eisenlohr U 1980 Flow field analysis of radial and backswept centrifugal compressor impellers, Part 2: Comparison of potential flow calculation and measurements. In: *Performance Prediction of Centrifugal Pumps and Compressors*, ASME.

Senoo Y 1984 Vaneless diffusers. In: *Flow in Centrifugal Compressors*, VKI Lecture Series 1984-07.

Senoo Y and Ishida M 1975 Behavior of severely asymmetric flow in a vaneless diffuser. *Trans. ASME, Journal of Engineering for Power*, 97, 375–387.

Senoo Y and Kawaguchi N 1983 Pressure recovery of collectors with annular curved diffusers, ASME 83-GT-35.

Senoo Y and Kinoshita Y 1977 Influence of inlet flow conditions and geometries of centrifugal vaneless diffusers on critical flow angle for reverse flow. *Trans. ASME, Journal of Fluids Engineering*, 99, 98–103.

Senoo Y and Kinoshita Y 1978 Limits of rotating stall and stall in vaneless diffuser of centrifugal compressors. ASME 78-GT-19.

Senoo Y and Nishi M 1977a Prediction of flow separation in diffusers by boundary layer calculation. *Trans. ASME, Journal of Fluids Engineering*, 99, 379–389.

Senoo Y and Nishi M 1977b Deceleration rate parameters and algebraic prediction of turbulent boundary layer. *Trans. ASME, Journal of Fluids Engineering*, 99, 390–395.

Senoo Y, Kinoshita Y and Ishida M 1977 Asymmetric flow in vaneless diffusers of centrifugal blowers. *Trans. ASME, Journal of Fluids Engineering*, 99, 104–114.

Senoo Y, Kawaguchi N and Nagata T 1978 Swirl flow in conical diffuser. *Bulletin of JSME*, 21(151).

Senoo Y, Hayami H and Ueki, M 1983 Low solidity tandem cascade diffusers for wide range centrifugal blowers. ASME 83-GT-3.

Shibata T, Yagi M, Nishida H, Kobayashi and Tanaka M 2010 Effect of impeller blade loading on compressor stage performance in high specific speed range. ASME GT-2010-2228 1.

Shouman AR and Anderson JR 1964 The use of compressor-inlet prewhirl for the control of small gas turbines. *Trans. ASME, Journal of Engineering for Power*, 86, 136–140.

Sideris M 1988 Circumferential Distortion of the Flow in Centrifugal Compressors due to Outlet Volutes. PhD Thesis, von Karman Insititute – Universiteit Gent.

Sideris M and Van den Braembussche RA 1986 Evaluation of the flow in a vaneless diffuser followed by a volute. Proceedings of the 3rd International Conference on Turbocharging and Turbochargers, IMechE C102/86.

Sideris M, Ayaz Y and Van den Braembussche RA 1987a Centrifugal impeller response to a circumferential variation of exit pressure. Proceedings of the International Gas Turbine Congress, Tokyo, 87-IGTC-10.

Sideris M and Van den Braembussche RA 1987b Influence of a circumferential exit pressure distortion on the flow in an impeller and diffuser. *Trans. ASME, Journal of Turbomachinery*, 109, 48–54.

Simon H and Bulskamper A 1984 On the evaluation of Reynolds number and relative surface roughness effects on centrifugal compressor performance based on systematic experimental investigations. *Trans. ASME, Journal of Engineering for Gas Turbines and Power*, 106, 489–501.

Simon H, Wallmann T and Mönk T 1986 Improvement in performance characteristics of single stage and multistage centrifugal compressors by simultaneously adjusting inlet guide vanes and diffuser vanes. ASME 86-GT-127.

Simon JS, Valavani L, Epstein AH and Greitzer EM 1993 Evaluation of approaches to active compressor surge stabilization. *Trans. ASME, Journal of Turbomachines*, 115, 57–67.

Sirakov B and Casey M 2011 Evaluation of heat transfer effects on turbocharger performance. ASME GT-2011-45887.

Sirakov B, Gong Y, Epstein AH and Tan CH 2004 Design and characterization of micro-compressor impellers. ASME GT-2004-53332.

Sivagnanasundaram S, Spence S, Early J and Nikpour B 2010 An investigation of compressor map width enhancement and the inducer flow field using various configurations of shroud bleed slot. ASME GT-2010-22154.

Sivagnanasundaram S, Spence S and Early J 2012 Map width enhancement technique for a turbocharger compressor. ASME GT-2012-69415.

Sivagnanasundaram S, Spence S, Early J and Nikpour B 2013 An impact of various shroud bleed slot configurations and cavity vanes on compressor map width and the inducer flow field. *Trans. ASME, Journal of Turbomachines*, 135, 041003 1–9.

Skoch G 2003 Experimental investigation of centrifugal compressor stabilisation techniques. *Trans. ASME, Journal of Turbomachines*, 125, 704–713.

Skoch G 2005 Experimental investigation of diffuser hub injection to improve centrifugal compressor stability. *Trans. ASME, Journal of Turbomachines*, 127, 107–117.

Smith AG 1957 On the generation of the streamwise component of vorticity for flows in a rotating passage. *Aeronautical Quarterly*, 8, 369–383.

Spakovsky ZS and Paduano JD 2000 Tip-clearance actuation with magnetic bearings for high-speed compressor stall control. ASME 2000-GT-0528.

Stanitz J 1952 One-dimensional compressible flow in vaneless diffusers of radial and mixed flow compressors including effects of friction, heat transfer and area change. NACA TN 2610.

Stanitz J 1953 Effect of blade thickness tapes on axial velocity distribution at the loading edge of an entrance rotor blade row with axial inlet and the influence of this distribution on alignment of the rotor blade for zero attack. NACA TN 2986.

Stanitz J and Prian V 1951 A rapid approximation method for determining velocity distribution of impeller blades of centrifugal compressors. NACA TN 2421.

Startsev A, Brailko I and Orekhov I 2015 Development of a centrifugal compressor outlet system in ESPOSA Project. Proceedings of the 12th International Symposium on Experimental and Computational Aerothermodynamics of Internal Flow, ISAIF12-003, Lerici, Italy.

Steglich T, Kitzinger J, Seume J, Van den Braembussche RA and Prinsier J 2008 Improved impeller/volute combinations for centrifugal compressors. *Trans. ASME, Journal of Turbomachinery*, 130, 011014-1(9).

Stepanoff AJ 1957 *Centrifugal and Axial Flow Pumps.* 2nd edition, pp. 116–123. John Wiley and Sons, Inc., New York.

Stiefel W 1965 Experimental investigations of radial compressors to evaluate the parameters leading to a variation of the design point. In: *Theoretical and Experimental Research on Limit Loading Radial Compressors*, VKI CN 53 Part B.

Stiefel W 1972 Experiences in the development of radial compressors. In: *Advanced Radial Compressors*, VKI LS 50.

Stodola 1924 *Dampf- und gasturbinen*, Springer Verlag; 1927 *Steam and Gasturbines*, McGraw-Hill, New York.

Strub RA, Bonciani L, Borer CJ, Casey MV, Cole SL, Cook B.B, Kotzur J, Simon H and Strite MA 1987 Influence of the Reynolds number on the performance of centrifugal compressors. *Trans. ASME, Journal of Turbomachinery*,109, 541–544.

Sugimura K, Jeong S, Obayashi S and Kimura T 2008 Multi-objective robust design optimization and knowledge mining of a centrifugal fan that takes dimensional uncertainty into account. ASME GT-2008-51301.

Sugimura K, Kobayashi H and Nishida H 2012 Design optimization and experimental verification of centrifugal compressors with curvilinear element blades. ASME GT-2012-69162.

Swain E 2005 Improving a one-dimensional centrifugal compressor performance prediction method. *Proceedings of the IMechE Part A Journal of Power and Energy*, 219, 653–659. DOI: 10.1243/095765005X31351.

Tamaki I 2011 Effect of recirculation device with counter swirl vane on performance of high pressure ratio centrifugal compressor. ASME GT-2011-45360.

Tamaki H, Zheng X and Zhang Y 2012 Experimental investigation of high pressure ratio centrifugal compressor with axisymmetric and non-axisymmetric recirculation device. ASME GT-2012-68219.

Tan J, Qi D and Wang, R 2010 The effects of radial inlet guide vanes on the performance of variable inlet guide vane in a centrifugal compressor stage. ASME GT-2010-22177.

Taylor ES 1964 The centrifugal compressor. Section J In: *The Aerodynamics of Turbines and Compressors*, Vol. X of *High Speed Aerodynamics and Jet Propulsion*. Princeton University Press, pp. 553–586.

Thévenin D and Janiga G. (Eds) 2008 *Optimization and Computational Fluid Dynamics.* Springer Verlag, Berlin, ISBN 978-3-540-72153-6.

Thomas JL and Salas MD 1986 Far-field boundary conditions for transonic lifting solutions to the Euler equations. *AIAA Journal*, 24, 1074–1080.

Tiow WT and Zangeneh M 2000 A three-dimensional viscous transonic inverse design method. ASME 2000-GT-0525.

Tiow WT, Yiu KFC and Zangeneh M 2002 Application of simulated annealing to inverse design of transonic turbomachinery cascades. Proceedings of the IMechE, Vol. 216, Part A: J. Power and Energy, pp. 59–73.

Tomica H, Wonsak G and Saxena S 1973 *Ausbildung und Untersuchung von Saugkrümmern für zweiflutige, radial angeströmte Kreiselpumpen.* Pumpentagung, Karlsruhe.

Toyama K, Runstadler PW and Dean RC 1977 An experimental study of surge in a centrifugal compressor. *Trans. ASME, Journal of Fluids Engineering*, 99, 115–131.

Traupel W 1962 *Die Theorie der Strömung durch Radialmaschinen.* Karlsruhe, Verlag Braun.

Trébinjac I, Kulisa P, Bulot N and Rochuon N 2008 Effect of unsteadiness on the performance of a transonic centrifugal compressor stage. ASME GT-2008-50260.

Trébinjac I, Ottavy X., Rochuon N and Bulot N 2009 On the validity of steady calculations with shock blade row interactions. Proceedings of the 9th International Symposium of Experimental and Computational Aerothermodynamics of Internal Flow, ISAIF9-sc-3, Gyeongju, Korea.

Tsujimito Y, Yoshida Y and Mori Y 1994 Study of vaneless diffuser rotating stall based on two-dimensional inviscid flow analysis. In: *Fluid Machinery*, Vol. 195, ASME FED.

Tun MT, Sakaguchi D, Numakura R and Wang B 2016 Multi-point optimization of recirculation flow type casing treatment in centrifugal compressors. ASME GT-2016-56610.

Uchida H, Inayoshi M and Sugiyama, K 1987 Effect of a circumferential static pressure distortion on small-sized centrifugal compressor performance. Proceedings of the International Gas Turbine Congress, Tokyo, 87-IGTC-11.

Ueda K, Kikuchi S, Inokuchi Y, Yamasaki N and Yamagata, A 2015 The effects of upstream bent pipe on the performance of a small centrifugal compressor. Proceedings of the 12th International Symposium on Experimental and Computational Aerothermodynamics of Internal Flows, ISAIF12-116, Paper 116, Lerici, Italy.

Van Bael J and Van den Braembussche RA 1994 Active control of centrifugal compressor surge. Proceedings of the 3rd Belgisch Nationaal Congres over Theoretische en Toegepaste Mechanica, Liège, Belgium.

Van den Braembussche RA 1994 Inverse design method for axial and radial turbomachines. In: *Numerical Methods for Flow Calculation in Turbomachines*, VKI LS 1994-06.

Van den Braembussche RA 2005 Micro-Gasturbines – A short survey of design problems. In: *Micro Gasturbines*, RTO-VKI Lecture Series, RTO-EN-AVT-131, ISBN: 92-837-1151-3.

Van den Braembussche RA 2010 Global optimization methods: Theoretical aspects and definitions. In: *Strategies for Optimization and Automated Design of Gas Turbine Engines*, RTO-MP-AVT-167, Chapter 3, ISBN 978-92-837-0124-8.

Van den Braembussche RA and Hände BM 1990 Experimental and theoretical study of the swirling flow in centrifugal compressor volutes. *Trans. ASME, Journal of Turbomachinery*, 112, 38–43.

Van den Braembussche RA, Frigne P and Roustan M 1980 Rotating non uniform flow in radial compressors. In: *Centrifugal Compressors, Flow Phenomena and Performance*, AGARD CP-282, paper 12.

Van den Braembussche RA, Sideris M and Soumoy V 1987 Comparative study of four prediction methods for radial vaneless diffusers. Proceedings of the International Gas Turbine Congress, 87-Tokyo-IGTC-6, pp. II-37-44.

Van den Braembussche RA, Demeulenaere A and Borges J 1993 Inverse design of radial flow impellers with prescribed velocity at hub and shroud. AGARD CP 537, paper 18.

Van den Braembussche RA, Ayder E, Hagelstein D, Rautenberg and Keiper R 1999 Improved model for the design and analysis of centrifugal compressor volutes. *Trans. ASME, Journal of Turbomachinery*, 121, 619–625.

Van den Braembussche RA, Prinsier J and Di Sante A 2010 Experimental and numerical investigation of the flow in rotating diverging channels. *Journal of Thermal Science* 19(2), 115–119, DOI: 10.1007/s11630-010-0115-4.

Van den Braembussche RA, Alsalihi Z, Verstraete T, Matsu A, Ibaraki S, Sugimoto K and Tomita I 2012 Multidisciplinary multipoint optimization of a transonic turbocharger compressor. ASME GT2012-69645.

Van den Braembussche RA, Aksoy HG and Hillewaert K 2015 About frozen rotor calculations. Proceedings of the 12th International Symposium on Experimental and Computational Aerothermodynamics of Internal Flow, ISAIF12-116, Lerici, Italy.

Van Laarhoven PJM and Aarts EHL 1987 *Simulated Annealing: Theory and Application*. Kluwer Academic Publishers.

Vavra M 1970 Basic elements for advanced design of radial flow compressors. In *Advanced Compressors*, AGARD LS 39.

Vavra M 1974 *Aero-thermodynamics and Flow in Turbomachines*. Robert E. Krieger Company, New York.

Verdonk G 1978 Vaned diffuser inlet flow conditions for a high pressure ratio centrifugal compressor, ASME 78-GT-50.

Verstraete T 2008 Multidisciplinary Turbomachinery Component Optimization Considering Performance, Stress, and Internal Heat Transfer. PhD thesis, von Karman Intsitute – Universiteit Gent.

Verstraete T 2016a Introduction to optimization and multidisciplinary design. In: *Radial Compressor Design and Optimization*, VKI LS 2017.

Verstraete T 2016b Multidisciplinary optimization of turbomachinery components using differential evolution. In: *Radial Compressor Design and Optimization*, VKI LS 2017.

Verstraete T, Alsalihi Z and Van den Braembussche RA 2007 Numerical study of the heat transfer in micro gasturbines. *Trans. ASME, Journal of Turbomachinery* 129, 835–841.

Verstraete T, Alsalihi Z and Van den Braembussche RA 2008 Multidisciplinary optimization of a radial compressor for micro gas turbine applications. *Trans. ASME, Journal of Turbomachinery*, 132(3), 03104-1 (7).

Verstraete T, Müller L and Müller JD 2017 Multidisciplinary adjoint optimization of turbomachinery components including aerodynamic and stress performance. AIAA 2017-4083, DOI: 10.2514/6.2017-4083.

Vignau H, Rodellar R and Silet J 1987 Intérêst de la géométrie variable pour les turbomoteurs à faible puissance. In: *Advanced Technology for Aero Gas Turbine Components*, AGARD-CP-421, paper 6.

Vilmin S, Lorrain E, Tartinville B, Hirsch Ch and Swoboda M 2006 Unsteady flow modeling across the rotor/stator interface using the nonlinear harmonic method. ASME GT2006-90210.

Vilmin S, Lorrain E, Tartinville B, Capron A and Hirsch Ch 2013 The nonlinear harmonic method: from single stage to multi-row effects. International Journal of Computational Fluid Dynamics, DOI:10.1080/10618562.2012.752074.

von Backström TW 2008 The effect of specific heat ratio on the performance of compressible flow turbomachines. ASME GT-2008-50183.

Wachter J and Rieder M 1985 Einfluss von Maschinenspezifischen grössen auf den Beginn und das Erscheinungsbild von rotierenden Ablösesrömungen in einem einstufigen radial Verdichter [Influence of design data on the onset and behavior of rotating stall in a single-stage centrifugal compressor]. Proceedings of VDI Tagung, Bochum, pp. 591–605.

Walitt L 1980 Numerical analysis of the 3D viscous flow field in a centrifugal impeller. In: *Centrifugal Compressors, Flow Phenomena and Performance*, AGARD CP-282, paper 6.

Watanabe H Konomi A and Ariga I 1994 Transient process of rotating stall in radial vaneless diffusers, ASME 94-GT-161.

Weber CR and Koronowski ME 1986 Meanline performance prediction of volutes in centrifugal compressors. ASME 86-GT-216.

White R and Kurz R 2006 Surge avoidance for compressor systems. Proceedings of the 35th Turbomachinery Symposium, Texas A&M University.

Whitfield A and Robert DV 1983 Alternative vaneless diffusers and collecting volutes for turbocharger compressors. ASME 83-GT-32.

Wiesner FJ 1967 A review of slip factor for centrifugal impellers. *Trans. ASME, Journal of Engineering for Power*, 89, 588–572.

Wiesner FJ 1979 A new appraisal of Reynolds number effects on centrifugal compressor performance. *Trans. ASME, Journal of Engineering for Power*, 101, 384–396.

Wilbur S 1957 An investigation of flow in circular and annular 90° bends with a transition in cross section. NACA TN 3995.

Will BC, Benra FK and Dohmen H-J 2011 Investigation of the flow in the side cavities of a centrifugal pump with volute casing. Proceedings of the 10th International Symposium on Experimental and Computational Aerothermodynamics of Internal Flow, ISAIF10-009, Brussels.

Worster RC 1963 An investigation of the flow in centrifugal pumps at low deliveries. BHRA RR 770.

Wu CH 1952 A general theory of 3D flow in subsonic and supersonic turbomachines of axial, radial and mixed flow types. NACA TN 2604.

Wunsch D, Hirsch Ch, Nigro R and Coussement G 2015 Quantification of combined operational and geometrical uncertainties in turbo-machinery design. ASME GT-2015-43399.

Xin J, Wang X, Zhou L, Ye Z and Liu H 2016 Numerical investigation of the flow field and aerodynamic load on impellers in centrifugal compressor with different radial inlets. ASME GT-2016-57180.

Xu C and Amano S 2012 Aerodynamic and structure considerations in centrifugal compressor design – Blade lean effects. ASME GT-2012-68207.

Yagi M, Kishibe T, Shibata T, Nishida H and Kobayashi H 2008 Performance improvement of centrifugal compressor impellers by optimizing blade-loading distribution. ASME GT-2008-51025.

Yagi M, Shibata T, Nishida H, Kobayashi H, Tanaka M and Sigimura K 2010 Optimizing a suction channel to improve performance of a centrifugal compressor stage. ASME GT-2010-203019.

Yamada K, Furukawa M, Fukushima H, Ibaraki S and Tomita I 2011 The role of tip leakage vortex breakdown in flow fields and aerodynamic characteristics of transonic centrifugal compressor impellers. ASME GT-2011-46253.

Yang M, Zheng X, Zhang Y, Bamba T, Tamaki H, Huenteler J and Li Z 2010 Stability improvement of high-pressure-ratio turbocharger centrifugal compressor by asymmetric flow control – Part I: Non-axisymmetric flow in centrifugal compressor. ASME GT-2010-22581.

Yang C, Zhao, Ma CC, Lao D and Zhou M 2013 Effect of different geometrical inlet pipes on a high speed centrifugal compressor. ASME GT-2013-94254.

Yi J and He L 2015 Space time gradient methods for unsteady blade row interaction – Part I: Basic methodology and verification. *Trans. ASME, Journal of Turbomachinery*, 137, 111008-(1-13).

Yoon SY, Lin Z, Jiang GW and Allaire PE 2012 Flow-rate observers in the suppression of compressor surge using active magnetic bearings. ASME GT-2012-70011.

Yoshida Y, Murakami Y, Tsurusaki H and Tsujimoto Y 1991 Rotating stalls in centrifugal impeller/vaned diffuser systems. In: *General Topics in Fluids Engineering*, Vol. 107, ASME FED.

Yoshida Y, Tsujimoto Y, Tateishi T and Tsurusaki H 1993 Active control of vaneless diffuser rotating stall. In: *Rotordynamic Instability Problems in High-Performance Turbomachinery 1993*, NASA CP 3239.

Yoshinaga Y, Gyobu I, Mishina H, Kosekli F and Nishida H 1980 Aerodynamic performance of a centrifugal compressor with vaned diffuser. *Trans. ASME, Journal of Fluids Engineering*, 102, 486–493.

Yoshinaga Y, Kaneki T, Kobayashi H, Hoshino M 1985 A study of performance improvement for high specific speed centrifugal compressors by using diffusers with half guide vanes. In: *Three-dimensional Flow Phenomena in Fluid Machinery*, ASME Winter Annual Meeting, Miami.

Yoshinaka T 1977 Surge responsibility and range characteristics of centrifugal compressors. Proceedings of the Tokyo Joint Gas Turbine Congress, pp. 381–390.

Zangeneh M 1991 A compressible three-dimensional design method for radial and mixed flow turbomachinery blades. *International Journal of Numerical Methods in Fluids*, 13, 599–624.

Zangeneh M 1993 Inviscid–viscous interaction method for 3D inverse design of centrifugal impellers. ASME 93-GT-103

Zangeneh M 1998 On 3D inverse design of centrifugal compressor impellers with splitter blades. ASME 98-GT-507.

Zangeneh M, Goto A and Harada H 1998 On the design criteria for suppression of secondary flows in centrifugal and mixed flow impellers. *Trans. ASME, Journal of Turbomachinery*, 120, 723–735.

Zangeneh M, Mendonça F, Hahn Y and Cofer J 2014 3D Multi-disciplinary inverse design based optimization of a centrifugal compressor impeller. ASME GT-2014-26961.

Zheng X and Lan C 2014 Improvements of the performance of a high-pressure-ratio turbocharger centrifugal compressor by blade bowing and self-recirculation casing treatment. *Proceedings of the IMechE, Journal Automotive Engineering*, 228, 73–84.

Zheng X, Zhang Y, Yang M, Bamba T and Tamaki H 2010 Stability improvement of high-pressure-ratio turbocharger centrifugal compressor by asymmetric flow control – Part II: Non-axisymmetric self recirculation casing treatment. ASME GT-2010-22582.

Zhou L, Xi G. and Cai YH 2007 Unsteady numerical simulation in a centrifugal compressor using the time-inclined operator. Proceedings of the IMechE, Part G: *Journal of Aerospace Engineering*, 221, 795–803, DOI: 10.1243/09544100JAERO194.

Ziegler KU, Gallus HE and Niehuis R 2003a A study on impeller–diffuser interaction – Part I: Influence on the performance. *Trans. ASME, Journal of Turbomachinery*, 125, 173–182.

Ziegler KU, Gallus HE and Niehuis R 2003b A study on impeller–diffuser interaction – Part II: Detailed flow analysis. *Trans. ASME, Journal of Turbomachinery*, 125, 183–192.

Index

Design and Analysis of Centrifugal Compressors, First Edition. René Van den Braembussche.
© 2019, The American Society of Mechanical Engineers (ASME), 2 Park Avenue, New York, NY, 10016, USA (www.asme.org).
Published 2019 by John Wiley & Sons Ltd.